Biomathematics

Volume 4

Edited by

K. Krickeberg · R. C. Lewontin · J. Neyman
M. Schreiber

M. Iosifescu · P. Tăutu

ACADEMY OF THE SOCIALIST REPUBLIC OF ROMANIA
Centre of Mathematical Statistics, Bucharest

Stochastic processes and applications in biology and medicine

II

MODELS

Editura Academiei
București

Springer-Verlag
Berlin · Heidelberg · New York

1973

This work is based on the book
"Procese stohastice şi aplicaţii în medicină şi biologie"

Editura Academiei, 1968
str. Gutenberg 3 bis, Bucureşti

AMS Subject Classifications (1970)
Primary 60J10, 60J15 60J80, 60J70, 92A05 92A10, 92A15
Secondary 60G05, 60J40, 60K05, 60K25, 92A25

ISBN-13: 978-3-642-80755-8 e-ISBN-13: 978-3-642-80753-4
DOI: 10.1007/978-3-642-80753-4

Softcover reprint of the hardcover 1st edition 1973

Preface

This volume is a revised and enlarged version of Chapter 3 of a book with the same title, published in Romanian in 1968. The revision resulted in a new book which has been divided into two parts because of the large amount of new material. The whole book is intended to introduce mathematicians and biologists with a strong mathematical background to the study of stochastic processes and their applications in biological sciences. It is meant to serve both as a textbook and a survey of recent developments.

Biology studies complex situations and therefore needs skilful methods of abstraction. Stochastic models, being both vigorous in their specification and flexible in their manipulation, are the most suitable tools for studying such situations. This circumstance determined the writing of this volume which represents a comprehensive cross section of modern biological problems on the theory of stochastic processes. Because of the way some specific problems have been treated, this volume may also be useful to research scientists in any other field of science, interested in the possibilities and results of stochastic modelling. To understand the material presented, the reader needs to be acquainted with probability theory, as given in a sound introductory course, and be capable of abstraction.

This work is the result of the collaboration between a mathematician and a physician (by training) and as such has both the virtues and the shortcomings of such an enterprise. We have attempted to keep a balance between length and depth and between omission and suggestion.

We draw the reader's attention to the numbering, because there are structural differences between the two volumes. In Part II there are Sections, Subsections, Paragraphs, and Subparagraphs. Thus the numbering a.b.c.d refers to Subparagraph d of Paragraph c of Subsection b of Section a. Theorems and lemmas are numbered a.n, n = = 1, 2, ..., where a indicates the section. In Part I there are Chapters, Sections, Subsections, Paragraphs, and Subparagraphs. Thus the numbering a.b.c.d.e refers to Subparagraph e of Paragraph d of Subs section c of Section b of Chapter a. Definitions, theorems, lemma- and propositions are numbered a.b.n, n = 1, 2, ..., where a indicates the chapter and b the section. In the present volume references to Part I are given by I. Thus I; a.b.c.d refers to Paragraph d of Subsection c of Section b of Chapter a of Part I.

We also wish to call the reader's attention to the bibliography: it exceeds the number of works quoted in the course of the volume in order not to overburden the text. Thus the bibliography should be consulted with the purpose of discovering historical sources, parallel researches and, in a general way, starting points for new investigations.

The authors desire to express their thanks to Dr. Klaus Peters of Springer-Verlag for his interest and patience in editing this new book, to Mrs. Sorana Gorjan of the Publishing House of the Romanian Academy, to Mrs. Ruxandra Goga, and to Mr. K. Wickwire of Springer-Verlag for their careful checking of the language.

March 1972

MARIUS IOSIFESCU

PETRE TĂUTU

Contents

5. Models in physiology and pathology

0. Prolegomenon

In the year 1873, Alphonse de CANDOLLE, one of the best known geo-botanists of his time, investigating the role of fertility in the struggle for life of the human species stated that the extinction of family surnames was unavoidable. In the same year, FRANCIS GALTON, interested in the decay of outstanding English families ("men of note": peers, judges and the like), published the famous 4001 problem, the first mathematical formulation of this extinction process in the *Educational Times* (subtitled "Journal of the College of Preceptors").

Its solution was given four months later (August 1, 1873) in the same journal by HENRY WILLIAM WATSON, FRS, clergyman, mathematician and alpinist (as he was depicted in KENDALL (1966)). One year later, this time in an anthropology journal, F. GALTON and H. W. WATSON presented their joint work *On the probability of the extinction of families* (published as an appendix to GALTON's book *Natural Inheritance* (1889)). GALTON mentioned in his book that "M. Alphonse de Candolle has directed attention to the fact that, by the ordinary law of chances, a large proportion of families are continually dying out, and it evidently follows that, until we know that proportion, we cannot estimate whether any observed diminution of surnames among the families whose history we can trace, is or is not a sign of their diminished fertility".

This is the beginning of the prehistory of the theory of branching processes and, by extension, of a large and important class of stochastic processes generated by problems in biology. Some essential vital processes, such as the multiplication of organisms or the development of species, have created empirical situations which have led to mathematical models of quite general character. The survival of family names appears to have been the first chain reaction studied by probability methods.

Stimulated by the problems of Darwinist biology, YULE (1924), FELLER (1939) and KENDALL (1948 a) created stochastic models for the numerical increase of populations, starting from synthetic nomologic statements in biology. It is known that the substitution of empirical statements by mathematical ones, the deduction of the corresponding mathematical consequences and the substitution of some of these consequences by empirical statements, represent the methodology of applied mathematics (see KÖRNER (1960)). G. U. YULE started from the problem of the birth of new species within a genus, while W. FELLER took as starting point the theory of the struggle for life and D. G. KENDALL set to study migration processes and the analysis of the stability of populations. For YULE, the species, rather than individuals, were the multiplying objects. In his now classical paper, published in 1924,be considered two types of mutations according to their occurrence within the species of the genus: *specific* and *generic* mutations: "Within any species, in any interval of time an 'accident' may happen that brings about a (viable) 'specific mutation', i.e., the throwing of a new form which is regarded as a new species, but a species within the same genus as the parent... Within any genus in any interval of time an 'accident' may happen that brings about the throwing of a (viable) 'generic mutation', i.e., a new form so different from the parent that it will be placed in a new genus". In another work, published in 1920 (jointly with Major GREENWOOD), G. U. YULE gave the solution for the general linear birth process, a solution somewhat different from that found by A. G. MCKENDRICK as early as 1914. We remind the reader that the linear birth-and-death process is one of the simplest branching processes in continuous time.

However, in addition to a discussion of the permanent, empirical sources of stochastic processes, a discussion *within* biology itself is very important. A series of theoretical conclusions concerning the structure and behaviour of living systems must evolve out of the structure and the results of abstract mathematical processes.

A characteristic example of this process is given by the GALTON-WATSON problem itself. Correcting the error of WATSON, AGNER KRARUP ERLANG (whose own maternal family name, KRARUP, was dying out), conjectured the basic theorem on the subject: extinction is almost certain for subcritical populations with $m < 1$ and for critical populations with $m = 1$, but there is always a positive chance of survival for supercritical populations with $m > 1$, where m is the mean number of offspring per individual. Moreover, according to I; Corollary to Theorem 1.2.9, the population dies out or suffers an explosive increase if $m > 1$.

This situation takes on particular significance if it is put into the framework of Darwin's theory of natural selection. Let us suppose that in a population with fixed size a mutant gene appears which will play the role of a zero-generation particle.

Assume that the distribution probability $\mathfrak{p}_k$, $k \in N$, that a mutant gene is present in exactly k organisms in the next generation is approximately Poisson with mean $m = 1 + \varepsilon$, $\varepsilon > 0$. If $\varepsilon = 0$, the assumption is based on the following analogy (FISHER (1922)): suppose a corn plant produces N male and N female gametes, each having probability 1/2 of possessing a mutant gene. Considering that the population size remains constant, these gametes are drawn as a random sample from the population to form the next generation. The probability $\mathfrak{p}_k$ is comparable to the probability of exactly k successes in $2N$ Bernoulli trials with probability $1/2N$.

If the mutant gene confers a small selective advantage ($\varepsilon > 0$), according to I; Theorem 1.2.9, the extinction probability q will be the non-unit root of the equation $x = e^{m(x-1)} (= \sum_{k \in N} \mathfrak{p}_x x^k)$. We obtain $q \approx \approx \exp\{2(1-m)/m\} = \exp\{-2\varepsilon/m\}$. In his pioneering work published in 1922 as well as in his book *The Genetical Theory of Natural Selection* (1930), R. A. FISHER calculated the probabilities of extinction at the nth generation of a mutant gene with $m = 1$ and $m = 1.01$ (selective advantage 1%).

In case of no advantage, the probability of extinction at the 63rd generation equals 0.9698, while in the case of selective advantage, it is 0.9591. After 64 additional generations, the situation becomes clearer. The chance that the 127th generation is nonempty is 0.0271 if $m = 1.01$; without selective advantage this chance is 0.0153. In the absence of selection the number of descendants from a single mutant cannot greatly exceed the number of generations produced since its occurrence. The odds in favour of the mutant persisting are greater if the population happens to be increasing in number when it appears, and less if the population is declining (hence the fundamental supposition on the constancy of the population). In general, however, the majority of new mutant genes produced by fresh mutations are likely to be eliminated in the course of the next 10 to 20 generations in a more or less random manner (H. HARRIS (1970 a)).

A clear and brief treatment of this example may be found in BAILEY (1964 a) and the question of the survival of a single mutant was expounded by KIMURA (1957, 1964) and MORAN (1961, 1962, Ch. V). See also ROBERTSON (1962) for the selection problem for heterozygotes in small populations.

The biological implications of these results are extremely important, and produce significant conclusions when the identification of the empirical situation is appropriately carried out. We shall give a single example: the survival of sickle-cell hemoglobin gene. This mutant gene determines the substitution of a valine for a glutamic acid residue in the sixth N (nitrogen) terminal position of the hemoglobin β-chain, an event having physico-chemical, structural, physiological and pathological consequences. During hypoxia, physico-chemical modifications of the abnormal hemoglobin take place and deform the red cell membrane into its characteristic sickle shape, also causing increased blood viscosity. As the most frequent iatrogenic cause of hypoxia is the anesthetic state, the *sickling phenomenon* also has specific importance in pharmacogenetics (MOTULSKY and STAMATOYANNOPOULOS (1968)).

The presence and the persistency of the sickle-cell hemoglobin gene in some populations (e.g. in Central Africa) is due to the action of strong selection forces offering a clear selective advantage to heterozygotes. Homozygotes will suffer in turn a severe and characteristic hemolytic anemia (sickle-cell anemia).

Because of the incidence of malaria observed in heterozygotes with a sickle-cell gene, it was supposed that *Falciparum malaria* was the selective agent (an assumption also valid in the case of β-thalassemia, a genetically determined syndrome reflected by a decreased synthesis of β-globin, with a consequent decreased synthesis of hemoglobin A). It was calculated that the selective advantage is 24%. The sickle-cell disease is clearly a molecular disease (ZUCKERKANDL and PAULING (1962)); yet, a molecular disease may be maintained in the species because certain agents in the environment render it innocuous, and, on the other hand, it may be maintained because it renders relatively innocuous certain agents present in the environment. Therefore, if through its eradication malaria does not remain a selective agent, the relative advantage of the heterozygote will be diminished. Instead of eradication one may suppose a change of environment by immigration from other regions (if the "founder-effect" does not appear). *Other things being equal*, the frequency of the sickle-cell gene will then become lower (CROW (1968)). Data obtained by ALLISON (1954) show that the frequency of the gene in African Negroes is 0.206 (in malaria stricken zones) but that it declines to 0.046 in American Negroes.

This remark calls for a new discussion about the exact significance of the word "population". Here, as in the case of those "inborn errors of metabolism" (an expression introduced in 1909 by A. E. GARROD; see H. HARRIS (1970 b, Ch.6)), not only more than one different abnormal allele producing a specific effect may occur among the members of any given population, but considerable differences are likely to occur both in the actual alleles that are present and in their relative frequencies from population to population *with different ancestries*. We can now explain the "founder effect". It is used to refer to situations where a relatively small number of individuals from one population have migrated elsewhere to found a new community, which has subsequently increased considerably in number, more or less in isolation. If by chance one of the founder members happens to be a heterozygous carrier of a particular rare mutant gene, then this might quite fortuituously come to have an extremely high frequency among the descendants. H. HARRIS (1970 b, p. 241) thus specifies the remarkable incidence of hereditary tyrosinaemia in an isolated French-Canadian population living in Northern Quebec.

We hope that our discussion and understanding of the problems of stochastic modelling in biology pays due respect to the tradition of the first biometricians. This volume is dedicated to their memory.

1. Preliminary considerations

1.1. Stochastic and deterministic models in biology

The concept of a stochastic model [στοχάζτομαι, as explained by IRWIN (1967), has two main meanings: (a) to aim at or shoot at, or to endeavour to; (b) to attempt to make out or guess anything, or to conjecture] is used to signify an abstract scheme of probabilistic nature, capable of representing any phenomenon whose evolution submits to random laws. We do not consider it redundant to stress the random evolution of the phenomenon since in this way we emphasize the idea that the phenomenon studied is itself random.

It is interesting to add the characterization of LEWONTIN (1963) concerning stochastic models of biological populations: Certainly real populations of plants and animals are not infinitely large and the number of effective parents of each generation may be much smaller than the total population, as in social units of bees. Moreover, the environment is not usually constant but undergoes a certain amount of random fluctuation. A more realistic group of models is then one in which the role of chance in determining which individuals will leave what sort of offspring is taken into account. These are, of course, the *stochastic* models.

1.1.1. Comparisons with deterministic models

1.1.1.1. The deterministic representation of a real process is based on the assumption that the development of the process is governed by dynamic laws. This leads to the consideration of the possibilities of each model separately and to an analysis of the conditions of its utilization. The mathematical representation is a stage between theory and experiment and therefore it is useful to know the descriptive, explanatory and predictive capacity of deterministic and stochastic models. As FELLER (1968) has said, in applications, abstract mathematical models serve as tools and different models can describe the same empirical situation. *The manner in which mathematical theories are applied does not depend on preconceived ideas*; *modeling is a technique depending on, and changing with, experience.*

Let us immediately observe that, in the case of a deterministic model, the representation of the empirical process is unique in the sense that the evolution of the process is completely determined. The increase of a cellular population submitting to an exponential deterministic law such as

$$x(t) = x_0 e^{bt}, \tag{1.1}$$

will have the same values whenever x_0 is the same, for fixed b and t. The predictive capacity of a deterministic model seems, therefore, to be limited.

1.1.1.2. The building of a stochastic model cannot lead to the prediction of a single realization. The development of the process is not uniquely determined. A graph of population growth would show appreciable fluctuation, and the question arises how this fluctuation is related to a probability distribution.

Thus, instead to the unique value of $x(t)$, the linear birth model leads to a general result such as

$$p_x(t) = \binom{x-1}{x_0-1} e^{-bx_0 t} (e^{bt} - 1)^{x-x_0}. \tag{1.2}$$

The meaning of (1.2) is as follows: if we imagine a large number of populations, all starting off simultaneously with x_0 individuals, then the proportion of these populations having size x at epoch t is theoretically $p_x(t)$ with average size $m(t) = x_0 e^{bt}$ and variance $\sigma^2(t) = x_0 e^{bt}(e^{bt} - 1)$. One immediately notes that $m(t)$ is identical to the deterministic value in (1.1). Evidently, if the population size is large enough, the deterministic treatment would be a successful substitute for any stochastic discussion that was primarily concerned with the mean value. Mean values are often used as a first step towards describing the behaviour of a stochastic process, so that it is clearly convenient when the more easily calculated deterministic value is the same (BAILEY (1967 b)).

1.1.1.3. A probabilistic model is a relation of the type

$$\mathrm{P}(\xi \leqslant x) = F(\boldsymbol{\theta};\ \mathbf{X};\ x), \tag{1.3}$$

where ξ is a random variable, F is a probability distribution with argument x and vector parameters $\boldsymbol{\theta}$ and $\mathbf{X}$, with $\boldsymbol{\theta}$ a vector invariant over a class $\mathcal{X}$ of situations of interest, and $\mathbf{X}$ a vector which varies over the class $\mathcal{X}$. LUCAS (1964) calls ξ the *dependent* variable (the response, the predictand or the output) and $\mathbf{X}$ the vector of *independent* variables (the treatment factors, the predictors, or the inputs), assuming that any probabilistic elements are distributed with at least the first two moments finite.

We may now write down the random variable as

$$\xi = m(\boldsymbol{\theta}; \mathbf{X}) + \varepsilon, \tag{1.4}$$

where

$$m(\boldsymbol{\theta}; \mathbf{X}) = \int_{-\infty}^{\infty} x \, \mathrm{d}_x F(\boldsymbol{\theta}; \mathbf{X}; x) = \mathsf{E}[\xi \mid \boldsymbol{\theta}; \mathbf{X}],$$

and

$$\varepsilon = \xi - m(\boldsymbol{\theta}; \mathbf{X}).$$

We also have

$$\mathsf{P}(\varepsilon \leqslant e) = F(\boldsymbol{\theta}; \mathbf{X}; x) = F(\boldsymbol{\theta}; \mathbf{X}; e + m(\boldsymbol{\theta}, \mathbf{X})),$$

where

$$e = x - m(\boldsymbol{\theta}; \mathbf{X}) \quad \text{and} \quad \mathsf{E}[\varepsilon \mid \boldsymbol{\theta}; \mathbf{X}] = 0.$$

H. L. LUCAS emphasizes the deterministic aspects of the model, e.g. the forms of F and m, and the values taken by the elements of $\boldsymbol{\theta}$ and $\mathbf{X}$. In (1.4), $m(\boldsymbol{\theta}; \mathbf{X})$ is deterministic and ε is random. By a deterministic model, he means, of course, a model developed with no randomness.

H. L. LUCAS points out that many biological problems can be set up in terms of input-output systems which embrace both stochastic and deterministic elements. He admits that it is convenient to start with an idealized basic vector relation, namely

$$\boldsymbol{\eta} = A(\boldsymbol{\zeta}),$$

where $\boldsymbol{\eta}$ is the vector of output, $\boldsymbol{\zeta}$ is the vector of input and A, a vector valued transformation. The variable $\boldsymbol{\eta}$ is uniquely determined by A and $\boldsymbol{\zeta}$, which is viewed by the author as a clearly deterministic relation; yet it can serve as the starting point for a stochastic treatment. One must think of A as a sort of universal transformation and must incorporate characteristics of the organism into $\boldsymbol{\eta}$ and $\boldsymbol{\zeta}$.

1.1.1.4. We emphasize that some of H. L. LUCAS' interpretations are due to his conception of probability. The concept of *randomness* may have applied simply as a device to cope with ignorance about the causes of events. Thus, he introduces the term *inherent unexplainability*, true randomness implying that there is no ultimate explanation or reason that a given event should occur instead of another. Truly random events should be defined as mutually independent with a distribution characterized by the form and the parameters of some probability distribution. Consequently, the occurrence of an event is not in any way conditioned by the previous or the concomitant occurrence of another event. If two variables are distributed by a bivariate normal distribution, for example, it seems possible to explain any correlation between the two variables in terms of an underlying cause common to both, or to regard as truly random only that part of the variation of one variable that is not explained by the value taken by the other variable. Thus, pseudo-randomness applies to the apparent randomness generated by ignorance.

However, our opinion is that this independence must be understood as being of causal not of stochastic nature. In the last case, two random events are dependent if the probability of one of them is influenced by the occurrence of the other. The mutation of an organism belonging to a cellular population depends on its birth, since it is evident that its genetic modification has no significance if it does not exist. But this mutation may be stochastically independent of the birth, death or mutation of another organism. E.g., the basic postulate of branching processes is the independent evolution of every new individual appearing in the population; they must not interfere with one another.

1.1.1.5. The implication of what we have written above is that the intrinsic random character of some processes may appear under at least three aspects:

(i) *Random perturbations on a deterministic trajectory*: the evolution of the process is governed by a deterministic law, but this evolution may be complicated by the random perturbations acting during the evolution of the process.

(ii) *Initial random conditions*: the evolution of the process is completely defined, starting from some initial conditions, but the process may be repeated under conditions modified at random.

(iii) *Integral (pure) randomness*: the evolution of the process is purely random, as in the random walk. (It will be observed later that to a certain degree, real processes not wholly random can nevertheless be approximated as such.)

1.1.2. Equipollent and conjunct models

1.1.2.1. There are many real situations which cannot be represented as purely stochastic or purely deterministic processes. RAMAKRISHNAN (1959) called "blended" a process whose trajectories are deterministic, with the exception of a finite number of points where random perturbations take place. Also, BHARUCHA-REID (1960) has called "mixed" a process in which the variable undergoes continuous modifications governed by a deterministic law as well as discrete modifications of random character. We will use the term *conjunct models* to designate all the processes in which the deterministic and the random characters are associated together in one form or another (*conjunctus* = joined together).

1.1.2.2. The great majority of demographic models is made up of deterministic models existing in two forms: those using a continuous time-variable and a continuous age-scale (thus continuing SHARPE and LOTKA's (1911) classical work) and those using a discrete time-variable and a discrete age-scale (continuing BERNARDELLI's (1941) and LESLIE's (1945) investigations). The integral equation due to LOTKA (1939) represents the starting point for the analysis of the continuous model, while for the discrete model this analysis is based upon the matrix approach, which is somewhat more recent and due to LESLIE (starting in 1945). We note together with POLLARD (1966) that the discrete formulation is closer to actuarial practice than the continuous model and is preferable when the age-specific birth and death rates are to be given on the basis of empirical data. Furthermore, the matrix gives the same results as the integral equation if the age intervals are made small enough and, with five-year age intervals common in practice,

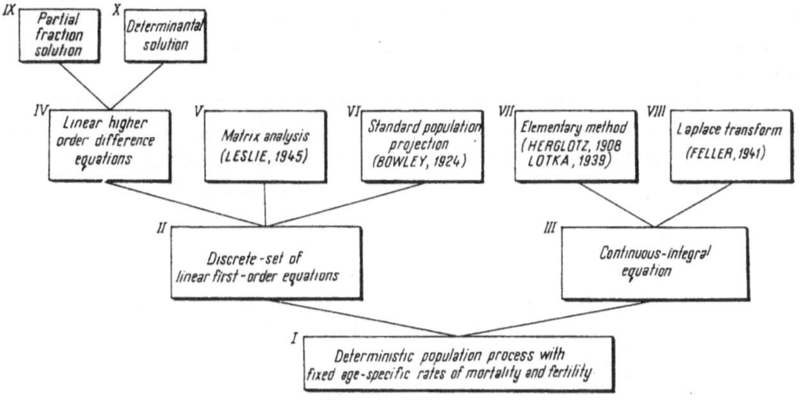

Fig. 1

gives a finite approximation to the same result (KEYFITZ (1967 b); see also GOODMAN (1967 a)). In Figure 1 we have shown (after KEYFITZ (1967 a)) methods specific to both models.

In 1969, Z. M. SYKES proposed three stochastic versions for the classical discrete deterministic models for population dynamics, which versions we consider as typical conjunct models because they respectively suppose that (i) the deterministic model is subject to additive random errors (recalling the "stochastization" stages proposed by LUCAS),

(ii) the transition rates are probabilities; (iii) the transition rates are random variables. Usually, the simple linear deterministic model is of the form

$$\mathbf{x}_{n+1} = \mathbf{A}^n \mathbf{x}_0, \qquad n \in N,$$

where $\mathbf{x}_0$ is the k-dimensional vector with components $x_{0i} \geqslant 0$ ($1 \leqslant i \leqslant k$) and $x_{0i} > 0$ for at least one class (type) i and $\mathbf{A}$ is a matrix with elements

$$a_{ij} = \begin{cases} b_j, & i = 1, \quad 1 \leqslant j \leqslant k-1 \\ s_j, & 2 \leqslant i \leqslant k, \quad j = i-1 \\ 0, & \text{otherwise} \end{cases}$$

where $b_j \geqslant 0$, $b_k > 0$ and $0 < s_j \leqslant 1$. Each b_j represents the number of individuals contributing to the first class at time $(n+1)$ by an individual of type j at time n, so that $\{b_j\}$ is in effect a set of "birth" rates defined in some appropriate way. Similarly, the requirement that $a_{ij} = 0$ $(2 \leqslant i \leqslant k,\ j \neq i-1)$ expresses the usual situation in closed biological populations, in which a member of an older class at time $(n+1)$ must have been in the next younger class at time n, and $\{s_j\}$ is thus the set of "survival" rates for the population, representing the proportions of individuals of type j who will survive to become members of class $j+1$.

The conjunct model considered satisfactory by Z. M. Sykes is that in which it is supposed that the population $\mathbf{x}_n$ results from the successive application of n sets of random birth and death rates to the initial population $\mathbf{x}_0$. For details and numerical example, see Sykes (1969 a).

1.1.2.3. A comparative investigation of deterministic and stochastic models caused us to call equipollent (*aequipollens* = practically equivalent) deterministic and stochastic models that reach analogous or equivalent results.

In the Romanian version of this book, we presented (under the title "frontier model") a deterministic model of cellular growth (Trucco (1965)) from which we deduced the so-called Von Foerster equation (1959), otherwise similar to those obtained by Scherbaum and Rasch (1957), Bergner (1965 a) or Hirsch and Engelberg (1966). Martinez (1966), however, showed that the Hirsch-Engelberg equation may be obtained, in certain conditions, from the von Foerster as well as the Bellman-Harris equation (I; (2.3.35)) for age-dependent branching processes (see I; 2.3.3.4.7). Without doubt, a close analysis of this situation will discover equipollent deterministic and stochastic models. (An excellent guide is the paragraph on the simple deterministic models of population growth with age distribution, in Karlin's (1966) book.)

RAPOPORT (1965) employed the term "equivalent mathematical models" specifying that within learning processes, Landahl's (1941) equation is a limiting case of the stochastic learning model of W. K. ESTES, and that the more general form of Landahl's equation is exactly equivalent to a certain approximation of the stochastic model, assuming that learning eventually proceeds to the fixation of the correct response.

1.2. The structure of biological populations

Stochastic models generally deal with the description and prediction of events taking place within populations, including events leading to the constitution of the population itself. We shall consider those populations which represent sets of objects with a determined and complex character, i.e., biologic objects, abbreviated to B-objects. Except for certain populations of a special type or in certain situations where intuitive considerations dictate otherwise, we shall denote as B-object populations taxons and clones as well as the population in a biotope, in a bacterial culture, etc.

1.2.1. Elements of bio-logic

1.2.1.1. In this paragraph we shall briefly present some aspects of the logic and algebra of biological objects — thus setting up a formalization of biological statements, a field whose pioneer works are due to J. H. WOODGER. This field of formal biology, ignored for a long time, grew considerably when the fundamental work of W. S. MCCULLOCH and W. PITTS was published in 1943; here the neural networks were described by means of the propositional calculus. Also worthy of mention are the works of JOHN VON NEUMANN between 1948 and 1952 (see VON FOERSTER (1952)). Models of behaviour can now be described in logical terms, independent of whether that logic be of a precise or a probabilistic form.

The logic of kinship relations has not been able to offer until now a worthy reply to the McCulloch-Pitts model. Some books on mathematical logic carry examples concerning formal expressions of parental relations. For instance, denoting by $\mathsf{S} < a, b >$, the relation "*a* is the grandfather of *b*", one shows that by means of the concatenation process two relations can be chained such that $\mathsf{S} < a, b >$ may be written as

$$\mathsf{S} < a, b > = a(\mathsf{Q} \mid \mathsf{R})\, b,$$

where $\mathsf{Q} < a, x >$ is the relation "*a* is the father of *x*", and $\mathsf{R} < x, b >$ is the relation "*x* is the parent of *b*". A formal definition of genealogy was given in GLEASON (1966).

WOODGER (1962) considers the two important parental relations, the sexual (Ps) and the asexual (Pa), each of these being the sum of two relations. We say that a life x stands in the female parental relation (Pf) to a life y iff x produces the egg which unites with a male gamete to form the beginning of y; and we say that z is the male parent (Pm) of y if it produces the male gamete. One can say that

$$Ps = Pf \cup Pm,$$

and that it has two forms Ps_1 and Ps_2, so that

$$Ps_1 = Pf \cap Pm,$$

$$Ps_2 = (Pf - Pm) \cup (Pm - Pf).$$

Thus Ps_1 is the product of the two relations Pf and Pm and it is the relation among hermaphrodites; Ps_2 is their symmetric difference and it is the relation among organisms in which sexes are separate.

1.2.1.2. There are, however, a series of difficulties inherent to biology, since, as BOYD (1968) wrote, dealing with the algebraic structure of parenthood, the first requirement of a model of complex behaviour is a *grammar*, or a set of rules which recursively generate all of the data. The reader interested in this type of problem may find in LOUNSBURY's (1964) book the idea of "kinship grammars" and in SALTARELLI and DURBIN's (1964) paper a semantic interpretation of the kinship system. According to F. G. LOUNSBURY a "formal account" of a collection of empirical data has been given when there have been specified (i) a set of primitive elements, and (ii) a set of rules for operating on these, such that by the application of the latter to the former, the elements of a "model" are generated. A formal account of the problem of biologic descendants for the asexual parental relation will be given in Paragraph 1.2.4.

1.2.1.3. The term of "kinship network" was used by CUISENIER (1968) to express the whole set of filiation, alliance and fraternity relations existing among individuals forming a certain social system. In such a network, each individual may be considered as a node, each liaison as an arc. If the universe of parents under examination is a community, village, urban district or fraction of a tribe, a graphic description of existing relations is possible. The majority of operations on oriented graphs and networks will then be possible, leading towards the discovery of some special configurations of the considered social system.

1.2.1.4. In his 1965 paper, J. H. WOODGER dealt with logic on random evolution, starting from parental relations of asexual type. By introducing sets of *environmentally random* B-objects, WOODGER obtains two theorems emphasizing the influence of the environment upon mutations and the growth of the number of mutants in the following generations if environmental conditions remain the same. The logical treatment shows that J. H. WOODGER implicitly admitted a subcritical population of standard individuals. We mention this work not only because of the problems of bio-logics it raises but also to simultaneously stress the fact that random environment variables have only recently been considered in the theory of stochastic processes. We point out in this sense some papers on branching processes with random environment: ATHREYA and KARLIN (1970), SMITH (1968), SMITH and WILKINSON (1969), WILKINSON (1967), and WEISSNER (1971).

1.2.1.5. We shall use here some particular symbols: I — objects for the mature individuals, Z — objects for the zygotes, G — objects for gametes, and C — objects for chromosomes. We shall thus easily be able to note a series of operations connected, for example, with the process of reproduction and replication: parthenogenesis, syngamy, meiosis, mitosis, binary division, etc. In the case of syngamy (sexual reproduction) we shall write

$$\mathsf{SYN}\,(\mathsf{G}_1 \cup \mathsf{G}_2 \mid \mathsf{Z}),$$

where G_1 and G_2 are gametes of different sexes (1 = male, 2 = female) while Z is the resulting zygote.

1.2.2. Distinguishability and indistinguishability

1.2.2.1. Let E be an equivalence relation in the basic set X and let Q be the quotient set of X by E. Then Q is a partition of X, that is, an exhaustive collection of mutually exclusive subsets of X. The equivalence classes thus formed, $X_1, X_2, \ldots, X_k$, contain equivalent elements. We shall call *alteridem objects* (= another of the same one) (LEITNER (1965)) those forming an equivalence class and *alterexter objects* (= another foreign one) those belonging to different equivalence classes.

With respect to the characteristic (property) that caused them to be grouped together into the same class, the objects in an equivalence class are *indiscernible*. However, equivalence classes are monothetic classes of objects, i.e., collections of objects which possess a certain property in equal measure. If the objects are only partially indiscernible, E does not have in this case the property of transitivity and the relation will be a relation of similitude or proximity.

1.2.2.2. Let us observe that we are dealing with a group of properties or characteristics $P = \bigcup_{i=1}^{k} P_i$, each class $X_1, \ldots, X_k$ being formed by exactly those objects which have a corresponding property $P_1, \ldots, P_k$.

In some cases one may distinguish a series of subproperties of a property P_i so that $P_i = \bigcup_{\alpha=1}^{r} P_{i_\alpha}$. This amounts to saying that another partition is operating within an equivalence class X_i: if $\mathcal{P}$ is any partition of X_i, then there is a unique equivalence relation F in X_i having $\mathcal{P}$ as its quotient set. Generalizing, we can assert that to obtain steadily more precise distinctions within a population, one must achieve a chain of finer and finer partitions in a lattice of partitions on X. We thus obtain, as in fundamental taxonomy, a chain $\mathcal{C} = (\mathcal{P}_0, \mathcal{P}_1, \ldots, \mathcal{P}_k, \mathcal{P}_{k+1})$, where $\mathcal{P}_0$ is the finest partition, that is, the partition for which each class holds precisely one element of X. The classes of $\mathcal{P}_{k+1}$ are obtained by union of classes from $\mathcal{P}_k$, and so on. The number of distinct partitions, i.e., the length of $\mathcal{C}$, is smaller or equal to n, the cardinal of X.

In this sense, indistinguishability and distinguishability have a precise significance related to the length of the chain $\mathcal{C}$.

1.2.2.3. A different approach to this problem has been made by WOODGER (1962) who started from the philosophical concept of "epistemic priority". When time objects, as WOODGER says, are distinguishable by some test, we ordinarily assume that at least one pair of corresponding parts of these objects are distinguishable, and this obviously leads to the characteristic atomicity of natural science, which takes us ever farther away from statements of greatest epistemic priority. Objects not on this level cannot be defined in terms of objects that *are* on it, but the relation of indistinguishability between such objects as belonging to levels of inferior epistemic priority can be defined, by successive steps, with the help of the notion of indistinguishability between objects belonging to the level of highest epistemic priority.

1.2.2.4. We shall conclude this paragraph with an example taken from immunogenetics about the degree of indistinguishability between two alleles at locus H-2 in mice. This gene locus on a chromosome is both a blood group locus and a histocompatibility locus controlling the fate of a tumor and tissue transplants. By serological studies it was shown that 33 antigenic determinants are related to this locus. If we note the antigenic specificities characteristic of alleles $H\text{-}2^a$ and $H\text{-}2^k$, we shall observe that the determinant 32 is the only one present in $H\text{-}2^k$ and not in $H\text{-}2^a$ (see ELANDT-JOHNSON (1969)):

$H\text{-}2^a$:	1,	3,	4,	5,	6,	8,	10,	11,	13,	14,	25,	27,	28,	29
$H\text{-}2^k$:	1,	3,		5,		8,		11,			25,			32.

1.2.3. Equivalence relations and graphs

1.2.3.1. We shall illustrate this paragraph with the classical problem of panmictic mating (*pan* — *mixis*, indiscriminate interbreeding;random mating). We know that in panmictic populations any individual has equal chances of mating with any other individual of the other sex belonging to the same population.

Let us consider a population of *beige*-fur mice (a mouse strain studied, for example, at the Brown University and Harwell laboratories of genetics). The colour of thc fur is determined by the action of a gene denoted by *bg*, which has as a correspondent the pair (homologous) chromosome, either also a *bg* gene, or an alternative form, the mutant *slt* (slate) gene. Thus, as regards this colour gene, an individual may be of the type *(bg, bg)*, *(bg, slt)* or *(slt, slt)*, corresponding to whether the gene is of type *bg* on both chromosomes of the particular homologous pair,or type *slt* or both, or of type *bg* on one and *slt* on the other. This particular gene combination is called the *genotype* of an individual.

In our example, since we are not concerned with any other genes on the other chromosomes, the pair order is indistinguishable and therefore the genotypes *(bg, slt)* and *(slt, bg)*. Also, because we are considering a character which is not sex-linked, it is nevertheless necessary to consider the genotype of each sex in order that this "equally likely" assumption be properly applied.

The process of transmission of the genotype takes place during the fecundation process. By using the notations introduced in Subparagraph 1.2.1.5 we shall consider a population of gametes $\{G_i\}$, $i=1, 2$ (1=male, 2=female) transmitting their genotypes to the resulting Z-objects. For reasons of generality, we shall denote the *bg* gene by a_1, and the *slt* gene by a_2. (In the case of genes, indices do not mean male-female, as in the case of objects denoted by capital letters). We shall thus symbolically write down the situations that may appear according to the Mendelian laws:

$$\text{(1)} \qquad \text{SYN}\,(G_1\colon a_1a_1 \cup G_2\colon a_1a_1 \mid Z\colon a_1a_1),$$

$$\text{(2)} \qquad \text{SYN}\,(G_1\colon a_1a_1 \cup G_2\colon a_1a_2 \mid Z\colon a_1a_1 \wedge Z\colon a_1a_2),$$

$$\text{(3)} \qquad \text{SYN}\,(G_1\colon a_1a_1 \cup G_2\colon a_2a_2 \mid Z\colon a_1a_2),$$

$$\text{(4)} \qquad \text{SYN}\,(G_1\colon a_1a_2 \cup G_2\colon a_1a_2 \mid Z\colon a_1a_1 \wedge Z\colon a_1a_2 \wedge Z\colon a_2a_2),$$

$$\text{(5)} \qquad \text{SYN}\,(G_1\colon a_1a_2 \cup G_2\colon a_2a_2 \mid Z\colon a_1a_2 \wedge Z\colon a_2a_2),$$

$$\text{(6)} \qquad \text{SYN}\,(G_1\colon a_2a_2 \cup G_2\colon a_2a_2 \mid Z\colon a_2a_2),$$

where $\wedge$ is the logical symbol "and"[1]). The six possible situations are in fact the six "types of mating". We shall thus also express the same situations in the terms of I-objects: if a male and female offspring of one of the six mating types are mated and offspring of this union are in turn

[1]) The reader is certainly familiar with the classical notations of the biologist, where gene a_1 is denoted by A and gene a_2 by a. Instead of the reunion symbol, the symbol $\times$ is used, such that the first type of mating is written $AA \times AA$.

mated, we have a sequence of events, namely matings in successive generations. The great majority of books on the theory of probability include this process as a simple example of a Markov chain with six states (the six types of mating) and with transition matrix

$$\mathbf{P} = \begin{pmatrix} 1 & 0 & 0 & 0 & 0 & 0 \\ 1/4 & 1/2 & 0 & 1/4 & 0 & 0 \\ 0 & 0 & 0 & 1 & 0 & 0 \\ 1/16 & 1/4 & 1/8 & 1/4 & 1/4 & 1/16 \\ 0 & 0 & 0 & 1/4 & 1/2 & 1/4 \\ 0 & 0 & 0 & 0 & 0 & 1 \end{pmatrix} \quad (1.5)$$

The calculation of these probabilities is made considering that each parental gene has probability 1/2 of being transmitted, and the successive trials are independent (see FELLER (1968, p. 133)). Thus, the probability to have a genotype y in the condition of random mating (4) is

$$\begin{array}{l} 1/4 \text{ if } y \text{ is } a_1a_1, \\ 1/2 \text{ if } y \text{ is } a_1a_2, \\ 1/4 \text{ if } y \text{ is } a_2a_2, \\ 0 \text{ otherwise.} \end{array}$$

1.2.3.2. One can easily associate an oriented graph (a digraph) with the stochastic matrix **P**, putting the set of states of the Markov chain in one-to-one correspondence with a set of points $\{v_1, \ldots, v_6\}$ and considering that a directed line v_iv_j, $(1 \leqslant i, j \leqslant 6)$ exists iff $p_{ij} > 0$. Such a digraph is said to be a *Markov digraph* (HARARY and LIPSTEIN (1962)). The digraph associated to the above-mentioned particular matrix is presented in figure 2, where the numbering of the digraph points $v_1, \ldots, v_6$ corresponds to the numbering of the states of the Markov chain thus represented.

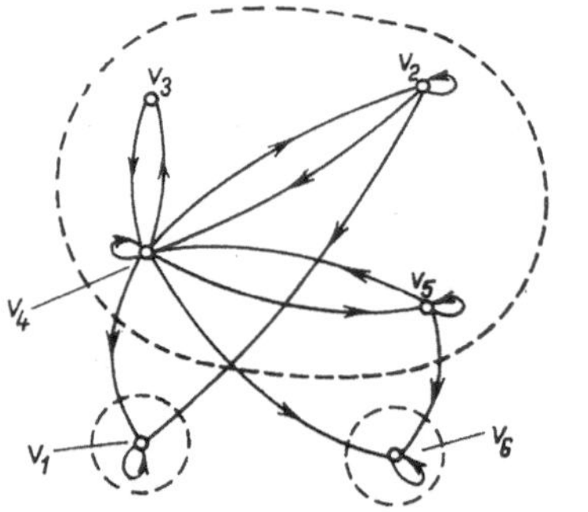

Fig. 2

By means of the Boolean matrices, well known in graph theory (see HARARY, NORMAN and CARTWRIGHT (1965)), e.g., the adjacency (or structure) matrix and the reachability matrix, one determines the three strong components of the considered graph, $S_1 = \langle v_1 \rangle$, $S_2 = \langle v_2, v_3, v_4, v_5 \rangle$ and $S_3 = \langle v_6 \rangle$. Thus one achieves an operational classification of the states of a Markov chain, the strong component S_2 includes all the nonrecurrent states, while the two other strong components represent the absorbing states of this genetic Markov chain.

1.2.3.3. The introduction of a relation of strict order $v_i > v_j$ in a graph, creates an "ancestor" of point v_j and a "descendant" of point v_i. A necessary and sufficient condition for a graph to admit such a relation is that it be acyclical and without loops.

The use of the terms "ancestor" and "descendant" leads to valuable associations with biological processes. In fact, there are two possibilities for using digraphs in the case of parental relations, according to the type of reproduction: asexual or sexual. In the first case, a special case of a latticeal graph is achieved, *a tree from a point*, which is a weak digraph with a source and no semicycle. It is often called *arborescence with root a*, considering that (i) every point v, $v \neq a$, has a degree of reception (or internal semi-degree) equal to one; (ii) the degree of reception of a is zero; (iii) the graph does not contain any circuit.

The second possibility is to carry out a genealogical tree where (i) there are two points of origin; (ii) every alternate circular path has a number of lines divisible by 4 (for details see ORE (1966)). These conditions are necessary to preserve the biological property of sex (one offspring has two parents of opposite sexes).

1.2.3.4. An arborescence is a set of divergent paths: we speak of the points with a degree of emission (or external semi-degree) greater than one as *branching*. A stochastic branching process is associated with such an arborescence. MYCIELSKI and ULAM (1969) have defined a *stochastic pairing process* schematizing the life process of a species with sexual reproduction. The associated tree is the second type in Subparagraph 1.2.3.3. The authors introduce here three different types of metrics ("genealogic distances").

1.2.4. Descendants, generations and families

1.2.4.1. One of the oldest assumptions about the forming of a biological population was that presuming its generation out of a single B-object. This generating element may be denoted by $<0>$ and is called an ancestor (= *antecessor*). It can give birth to a number of i descendants (= *posteri*): $<1>$, $<2>$, ..., $<i>$ forming a *generation* (denoted by biologists by F, from the Latin *filius*).

The problem of forming populations is not changed even when there are initially m ancestors, because a population starting at time $t = 0$ by m objects, coincides at another moment $t \neq 0$ with the sum of m independent populations, each started by a single ancestor. This property forms the basis of the theory of branching chains and processes. The assumption of the single ancestor was associated with the assumption of the *simultaneous* appearance of the offspring; the death of a parent and the birth of offspring must occur simultaneously. According to this assumption, populations do not overlap. The indications "first child", "second child" and so on, do not have here any temporal physical meaning.

The assumption of a very small number of ancestors has a biological foundation in the supposition of DARWIN (1869, p. 572): "I believe that animals are descended from at most only four or five progenitors, and plants from an equal or lesser number". One of the implications of this assumption refers to Eve, the *antica matre* (DANTE). It may be presumed that the so-called human population started at time $t = 0$ with a number m of Eves, each of them generating one *tribe*. A supplementary assumption considers the appearance of these Eves as random. Thus, PURI (1967 b) put forth the following problem: given that the human race does not eventually become extinct as time goes on, the progeny of how many Eves out of the original m would we expect to be present in the ultimate population, and in what proportion would the progeny of each be present? The answers to such questions are obtained in the form of limit theorems for Markov branching processes.

1.2.4.2. In the following pages we shall present the method of building descendancy, according to HARRIS' theory (1963), generalized by CRUMP and MODE (1968, 1969). An individual shall be denoted by $<i_1 \ldots i_n>$, the sequence $i_1 \ldots i_n$ indicating its line of descent. For example, $<I> \equiv <23>$ is the third child of the second child of the initial individual denoted by 0. The individual 0 comprises the 0-th *generation* and otherwise an individual of the form $<i_1, \ldots, i_n>$ is said to belong to the nth generation.

For each positive integer n, let $\mathcal{J}_n$ be the set of all n-tuples $i_1 \ldots i_n$ of positive integers and let $\mathcal{J} = \{0\} \cup \bigcup_{n=1}^{\infty} \mathcal{J}_n$.

Each individual I appearing in the population lives for a random length of time, ℓ_I, and at random times during its life this individual gives birth to offspring that behave in a similar fashion, the behaviour of each individual being independent of all others. The random function $N_I(t)$ which represents the number of offspring born to I in the age interval $[0, t]$ is determined by randomly stopping an arbitrary counting process K_I at ℓ_I; i. e.,

$$N_I(t) = \begin{cases} K_I(t) & \text{if} \quad t \leqslant \ell_I, \\ K_I(\ell_I) & \text{if} \quad t > \ell_I. \end{cases} \tag{1.6}$$

For each $I \in \mathcal{J}$, let T_I denote a sequence of the form ℓ_I, $t_I^{(1)}$, $t_I^{(2)}$, ..., where each element of the sequence is an extended nonnegative real number and $t_I^{(1)} \leqslant t_I^{(2)} \leqslant \ldots$. The quantity $t_I^{(j)}$ represents the time that elapses from the birth of I until the birth of Ij, the jth child of I. The distribution function of ℓ_I is usually denoted by G, $G(0) < 1$ and $G(\infty) = 1$.

By a *family history* we mean a sequence $\omega = \{T_0, T_1, T_{11}, T_2, \ldots\}$ where the subscripts run over all elements of $\mathcal{J}$ in some specified order.

Let Ω be the set of all such family histories. If for each $I \in \mathcal{J}$, $\mathcal{G}_I$ represents the set of all sequences T_I, then $\Omega = \prod_{I \in \mathcal{J}} \mathcal{J}_I$. Each $\omega \in \Omega$ represents a complete development of the family composed of 0 and its descendants, but each ω contains much additional information, since it is possible that $t_I^{(j)} > l_I$ for some I and j, and hence "births" may occur after the death of the parent. (However, the inclusion of such "births" enables one to represent Ω as a simple product space which is quite convenient from a mathematical viewpoint, so that when necessary we will simply ignore these "births".)

For each $I \in \mathcal{J}$ let P_I be a probability on $\mathcal{G}_I$ with all P_I having a common law. The probability P on Ω is taken to be the one determined uniquely by the product probability theorem (see Loève (1963, p. 91)). This assignment of probabilities implies, of course, that individuals live and reproduce independently. It is easily observed that these formal considerations are nothing other than the construction of the probability space over which one defines an age-dependent branching process.

1.2.4.3. Now, let $K_I(t)$ represent the number of points $t_I^{(j)}$ in the interval $[0, t]$. Each realization of $K_I(t)$ is non-decreasing, right continuous and assumes nonnegative integer values only, including possibly infinity. With this definition of $K_I(t)$, equation (1.6) defines $N_I(t)$ also as a function on Ω. It should be noted that (1.6) implies that a birth is counted even if it occurs at the exact instant of the death of the parent. $N_I(\infty)$, which represents the total number of offspring born to I, will be simply denoted N_I.

1.2.4.4. Considering now that the main object of the construction of such structures is the wish to express the numerical evolution in time of the population, we shall first make the following specifications:

The individual $<i_1 \ldots i_k>$ in the kth generation appears in the population iff

$$(A):\ i_1 \leqslant N_0,\ i_2 \leqslant N_{\langle i_1 \rangle}, \ldots, i_k \leqslant N_{\langle i_1 \ldots i_{k-1} \rangle};$$

The individual $<i_1 \ldots i_k>$ is born by time t iff (A) and

$$(B):\ t_0^{(i_1)} + t_{\langle i_1 \rangle}^{(i_2)} + \ldots + t_{\langle i_1 \ldots i_{k-1} \rangle}^{(i_k)} \leqslant t;$$

The individual $<i_1 \ldots i_k>$ has died by time t iff (A) and

$$(D):\ t_0^{(i_1)} + t_{\langle i_1 \rangle}^{(i_2)} + \ldots + t_{\langle i_1 \ldots i_{k-1} \rangle}^{(i_k)} + l_{\langle i_1 \ldots i_k \rangle} \leqslant t,$$

and $<i_1 \ldots i_k>$ is alive at time t iff A, B, and not D.

Denoting by $\xi(t, \omega)$, $b(t, \omega)$ and $d(t, \omega)$ the random variables representing the number of individuals alive at time t, the number of individuals born by time t and the number of individuals which have died by time t, respectively, we have

$$\xi(t, \omega) \leqslant b(t, \omega), \qquad d(t, \omega) \leqslant b(t, \omega),$$

and

$$\xi(t, \omega) = b(t, \omega) - d(t, \omega).$$

For each $\omega = \{T_0, T_1, T_{11}, T_2, \ldots\} \in \Omega$, let $\omega_i = \{T_i, T_{i1}, T_{i11}, T_{i2}\}$, $i \in N^*$.

We have

$$\xi(t, \omega) = \sum_{j=1}^{N_0(t)} \xi(t - t^{(j)}, \omega_j) + 1 - \delta(t - l), \tag{1.7}$$

where $\delta(t)$ equals 1 iff $t \geqslant 0$ and 0 otherwise, and the sum is zero if $N_0(t) = 0$. Intuitively, (1.7) expresses the fact that if k offspring have been born to 0 by time t then the size of the population at time t is the sum of the numbers of individuals alive that descend from each of these k offspring plus a term that takes into account the possibility that 0 may still be alive.

For the structure of a branching process with an arbitrary set of particle types, see HARLAMOV (1968) who also studies, among other things, the form of a genealogical tree section and the relationship distance.

1.2.5 The temporal structure of populations and biological objects

1.2.5.1. In Paragraph 1.2.4.2, we have shown that in the development of B-object populations there are two temporal dimensions connected between them, l_I the lifespan of the object I (a fundamental time distance) and $t_I^{(j)}$ the time distance from the birth of I to the birth of his jth child.

If $l_I \equiv t_I^{(j)}$, $1 \leqslant j \leqslant r$, for some $r \in N^*$ and $t_I^{(j)} = \infty$ for $j > r$, then we are in the classical situation when the death of the parent coincides with the birth of his r offspring; in this case, l_I represents the division time, that is, the time which elapses between the "birth" of an individual and the instant of its own subdivision into "new" individuals (see for example KENDALL (1953)).

1.2.5.2. The lifetime is often split into parts bearing various names ("ages", "stages", "phases", "cycles", "periods"), separated by more or less well-defined transitions from newborn to adult stage or through various metamorphosis steps. Since most individuals' characteristics change with growth, subclassification with regard to age is common. The simplest variable indicating the change of a B-object's intrinsic properties is, therefore, age. The immediate consequence at the level of the whole population is the creation of a nonhomogeneous time structure. Even in the case of populations with asexual (binary) reproduction the causes of inhomogeneity will be (i) the asynchronous division of objects and (ii) the possibility that both offspring may be of the zero age. These situations, as well as the case $\ell_I \neq t_I^{(r)} < \infty$, for some $r \in N^*$, determine the appearance of overlapped generations.

1.2.5.3. The age-dependent configuration of a population is revealed by the ages of the objects present after a long time. For example, in an age-dependent branching process with lifetime distribution G there is a limiting distribution of ages whenever the population does not die out, given by

$$A(x) = \frac{\int_0^x e^{-\alpha t} [1 - G(t)] \, dt}{\int_0^\infty e^{-\alpha t} [1 - G(t)] \, dt}, \tag{1.8}$$

where α is the so-called *Malthusian parameter* determined by

$$1 = m \int_0^\infty e^{-\alpha x} \, dG(x). \tag{1.9}$$

Here m is the mean number of offspring per individual and the parameter α must be considered as a measure of population increase or decrease. These facts can be deduced by an argument due to FISHER (1930, Ch. II) that permits one to find the form of $A(x)$ and also the value of α. See also HARRIS (1963, p. 152) and MORAN (1962, p. 59).

Starting from the age distribution problem, KENDALL (1949) proposed as a working instrument the characteristic functional (see also BARTLETT and KENDALL (1951)).

1.2.5.4. If the lifespan interval ℓ can be divided into several subintervals with a biological significance, these will be called "phases" (or "stages"). It is well known that according to present knowledge the multiplication process of the somatic cells takes place in four phases, denoted by G_1, S, G_2 and M, in successive order. M-phase, mitosis, consists in turn of four other well-studied phases: the prophase, metaphase, anaphase and telophase. Phases G_1, S and G_2 formed for a long time a single phase: the interphase, because until radioactive isotope techniques (autoradiography, for instance) were used, it was considered as a physiologically inert stage (interkinesis). However, it actually represents an essential moment of preparation of the division in which the syntheses of DNA, RNA and histones take place. DNA is synthetized during a specific and delimited period, known as the S-phase. This period is preceded and followed by two non-DNA synthetic periods of gaps called G_1-phase (postmitotic presynthetic gap) and G_2-phase (postsynthetic premitotic gap) (see fig. 3).

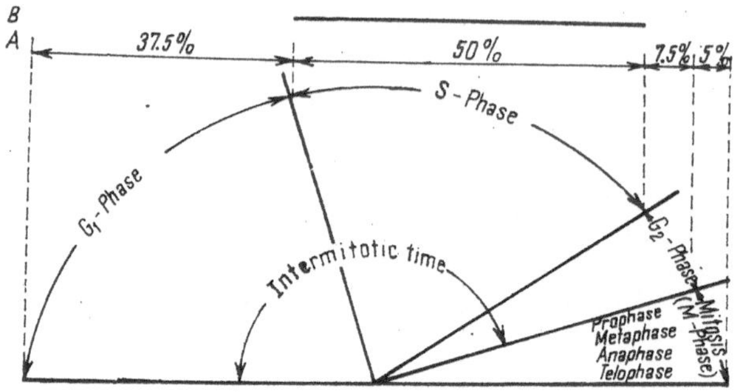

Fig. 3. — A: Phases duration (percent); B: DNA S-phase.

If we divide l depending on the cellular event of mitosis in a premitotic phase l_1 and a mitotic phase l_2 we shall call *mitotic index* the fraction of B-objects in l_2 — (or M) — phase. Going further, if a B-object of age u has probability $p(u)$ of being in M-phase, then the mitotic index — in a population with a limiting age distribution A, defined by (1.9) — is approximately $\int_0^\infty p(u)\,\mathrm{d}A(u)$. This quantity is called the limiting mitotic index (LMI).

Let us now suppose that l_1 and l_2 are random quantities independent of one another and moreover that the premitotic period is distributed as the sum of $k(k > 0)$ independent periods, each distributed like the mitotic period.

If so, we shall have the following formula for LMI (Harris (1963, p. 157)):

$$\mathrm{LMI} = \frac{m^{1/(k+1)} - 1}{m - 1}; \tag{1.10}$$

thus in case of multiplication by binary fission (Harris (1959)), LMI is $2^{1/(k+1)} - 1$. If k is large, we have $\mathrm{LMI} = 0.693/k + 1$; the factor 0.693 represents, therefore, the effect of the expansion of the population, since in a population of constant size this limiting index becomes $1/k + 1$ (see the use in Hoffman (1949), Walker (1954) and Hoffman, Metropolis and Gardiner (1956)).

1.2.5.5. Various biological time measurements (biological chronometry) should also be mentioned: life expectancy, survival time, turnover time, half-time, periodicity measures, etc. For periodical phenomena one may adopt the term "cycle", e.g., light-dark cycles, lunar developmental cycles, migratory cycles in birds, cyclic variations in population size, etc. (see SOLLBERGER (1965), BÜNNING (1967)). For the same reason, some authors consider that phases G_1, S, G_2 and M constitute the cell renewal cycle (LIPKIN (1965)). However, by observing that the duration of these phases differs even within a morphologically identical cell population (i.e., mouse epidermal cell populations), one may thus distinguish types of cell populations on the basis of temporal characteristics: G_1-population and G_2-population, with a relatively long G_2-phase (GELFANT (1963)). For more recent references, see SISKEN and MORASCA (1965) on intrapopulation kinetics of the mitotic cycle, as well as BARRET's (1966) paper.

We shall also observe the variability of another temporal characteristic of cell populations, the mean cell-doubling time, $d = \log 2\,(T_1 - T_2) / (\log n_2 - \log n_1)$, where $T_2 - T_1$ is the growth time, n_1 is the starting cell number and n_2 is the total cell number recovered at the end of the observation period. Because of continuous lengthening of d, the Ehrlich ascites carcinoma did not grow at a constant exponential rate; it follows a linear cube-root increase of cell number with time (HAUSCHKA et al. (1957)). For the variability of survival times and of the generation time see SCHACH and SCHACH (1970). The role of variable generation time in the development of a stochastic birth process was studied by KENDALL (1948 c). Cell populations with identical generation time were called "equivivants" (VON FOERSTER (1959)).

1.2.5.6. The temporal non-homogeneity of B-objects has recently raised the question of differences between *deme* and *compartment*. It was assumed (TRUCCO (1965)) that each cell generation constitutes a deme so that, formally, the birth of new cells is reduced to immigration from a preceding deme, with the 0-th generation coming from a precursor population.

The term compartment is prefered by biochemists, According to HEARON's (1963) definition, a compartment may be a physical region, perhaps bounded by a membrane, in which case it is assumed to be homogeneous and uniform. But chemical species can be formally treated as compartments with the understanding that "flow from one compartment to another" means chemical conversion. In a new assessment of these terms (FIRESCU and TAUTU (1967)), it was considered that a compartment may be defined not only as an indicator of the state of location, but also as the indicator of the change of state of a material object: the "transport" of such an object includes its *translocation* (change of position) as well as its *transfer* (change of state).

By applying these concepts to cell populations, we shall assume that phases G_1, G_2, S or M are transfer compartments, under the essential condition that all objects within a compartment should be alteridem ones. A translocation compartment may, for example, be the blood flow for red blood cells leaving the bone marrow in which they have passed through several transfer (maturation) compartments. In this way, the definition of the turnover rate as the number of cells entering or leaving a compartment per cell in the compartment, per unit time, gets a wider application. The oscillation between the "anatomical" and "dynamic" sense of the compartment may be observed from the following two examples: Studying the cell proliferation in the gastrointestinal tract of man, LIPKIN (1965) considers that the anatomical unit used to measure the number and location of (radioactive) labelled cells is the "crypt cell column" and the area within the crypt cell column in which cells proliferate is the "proliferative compartment". Yet, for LAJTHA and OLIVER (1962), a compartment is a defined volume or mass of cells with a uniform cell cycle.

The two terms, deme and compartment, may be considered equivalent if one follows in every case the rule of their homogeneous composition (alteridem objects). See LIČKO (1965) for a discussion on the definition of the compartment on the basis of the class equivalence notion.

1.2.5.7. What we presented until now stresses the need to classify cell populations in relation to factors characterizing and deciding their power of replication. We shall show here a classification according to LAJTHA and OLIVER (1962) which considers the time characteristics studied here.

(i) Type I is a cell population in which every cell divides regularly with the same time interval between divisions. Therefore populations with type I cells will be equivivant. Such a cell population is in the process of logarithmic growth (type Ia). If, however, in such a population there exists a mechanism by which a number of cells equal to the number of divisions is removed, the population will remain in a steady state (type Ib). There is, consequently, a balance between cell divisions and cell removal (e.g. for differentiation, or in other words, brought about by passage into another compartment). These are the type of populations existing in tissues with cell renewal (for rat intestinal epithelium, the mean renewal time is 1.4 days: see VON BERTALANFFY (1960).

(ii) Type II population is a cell population with a nonuniform cell cycle. It is to be assumed that there is a minimum possible period between mitoses, but that the cycle for any individual cell may have any value greater than this minimum. An t_1-phase of excessive length (e.g. several months) may represent a state of "dormancy". In type IIa population the nonuniformity is considered to be anatomical (the cells form a compartment with poor nutrient supply, for example); in type IIb population, the nonuniformity is a result of a balance between feedback system controlling compartment size and a mechanism for random removal of cells from the compartment. The time characteristics of these populations change. In tumors experimentally cultivated *in vivo* (type IIa), the cell cycle time and the population doubling time are not equivalent terms; in regenerating organs, after removal (type

IIb), the cell population shows changes in mitotic index and a shortening of the population turnover time.

(iii) Type III population is not a self-maintaining cell population. Type IIIa populations are considered to have no precursor compartment (e.g. adult central nervous system or adult ovary), and, therefore, are incapable of proper regeneration. This is the type of population existing in tissues described by VON BERTALANFFY "with no cell addition", to which, besides the two above mentioned examples, the adrenal medulla is added. Type III populations depend for their proper maintenance on a precursor compartment of type Ib (or, as is more likely in the case of bone marrow cells, of type IIb), from which regeneration—if and when demanded — may start. VON BERTALANFFY considers the type of tissues "with cell addition but no renewal" (e.g. liver, kidney, thyroid, etc.).

These three types of cell populations react differently to some physiological or pathological agents (see details in LAJTHA and OLIVER (1962) for response to irradiation).

1.2.6. Characteristic distributions for some biological populations

1.2.6.1. This paragraph is based primarily on BARTLETT's (1969) paper on distributions associated with cell populations. Although our book does not deal specifically with the statistical problems of stochastic processes, we consider it absolutely necessary to present some particular aspects related to the structure and behaviour of B-object populations. Modern research works on cell kinetics (concerning growth, division and synchronisation) largely carried out with radioactive isotopes as well as radiobiological researches, stress the need to know the phases of the multiplication process and implicitly, the age of the studied population. However, in many cases one must consider the capacity of object discrimination, therefore, the homogeneity of the population. SAMPFORD (1965) has shown that cells not susceptible to breakage by irradiation cannot be distinguished from susceptibles in which no breaks occur and because of this, the distribution of breaks in irradiated cells which are at a particular stage of the mitotic cycle is a truncated negative binomial (Pascal) distribution.

1.2.6.2. However, a good introduction into this question would be RAO's (1965) paper on discrete distributions encountered in certain situations with incomplete information. When an investigator collects a sample of observations produced by nature, according to a certain model, the original distribution may not be reproduced due to various reasons. The author refers to the following situations:

(i) *Non-observability of events.* Certain types of events, may (although they occur) be unascertainable and the observed distribution would therefore be *truncated.* An edifying example is given above by Sampford's experiment. Rao's example is the frequency of families with both parents heterozygous for albinism and having no albino children. There is no evidence that the parents are both heterozygous unless at least one albino is born and the families of parents having no albino children get confounded with normal families. The actual frequency of the event "zero albino children" is thus not ascertainable (as well as the event "zero susceptibility" in case of irradiated cells).

(ii) *Partial destructions of observations.* Observations of chance phenomena (such as number of eggs layed, number of accidents, etc.) may be partially destroyed or may be only partially ascertained. In such a case the original distribution will be distorted.

(iii) *Samples with unequal chances of observations.* Such situations may be found in genetics (also pharmacogenetics). For instance, if we wish to study the distribution of albino children in families capable of producing such children, we may contact a large number of families and ascertain the number of albinos from each family. The families with at least one albino child provide a truncated binomial distribution, and the probability of a child being an albino can be estimated from such a distribution. Yet this method of investigation is wasteful since the proportion of abnormal families is generally small and the actual number of families providing the information desired will be small unless a very large number of families is investigated. A convenient method in such a case is first to discover an albino child and through it obtain information about the family to which it belongs. Without doubt this mechanism of observation — often encountered in laboratories and clinics — may not give *equal chances* to all subgroups in which the event occurred. The chance for a family with a albinos is that of detecting at least one of its albino children, which may be a function of a. Here is a similar example in pharmacogenetics: the analysis of sleeping time after hexobarbital administration demonstrated differences among some genetic strains of mice (Jay (1965)). Moreover, male mice sleep longer than females, and this sex difference is maximal in inbred strains and diminishes as the strains are outbred. Only in small populations, with a homogeneous structure, does the phenomenon have equal chances of appearance.

The second difficulty in determining a real distribution was studied by Sethuraman (1965) for continuous distributions. The general case is the following: when the statistician starts to measure ξ, it is possible that an independent random variable η, with a continuous distribution function, may influence the experiment, and he is forced to observe a new random variable ζ, which is min (ξ, η). We assume that the statistician can also examine his experiment and determine whether $\zeta = \xi$ (i.e., the observation is an undamaged one) or whether $\zeta = \eta$ (i.e., the observation is a damaged one). The case $\xi = \eta = \zeta$ is excluded by the assumption that ξ and η have continuous distribution function. Let us suppose that ξ is the life span of a bacterial cell. What the scientist sometimes observes when the bacterium is destroyed can actually be $\zeta = \min(\xi, \eta)$, where η is the duration of the action of a substance having a bacteriostatic role or the effect of lysogeny (latent infection with a temperate bacteriophage), when a small number of bacterial cells always suffers spontaneous lysis.

It may sometimes happen that distributions obtained in various situations should have the same form. This fact was observed by NEYMAN (1965) while investigating the Pólya scheme of contagion and the mixture scheme on Greenwood and Yule concerning accident proneness. The conclusion is that different chance mechanisms may produce identical distributions of certain random variables (case of non-identifiability) and, therefore, the study of the distribution of a particular random variable (possibly a vector) need not provide the crucial information about the underlying chance mechanism.

1.2.6.3. Considering the autoradiographic grain count distributions as compound distributions, BARTLETT (1969) differentiates two broad classes: (i) distribution mixtures of a few (possibly only two) components, and (ii) compound distributions where the components are too numerous to be specified separately.

We shall deal here very briefly with the second class of distributions. The classical example of compound distribution is a compound Poisson distribution with variation in the Poisson components specified by a moment generating function $M(\theta)$ for the distribution of means m. Then if $G(z)$ is the probability generating function of the discrete compound distribution, we have

$$G(z) = M(z - 1). \qquad (1.11)$$

The best known case of (1.11) is

$$M(\theta) = (1 - \alpha\theta)^{-\nu},$$

whence

$$G(z) = [1 - \alpha(z - 1)]^{-\nu},$$

corresponding to a negative binomial distribution with mean $\alpha\nu$ and variance $\alpha(\alpha + 1)$. The second theoretical distribution noted by M. BARTLETT is the convolution of a Poisson and geometric distribution checked on experimental data of thymidine incorporation into mouse DNA, obtained by STEWART et al. (1965).

We shall stress on this occasion the fact that modern techniques of cytophotometry or automatic analysis often raise complicated problems of statistical interpretation. M. BARTLETT mentions some of the biological factors that may influence the type of the distributions: variability of cells in S-phase in their incorporation of thymidine due to their different stages of DNA development, halving of grain-count on division, or doubling of grain-count due to overlap of two cells. In the case of automatic autoradiographic grain counting one must not neglect error sources of the system: grain-images too small to quantize, cellular constituents counted as grains or background grain count (PRESTON and NORGREN (1969)).

1.2.6.4. Many works dealt with the utilization of the negative binomial distribution in counting processes such as, for example, the number of surviving insects after applying pesticides (ANSCOMBE (1949)). KATTI and GURLAND (1961) have shown that the Neyman type A, the negative binomial and the Poisson binomial distributions, which combine two of the elementary distributions through the processes of compounding and generalizing have been fitted with varying degrees of success to data from a number of biological populations. SKELLAM (1958) analyzed the applicability of Neyman's type A distribution in the study of population dispersion on an areal. It was supposed that a number of "centres" with B-objects were distributed at random in a large field, that each centre gave rise to a number of offspring and that the offspring from any centre were distributed in the space around that centre independently of one another. In the simplest case, each centre gives rise to a single compact "cluster" of offspring located at an arbitrary finite distance from the centre of origin. But the migration of the new descendants around their centre is not always independent: a group of lepidopterous larvae arising from an egg batch could persist as a gregarious band.

Starting from the same facts, S. K. KATTI and J. GURLAND suppose that the egg masses in a field have a Poisson distribution with probability generating function exp $(\lambda(z-1))$ and that the survivors within an egg mass have a negative binomial distribution with probability generating function $(q-pz)^k$, $p>0$, $q=1-p$, $k>0$. For the distribution of the survivors in the field, they thus obtain the Poisson Pascal distribution with probability generating function exp $(\lambda[q-pz]^{-k}-1)$. See also KHATRI and PATEL (1961).

1.2.6.5. HOLGATE (1969) deals with the distribution of species abundance. The term "abundance" of a given species is used to express the expected number of B-objects of that species caught with the expenditure of unit effort. The biological populations considered are ecological groups or ecosystems such as, for example: permeant ecological groups (a vast population of highly mobile insects) or pelagial ecosystems, e.g., the nekton (*nektós* = swimmer), the system of species actively swimming in a large homogeneous ocean. The method used to measure abundance (e.g. light trap for insects, random trawl for fish) must not affect the structure of the system. The problem of the distribution and the abundance of animals is considered by ANDREWARTHA and BIRCH in their textbook (1954), as a fundamental problem of animal ecology.

The relation between the number of species and the number of individuals in a random sample, studied in 1943 by R. A. FISHER together with A. S. CORBET and C. B WILLIAMS, led to the introduction of the logarithmic series distribution. R. A. FISHER supposed the number of individual members of a species present in a sample has a Poisson distribution with mean m, and that m has a gamma-type distribution. The log distribution implies that the number of species of which n members have been caught is proportional to c^n/n, where c is a positive constant ($c < 1$). This constant depends on the time spent in sampling and the volume of the sample (e.g. the size of the light trap). The coefficient of proportionality is called the index of diversity.

If we admit that the set of species is infinite, the set of abundances can be replaced by a continuous density, often by Pearson type III density function. Otherwise one prefers a log normal density, as shown in HOLGATE's (1969) paper, or a discrete (Poisson) log normal (see ANSCOMBE (1950) and CASSIE (1962)). For this mixed distribution see HAIGHT (1967, Ch. 3) and for applications to ecology, WILLIAMS (1944) and J. H. DARWIN (1960). Many studies on the statistical measurement of spatial patterns have been recently published, we refer the reader to BOSWELL and PATIL (1970 a, b) and STITELER (1970).

One may thus generally affirm that the basic process for which the log normal distribution is suitable is a *comminution process*; others for which it is suitable are some modes of a growth process, as suggested by some works of biorheology (BRAZEE et al. (1969)). The idea is the same as that in KENDALL's (1948 b) work where he studies population growth and immigration models.

1.2.6.6. We conclude this paragraph by calling attention to the interest existing at present for mixtures of distributions. The general case is the following: in a population of B-objects certain characteristics are measured for each object separately. However, many characteristics vary markedly with the age of the B-object. Then the trait has a distinct distribution for each age group so that the whole population has a mixture of distributions (CHOI and BULGREN (1968)). One should mention in this field TEICHER's recent works (starting in 1960).

2. Population growth models

Under the term of models of *homoblasty* (*homoios* — like, *blastos* — bud) we mean stochastic models describing the growth of populations with alteridem (one-type) B-objects. We used a term of biological origin for the simplest type of population process called by MOYAL (1962 a) the finite univariate population process, where there is only one kind of individual and where the total size of the population is finite with probability unity.

However, we must specify that the indiscernability of objects is relative and that within a general theory of stochastic population processes this indiscernability is a limiting case. In fact, a biological population consists of objects characterized by their age, weight, location, etc. as a population of stars is characterized by brightness, mass, position, velocity, etc. Nevertheless, we shall try to present in the first part of this section models of homoblasty, in the above defined sense.

2.1. Stochastic population processes

2.1.1. Point processes as models of stochastic populations

2.1.1.1. The foundations of a general theory of stochastic population processes were laid by MOYAL (1962 a) to whom we refer the reader for details and proofs (for references to earlier papers see BARTLETT (1966)).

By *population* we shall mean a collection of objects, each of which may be found in any one state x of a fixed set X (the *individual state space*). If the B-objects are *distinguishable* from each other, then a *state of the population* is defined as an *ordered set* $x^n = (x_1, \ldots, x_n)$ of individual states: i.e., it is the state where the population has n objects, the ith being in the individual state x_i, $1 \leqslant i \leqslant n$. The *population state space* $\mathfrak{X}$ is the class of all such states x^n, $n \in N$. Conventionally, x^0 denotes the state where the population is empty.

If the B-objects are *indistinguishable* from each other, then a state of the population is defined as an *unordered set* $x^{(n)}=[x_1, \ldots, x_n]$ of individual states, i.e., a state where the population has n objects with one each in the states $x_1, \ldots, x_n$. The population state space $\mathfrak{X}$ is now the set of all such $x^{(n)}$, $n \in N$, x^0 denoting again an empty population.

A triplet $(\mathfrak{X}, \mathcal{B}, \mathsf{P})$, where $\mathfrak{X}$ is a population state space, $\mathcal{B}$ a σ-field of sets in $\mathfrak{X}$ and P a probability on $\mathcal{B}$ constitutes a model of a stochastic population and will be called a (single-variate) *point process*.

If we wish to differentiate between the two cases of distinguishable and indistinguishable objects, we shall denote the population state space in the latter case by $\mathfrak{X}_a$, the set of all points $x^{(n)}$ by $X^{(n)}$, subsets of $X^{(n)}$ by $A^{(n)}$, and so on. Let π be the permutation $i_1, \ldots, i_n$ of $1, \ldots, n$; let $A_\pi^{(n)}$ be the subset of X^n (= the Cartesian product $X \times \ldots \times X$, n times) obtained from $A^{(n)}$ by the permutation $(x_{i_1}, \ldots, x_{i_n})$ of the *coordinates* $x_1, \ldots, x_n$ of each $x^{(n)}$ in $A^{(n)}$. A in $\mathfrak{X}$ is *symmetric* if $A^n = A \cap X^n = A_\pi^{(n)}$ for all permutations π and $n \in N^*$. The *symmetrization* of A is the symmetric set $\bigcup_{n \in N^*} \bigcup_{\pi} A_\pi^{(n)}$.

The relation between the two cases of distinguishable and indistinguishable objects is that, given X, we have the transformation $\mathfrak{T}$ from $\mathfrak{X}$ into $\mathfrak{X}_a$ which maps every x^n into $x^{(n)}$. Clearly, the inverse image $\mathfrak{T}^{-1}[x^{(n)}] = \bigcup_{\pi} x_\pi^{(n)}$, and hence there is a one-to-one correspondence between subsets of $\mathfrak{X}_a$ and *symmetric* subsets of $\mathfrak{X}$; thus we can *identify* sets of states in a population of indistinguishable objects with symmetric subsets of the state space $\mathfrak{X}$ of a population with distinguishable objects.

2.1.1.2. Let $\mathcal{B}_X$ be a σ-field of sets in X; the pair $(X, \mathcal{B}_X)$ is called the *individual measure space*. Let $\mathcal{B}_X^n$ be the minimal σ-field of sets in X^n containing all product sets $A_1 \times \ldots \times A_n$ such that $A_i \in \mathcal{B}_X$, $1 \leqslant i \leqslant n$, and let $\mathcal{B}$ be the class of all sets A in $\mathfrak{X}$, such that $A^n \in \mathcal{B}_X^n$, $n \in N$. It is easily proved that $\mathcal{B}$ is the minimal σ-field of sets in $\mathfrak{X}$ containing all sets $A^n \in \mathcal{B}_X^n$, $n \in N$.

The pair $(\mathfrak{X}, \mathcal{B})$ is called the *population measure space*. In the case of a population with indistinguishable objects, starting from $(X, \mathcal{B}_X)$, we define $(\mathfrak{X}, \mathcal{B})$ as above, and then define the σ-field $\mathcal{B}_a$ to be the class of all sets A in $\mathfrak{X}_a$, such that $\mathfrak{T}^{-1} A \in \mathcal{B}$ thus $\mathcal{B}_a$ may be identified with the σ-field of all symmetric sets of $\mathcal{B}$.

Next, as to the probability P on $\mathcal{B}$, let P_n be a measure on $\mathcal{B}_X^n$, $p_n = P_n(X^n)$ being the probability that the size of the population is n, and $\sum_{n \in N} p_n = 1$. It can be shown that the function P on $\mathcal{B}$ whose value at $A = \sum_{n \in N} A^n$ is

$$\mathsf{P}(A) = \sum_{n \in N} P_n(A^n) \tag{2.1}$$

is the unique probability on $\mathcal{B}$ whose restriction to $\mathcal{B}_X^n$ agrees with P_n for all $n \in N$.

In the case of indistinguishable objects, a probability P on $\mathcal{B}$ determines a unique probability distribution P_a on $\mathcal{B}_a$ such that for every $A \in \mathcal{B}_a$,

$$\mathsf{P}_a(A) = \mathsf{P}(\mathfrak{T}^{-1}A). \tag{2.2}$$

It can be shown that a probability P_a on $\mathcal{B}_a$ determines a unique symmetric probability P on $\mathcal{B}$ to which it is related by (2.2). By symmetric probability we mean a distribution on $\mathcal{B}$ which is invariant under coordinate permutations.

Thus we can identify a distribution on $\mathcal{B}_a$ with the corresponding symmetric distribution on $\mathcal{B}$, and one can use the same symbol P_a for both. It follows that a point process $(\mathfrak{X}, \mathcal{B}, \mathsf{P}_a)$ with distinguishable B-objects and a symmetric distribution, and the corresponding process $(\mathfrak{X}_a, \mathcal{B}_a, \mathsf{P}_a)$ with indistinguishable B-objects are the same thing for all intents and purposes.

2.1.1.3. There is an alternative method of characterizing stochastic populations, namely that of assigning to subsets A of X the number of individuals which are in state $x \in A$. Clearly, if in a population of total size n there is one object in each of the states $x_1, \ldots, x_n$, then the number of individuals with states in a given arbitrary subset A of X is given by the expression

$$N(A \mid x^{(n)}) = \sum_{i=1}^{n} \delta_{x_i}(A), \tag{2.3}$$

where $\delta_\cdot(A)$ is the characteristic function of the set A, $\delta_x(A) = 1$ iff $x \in A$ and zero otherwise. For each fixed $A \subset X$, $N(A \mid \cdot)$ is a function on the symmetric population state space $\mathfrak{X}_a$, while for each fixed $x^{(n)} \in \mathfrak{X}_a$, $N(\cdot \mid x^{(n)})$ is a function on the class $\mathcal{P}(X)$ of all subsets of X. Each such function on $\mathcal{P}(X)$ will have the following properties: it is (i) non-negative, (ii) finite, (iii) integral-valued, and (iv) completely additive, in the sense that if $(A_t)_{t \in \Theta}$ is an arbitrary indexed collection of mutually disjoint sets in X, then at most a finite number of sets, say $A_{t_1}, \ldots, A_{t_n}$ can be such that $N(A_t) \geqslant 1$ and $N(\sum_{t \in \Theta} A_t) = \sum_{i=1}^{n} N(A_{t_i})$.

The converse is also true: call *counting measure* a function on $\mathcal{P}(X)$ which has the properties (i) — (iv) above, and let $\mathfrak{N}$ be the *class of all counting measures* on $\mathcal{P}(X)$. Then, it can be proved that relation (2.3) defines a one-to-one correspondence between $\mathfrak{X}_a$ and $\mathfrak{N}$; let us denote this mapping by U.

If we are now given the symmetric point process $(\mathfrak{X}_a, \mathcal{B}_a, \mathsf{P}_a)$, let $\mathcal{B}_{\mathfrak{N}} = U^{-1}\mathcal{B}_a$, i.e., $\mathcal{B}_{\mathfrak{N}}$ is the σ-field of all sets of functions $N \in \mathfrak{N}$ whose inverse image belongs to $\mathcal{B}_a$. Let $\mathsf{P}_{\mathfrak{N}}$ be the probability on $\mathcal{B}_{\mathfrak{N}}$ such that $\mathsf{P}_{\mathfrak{N}}(A) = \mathsf{P}_a(U^{-1}A)$ for every $A \in \mathcal{B}_{\mathfrak{N}}$. Then, the triplet $(\mathfrak{N}, \mathcal{B}_{\mathfrak{N}}, \mathsf{P}_{\mathfrak{N}})$ denoted briefly by **N** is a probability space which is called a *counting process*. In other words U defines a one-to-one measure-preserving transformation from $\mathcal{B}_a$ onto $\mathcal{B}_{\mathfrak{N}}$.

2.1.1.4. We shall now define a stochastic population process. Let $(\Omega, \mathcal{B}_\Omega, \mathsf{P}_\Omega)$ be a given probability space, let $(X_t, \mathcal{B}_{X_t})_{t \in T}$ be an indexed family of individual measure spaces and $(\mathfrak{X}_t, \mathcal{B}_t)_{t \in T}$ the associated family of population measure spaces. Suppose that for each $t \in T$ we are given a random variable $x(t, \omega)$ on Ω taking values in $\mathfrak{X}_t$. Let $\mathfrak{X}_T = \prod_{t \in T} \mathfrak{X}_t$ and let $\mathcal{B}_T = \prod_{t \in T} \mathcal{B}_t$. Then the measurable transformation $\omega \to x(t, \omega)$ yields a probability P_T on $\mathcal{B}_T$ and we may call $(\mathfrak{X}_T, \mathcal{B}_T, \mathsf{P}_T)$ a *stochastic population process* (in the point process formulation).

For each $t \in T$ and $\omega \in \Omega$, the transformation (2.3) defines a counting measure $N(\cdot\,; \omega)$ on $\mathcal{B}_{\mathfrak{X}_t}$. Let $(\mathfrak{N}_t, \mathcal{B}_{\mathfrak{N}_t})_{t \in T}$ be the family of counting process measure spaces generated in this way; the transformation then yields the stochastic population process $(\mathfrak{N}_T, \mathcal{B}_{\mathfrak{N}_T}, \mathsf{P}_{\mathfrak{N}_T})$ in the counting process formulation, where $\mathfrak{N}_T = \prod_{t \in T} \mathfrak{N}_t$ and $\mathcal{B}_{\mathfrak{N}_T} = \prod_{t \in T} \mathcal{B}_{\mathfrak{N}_t}$.

2.1.1.5. Of the examples of population processes given by MOYAL (1962 a) we mention the multiplicative population processes (equivalent to branching processes) (see also MOYAL (1962 b, 1964)) and the cluster processes. A cluster process is a population process in which the individuals are grouped in independent clusters. Each cluster is itself a population whose state is characterized by the ordered pair (x, y^n) where x is characteristic of the cluster as a whole and y^n means that the cluster contains n individuals in states $y_1, \ldots, y_n$.

2.1.1.6. In Paragraph 2.1.2 we shall deal with cases in which the individual state space consists of a single point. Consequently, the only variable of interest will be the size of the population. Paragraph 2.1.4 will be concerned with another extreme situation, the individual state space consisting of the real line.

2.1.2. Homogeneous birth-and-death processes

2.1.2.1. Birth-and-death processes are frequently used as models of the growth of biological populations. They were presented in I; 2.3.4.7 within the framework of discrete state space homogeneous Markov jump processes.

2.1.2.2. *The Yule-Furry*[2] *process* was introduced in I; 2.3.4.7.6. It is the linear pure birth (or "escalator" process) with birth rates $b_i = bi, i \geq 1$.

G.U. YULE studied this type of process in 1924, starting from the assumption that if the number of species existing at time t is i, then the probability that a new species originate by a mutation in the time interval $(t, t + \Delta t)$ equals $bi\,\Delta\,t$. The creation of new species at a rate proportional to the number of existing species and independent of their ages may appear somewhat plausible if we suppose that (i) a species rather quickly reaches a maximum size determined by its environment and thereafter does not grow in size, and (ii) different species interfere only slightly with one another. These assumptions are based on Darwinist ideas: DARWIN specified that the process of species modification "will generally affect only a few species at the same time; for the variability of each species is quite independent of that of all others" (1869, p. 386).

At the same time, G.U. YULE dealt also with a particularly interesting problem from the biological point of view, i.e., the estimation of some parameters of the process under conditions of incomplete information from the remote past (see HARRIS (1963, p. 195)).

W. H. FURRY used the same model independently in 1937 in order to describe the multiplication of the particles in cosmic-ray showers.

2.1.2.3. *The multiple-phase birth process* will be presented in Paragraph 2.3.1.

2.1.2.4. *The Feller-Arley process* was introduced in I; 2.3.4.7.5. It coincides with the linear birth-and-death process, the most familiar homogeneous Markov branching process. FELLER (1939) proposed the model as an extension of deterministic models for the growth of biological populations. In 1943, in an unpublished paper, C. PALM first gave the probabilities explicitly. ARLEY (1943) treated the nonhomogeneous case $b =$ const., $d =$ const. $\times\ t$, where the increase of d with t represents the declining energy of a cosmic-ray particle as it suffers collisions. It would seem that the biological aging process could be treated in a similar manner.

2.1.2.5. *The Kendall process* is a birth-and-death process, for which $b_i = bi + \nu$, $d_i = di$, $i \in N$, where $\nu > 0$ is to be interpreted as the immigration rate per object from another population (or another region). If $b = d = 0$, the Kendall process reduces to the Poisson process. When $b = 0$, $d \neq 0$, we obtain the *immigration-death process* (or the *Chandrasekar-Rotschild process*). It can represent the fluctuations in

[2] We consider it legitimate to add A. G. MC KENDRICK's name, who first discussed the model in a remarkably early paper (1914) and who presented more flexible models of the same type in 1926.

the number of particles in a small volume under direct observation, entries and exits being independent of one another (see, e.g., Cox and MILLER (1965, p. 168)). Immigration from the surrounding medium may be regarded as a homogeneous Poisson process with parameter ν, while death (or emigration) is taken to have a rate proportional to the population size. This representation has been used in the study of colloidal particles in suspension (CHANDRASEKAR (1943)) and to investigate the movements of spermatozoa (ROTSCHILD (1953)).

If $b \neq 0$, $d = 0$, the Kendall process reduces to the so-called *contagion process* (see BARTLETT (1966, p. 57)).

2.1.3. A random walk example

2.1.3.1. When problems which essentially involve counting individuals themselves are posed, it is difficult or impossible to proceed by an analysis of the continuous-time situation, say by means of a Feller-Arley process $(\xi(t))_{t\geqslant 0}$. The answer to these problems can be found, as shown by SENETA (1967 a), by utilizing the imbedded Markov chain procedure, which reduces these problems to equivalent ones for a Bernoulli (discrete) random walk with absorbing barrier at the origin.

The corresponding discrete random walk $(\eta_k)_{k\in N}$ can be described by the following scheme, where by "transition" we mean a birth-death event:

Transition	Probability	
$i \to i+1$	$b/(b+d) = p$	$\left.\right\} i \in N^*$
$i \to i-1$	$d/(b+d) = q$	
$0 \to 0$	1	

See I; 2.3.4.6.3). The time units between transition are intervals between birth-death events.

Using $(\eta_k)_{k\in N}$ we can deduce the distribution of the maximum value of the population,

$$S(\xi) = \max_{0\leqslant t<\infty} \xi(t) = \max_{k\in N} \eta_k .$$

We have only to notice that the probability

$$\mathsf{P}(S(\xi) \geqslant l \mid \xi(0) = i)$$

equals the absorption probability $\alpha_i(\{l\})$ at l for the discrete random walk, starting in i, with an (imagined) absorbing barrier at $l \geqslant i$. According to I; 1.2.1.3.3 we have

$$\alpha_i(\{l\}) = \begin{cases} \dfrac{1-(q/p)^i}{1-(q/p)^l} & \text{if} \quad p \neq q \text{ (i.e. } d \neq b) \\ \\ \dfrac{i}{l} & \text{if} \quad p = q \end{cases} \tag{2.4}$$

Note that if we are concerned with results conditional on eventual extinction, it is sufficient to treat only the case $q \geqslant p$, for which the condition is automatically satisfied (see I; 1.2.1.3.2). The reason for this is that if $p > q$, in which case extinction is no longer certain, results conditional on extinction occurring may be obtained by substituting p for q in the results for $q > p$ (see e.g. WAUGH (1958)). Thus if $p > q$,

$$\mathsf{P}(S(\xi) \geqslant l \mid S(\xi) < \infty, \quad \xi(0) = i) = \frac{1-(p/q)^i}{1-(p/q)^l}.$$

2.1.3.2. In the case $q \geqslant p$, the search for the mean number of birth-death events occurring up to attainment of a maximum becomes, in random-walk language, the search for the mean number of steps required for the corresponding random walk to attain its maximum position before absorption[3].

Let

$$\theta(\eta) = \min\{n : \eta_n = S(\xi)\}.$$

Then we have, for $1 \leqslant s \leqslant l$, $1 \leqslant i < l$,

$$\mathsf{P}(\theta(\eta) \geqslant s, \quad S(\xi) = l \mid \eta_0 = i) =$$

$$= \mathsf{P}(\max_{0 \leqslant n < s} \eta_n < l, \ \max_{n \geqslant s} \eta_n = l \mid \eta_0 = i) =$$

$$= \sum_{r=1}^{l-1} \mathsf{P}(\max_{0 \leqslant n < s} \eta_n < l, \ \eta_{s-1} = r, \ \max_{n \geqslant s} \eta_n = l \mid \eta_0 = i) =$$

$$= \sum_{r=1}^{l-1} \mathsf{P}(\max_{n \geqslant s} \eta_n = l \mid \eta_{s-1} = r)\, \mathsf{P}(\eta_{s-1} = r, \ \max_{0 \leqslant n < s} \eta_n < l \mid \eta_0 = i) =$$

$$= \sum_{r=1}^{l-1} \mathsf{P}(S(\xi) = l \mid \eta_0 = r)\, \mathsf{P}(\eta_{s-1} = r, \ \max_{0 \leqslant n < s} \eta_n < l \mid \eta_0 = i).$$

[3] For a direct derivation see URBANIK (1956).

Now, for $s \in N, n \geqslant i,\ l \geqslant i,$

$$\mathsf{P}(\theta(\eta) \geqslant s | S(\xi) \leqslant n,\ \eta_0 = i) = \sum_{l=i}^{n} \frac{\mathsf{P}(\theta(\eta) \geqslant s,\ S(\xi) = l | \eta_0 = i)}{\mathsf{P}(S(\xi) \leqslant n | \eta_0 = i)} =$$

$$= \begin{cases} 1, & \text{if } s = 0, \\ \dfrac{\sum_{l=i}^{n} \sum_{r=1}^{l-1} \mathsf{P}(S(\xi) = l | \eta_0 = r) \mathsf{P}(\eta_{s-1} = r,\ \max_{0 \leqslant n < s} \eta_n < l | \eta_0 = i)}{\mathsf{P}(S(\xi) \leqslant n | \eta_0 = i)}, & \text{if } s \in N^* \end{cases} \tag{2.5}$$

from the above.

The only unknown quantity in this expression is

$$\mathsf{P}(\eta_{s-1} = r,\ \max_{0 \leqslant n < s} \eta_n < l | \eta_0 = i),$$

for $l \geqslant i + 1, s \geqslant 1, r \leqslant l - 1$. One may obtain this as follows: for $s \geqslant 2, 0 \leqslant h \leqslant s - 2,$

$$\mathsf{P}(\eta_{s-1} = r,\ \eta_h = l,\ \max_{h+1 \leqslant n \leqslant s-1} \eta_n < l | \eta_0 = i) =$$

$$= \mathsf{P}(\eta_{s-h-2} = r,\ \max_{0 \leqslant n \leqslant s-h-2} \eta_n < l | \eta_0 = l - 1) \times$$

$$\times\ \mathsf{P}(\eta_{h+1} = l - 1 | \eta_h = l) \mathsf{P}(\eta_h = l | \eta_0 = i) =$$

$$= q p_{il}^{(h)} \mathsf{P}(\eta_{s-h-2} = r,\ \max_{0 \leqslant n \leqslant s-h-2} \eta_n < l | \eta_0 = l - 1).$$

Thus,

$$\mathsf{P}(\eta_{s-1} = r,\ \max_{0 \leqslant n < s} \eta_n < l | \eta_0 = i) =$$

$$= p_{ir}^{(s-1)} - q \sum_{h=0}^{s-2} \mathsf{P}(\eta_{s-h-2} = r,\ \max_{0 \leqslant n \leqslant s-h-2} \eta_n < l | \eta_0 = l - 1) p_{il}^{(h)},$$

where the sum is to be replaced by 0 if $s = 1$. Summing over $s \in N$, we have after some arrangements:

$$\sum_{s \in N} \mathsf{P}(\eta_s = r,\ \max_{0 \leqslant n \leqslant s} \eta_n < l | \eta_0 = i) =$$

$$= \sum_{s \in N} p_{ir}^{(s)} - q \Big[\sum_{n \in N} p_{il}^{(n)}\Big]\Big[\sum_{s \in N} \mathsf{P}(\eta_s = r,\ \max_{0 \leqslant n \leqslant s} \eta_n < l | \eta_0 = l - 1)\Big] =$$

$$= \sum_{n \in N} p_{ir}^{(n)} - \frac{q \sum_{n \in N} p_{il}^{(n)} \sum_{n \in N} p_{l-1,r}^{(n)}}{1 + q \sum_{n \in N} p_{l-1,r}^{(n)}}.$$

2.1.3.3. Put

$$m_{lj} = \sum_{n \in N} p_{lj}^{(n)} .$$

Clearly, m_{lj} is the mean time spent in state j before absorption, tarting from state l. It is known (see BARNETT (1964)), that

$$m_{lj} = \begin{cases} \dfrac{(p/q)^j - 1}{p - q} & \text{if} \quad j \leqslant l \\[2ex] \dfrac{(p/q)^j - (p/q)^{j-l}}{p - q} & \text{if} \quad j \geqslant l \end{cases}$$

when $q > p$, and when $q = p = 1/2$,

$$m_{lj} = \begin{cases} 2j & \text{if} \quad j \leqslant l \\ 2l & \text{if} \quad j \geqslant l. \end{cases}$$

We deduce for $i \leqslant r \leqslant l - 1$, that

$$\sum_{s \in N} \mathsf{P}(\eta_s = r, \max_{0 \leqslant n \leqslant s} \eta_n < l \mid \eta_0 = i) = \begin{cases} \dfrac{(\alpha^i - 1)(\alpha^{l-r} - 1)}{(q - p)(\alpha^l - 1)} & \text{if } \alpha = \dfrac{q}{p} > 1 \\[2ex] \dfrac{2i(l - r)}{l} & \text{if } \alpha = 1. \end{cases}$$

Since

$$\sum_{s \in N} \mathsf{P}(\theta(\eta) \geqslant s \mid S(\xi) \leqslant n, \eta_0 = i) = \mathsf{E}[\theta(\eta) \mid S(\xi) \leqslant n, \eta_0 = i],$$

taking $i = 1$ for convenience, for $n \geqslant 2$, we obtain finally

$$\mathsf{E}[\theta(\eta) \mid S(\xi) \leqslant n, \eta_0 = 1] =$$

$$= \begin{cases} \dfrac{(1 - \alpha)^2 (1 - \alpha^{n+1})}{(\alpha - \alpha^{n+1})(q - p)} \displaystyle\sum_{l=2}^{n} \sum_{r=1}^{l-1} \dfrac{\alpha^l (1 - \alpha^r)(\alpha^{l-r} - 1)}{(1 - \alpha^{l+1})(1 - \alpha^l)^2}, & \alpha > 1 \\[2ex] \dfrac{2(n + 1)}{n} \displaystyle\sum_{l=2}^{n} \dfrac{l - 1}{6l}, & \alpha = 1. \end{cases} \tag{2.6}$$

It follows that

$$\mathsf{E}[\theta(\eta)\,|\,\eta_0 = 1] =$$

$$= \begin{cases} \dfrac{(1-\alpha)^2(d+b)}{d-b} \displaystyle\sum_{l\geqslant 2}\sum_{r=1}^{l-1} \dfrac{\alpha^l - (1-\alpha^r)(\alpha^{l-r}-1)}{(1-\alpha^{l+1})(1-\alpha^l)^2}, & \text{f} \quad \alpha > 1 \\ \infty, \quad \text{if} \quad \alpha = 1. & \end{cases}$$

2.1.4. Birth, death and diffusion processes

2.1.4.1. Continuing the series of models built by SKELLAM (1951), NEYMAN and SCOTT (1957) and SEVAST'JANOV (1958), S. R. ADKE and J. E. MOYAL presented in 1963 a simple model of a stochastic population diffusing in space. The population considered is formed by alteridem B-objects moving without restrictions in a linear biotope. The fundamental assumptions regard the behaviour of each B-object as well as the structure of the process. Thus, let us suppose that each such object diffuses independently of all the others and that the diffusion is a homogeneous Wiener process: if the object is at the point x at any given time, then after a time interval t, and conditioned on his survival in that interval, its position is specified by the probability density

$$p(t; x, y) = [2\pi\sigma^2 t]^{-1/2} \exp\left(-\frac{(y-x)^2}{2\sigma^2 t}\right),$$

where $\sigma^2/2$ is the constant coefficient of diffusion. Without any loss of generality, one can take $\sigma = 1$.

Assume also constant birth and death rates: the probability for a B-object at position x at time t to give birth to another object at the same position x in a small time interval Δt, is $b\Delta t$ ($b =$ const.), and the probability for a B-object to die at the same position x in the time interval $(t, t+\Delta t)$, is $d\Delta t$ ($d =$ const.). The probability of more than one birth and/or death in the time interval $(t, t+\Delta t)$ is $o(\Delta t)$.

2.1.4.2. The model as described above forms a multiplicative point process (see Subparagraph 2.1.1.5). The population state space $\mathfrak{X} = \sum_{n\in N} R^{(n)}$ is the set of all unordered sets $x^{(n)} = [x_1, \ldots, x_n]$, $n \in N^*$ of real numbers. Here, $R^{(0)}$ denotes conventionally the state of an extinct population.

The assumptions we have made on the model imply that the process is a multiplicative homogeneous Markov process: its temporal evolution, given indistinguishability, is completely specified by the symmetric (i.e. invariant under coordinate permutations; see Subparagraph 2.1.1.2) transition density function $p(t; x^k, y^n)$, $k, n \in N^*$, of a transition from x^k to y^n in the time interval t. Since the process is multiplicative, $p(t; x^k, \cdot)$ is also multiplicative, i.e., it is obtained by compounding the independent transition probabilities $p(t; x_i, \cdot)$, $1 \leqslant i \leqslant k$, corresponding to the k ancestors with positions $x_1, \ldots, x_k$ at $t = 0$.

Consequently, the generating function

$$G(z, \boldsymbol{\theta}; x^k, t) = \sum_{n \in N} z^n G_n(\theta^n; x^k, t), \tag{2.7}$$

where

$G_0(\theta^0;\ x^k,\ t)$ = extinction probability by time t given k ancestors with positions $x_1, \ldots, x_k$ at $t = 0$;

$$G_n(\theta^n;\ x^k,\ t) = \int_{R^n} \exp\left(\mathrm{i} \sum_{j=1}^{n} \theta_j y_j\right) p(t; x^k, y^n)\, \mathrm{d}y^n, \qquad n \in N^*$$

will have conspicuous properties.

In (2.7), z is a complex number such that $|z| \leqslant 1$, $\theta^n \in R^n$, $n \in N$, and $\boldsymbol{\theta}$ denotes the sequence $(\theta^n)_{n \in N}$. Indistinguishability implies that G_n is a symmetric function of θ^n and x^k. Combined with the multiplicative character of process, this implies that $G(z, \boldsymbol{\theta}; x^k, t)$ is equal to the symmetrized product of the generating functions $G(z, \boldsymbol{\theta}; x_i, t)$, $1 \leqslant i \leqslant k$. Thus, in particular

$$G_n(\theta^n; x_1, x_2, t) =$$

$$= \frac{1}{n!} \sum_{\pi} \sum_{j=0}^{n} G_j(\theta_{i_1}, \ldots, \theta_{i_j}; x_1, t) G_{n-j}(\theta_{i_{j+1}}, \ldots, \theta_{i_n}; x_2, t),$$

where the right hand side is summed over all permutations π of $\theta_1, \ldots, \theta_n$. We shall denote such symmetrized sums simply by

$$G_n(\theta^n; x_1, x_2, t) = \sum_{j=0}^{n} G_j(\theta^j; x_1, t)\, G_{n-j}(\theta^{n-j}; x_2, t). \tag{2.8}$$

2.1.4.3. It was shown by Moyal (1962 a) that $G(z, \boldsymbol{\theta}; x, t)$ satisfies the equation

$$\begin{aligned} \frac{\partial}{\partial t} G(z, \boldsymbol{\theta}; x, t) &= -(b + d) G(z, \boldsymbol{\theta}; x, t) + d + \\ &+ b G^{(2)}(z, \boldsymbol{\theta}; x, t) + \frac{1}{2} \frac{\partial}{\partial x^2} G(z, \boldsymbol{\theta}; x, t), \end{aligned} \tag{2.9}$$

where $G^{(2)}(z, \boldsymbol{\theta}; x, t)$ is the generating function conditional on two ancestors, both at x, and that the solution of (2.9) is unique and stochastic. The desired solution can be obtained by an iterative procedure; to this purpose it is necessary to expand G and $G^{(2)}$ as in (2.7), and equate the coefficients of successive powers of z. This yields the system of equations

$$\frac{\partial G_0}{\partial t} = -(b+d)G_0 + d + bG_0^{(2)} + \frac{1}{2}\frac{\partial^2 G_0}{\partial x^2}, \tag{2.10}$$

$$\frac{\partial G_n}{\partial} = -(b+d)G_n + bG_n^{(2)} + \frac{1}{2}\frac{\partial^2 G_n}{\partial x^2}, \quad n \in N^*. \tag{2.11}$$

Because of the nature of the diffusion process, one surmises that for all t, the transition density function $p(t; x, y^n)$ will depend only on the distances of the n survivors from the position x of the ancestor, i.e., that $p(t; x, y^n)$ will be a function of $(y_i - x)$, $1 \leqslant i \leqslant n$, and t.

If we assume this, we deduce that:

(a) G_0 is independent of x and hence equal to the probability $p_{10}(t)$ in the linear birth-and-death process,

$$p_{10}(t) = \begin{cases} d(1 - e^{-(b-d)t})(b - de^{-(b-d)t})^{-1} & \text{if} \quad b \neq d \\ bt(1-bt)^{-1} & \text{if} \quad b = d \end{cases}$$

which is, in fact, the solution of (2.10) with $\partial^2 G_0/\partial x^2 = 0$, $G_0^{(2)} = G_0^2$, and the initial condition $G_0(0) = 0$.

(b) The expression of $G_n^{(2)}$ takes from the above the following form:

$$G_n^{(2)}(\theta^n; x, t) = 2p_{10}(t)\,G_n(\theta^n; x, t) + \sum_{j=1}^{n-1} G_j(\theta^j; x, t)\,G_{n-j}(\theta^{n-j}; x, t).$$

(c) Formally,

$$\frac{\partial^2}{\partial x^2} G_n(\theta^n; x, t) = \frac{\partial^2}{\partial x^2}\int_{R^n} \exp\left(\mathrm{i}\sum_{j=1}^{n} \theta_j y_j\right) p(t; x, y^n)\,\mathrm{d}y^n =$$

$$= -\left(\sum_{j=1}^{n} \theta_j\right)^2 G_n(\theta^n; x, t).$$

Let us assume that the solution of (2.9) is such that the above is correct. We can then rewrite (2.11) as

$$\frac{\partial G_n}{\partial t} = -\left[b + d - 2bp_{10}(t) + \frac{1}{2}\left(\sum_{j=1}^{n}\theta_j\right)^2\right]G_n + b\sum_{j=1}^{n-1} G_j G_{n-j}, \qquad n \in N^*, \tag{2.12}$$

We shall now consider successive values of n. For $n = 1$, the last term in (2.12) vanishes and the solution with the initial condition $G_1(\theta; x, 0) = \exp(i\theta x)$ is

$$G_1(\theta; x, t) = p_{11}(t) \exp\left(i\theta x - \frac{\theta^2 t}{2}\right), \tag{2.13}$$

where

$$p_{11}(t) = \begin{cases} \left(\dfrac{b-d}{b-d\mathrm{e}^{-(b-d)t}}\right)^2 \mathrm{e}^{-(b-d)t}, & \text{if} \quad b \neq d \\ (1+bt)^{-2}, & \text{if} \quad b = d \end{cases}$$

For $n \geqslant 2$, equation (2.13) is a solution of the homogeneous part of (2.12), if we substitute

$$\Theta_n = \sum_{j=1}^{n} \theta_j$$

for θ. Hence the solution of (2.12) with the initial condition $G_n(\theta^n; x, 0) = 0$ yields the iteration relation

$$G_n(\theta^n; x, t) = b \int_0^t \frac{p_{11}(t)}{p_{11}(\tau)} \exp\left(-\frac{1}{2}\Theta_n^2(t-\tau)\right) \times$$

$$\times \sum_{j=1}^{n-1} G_j(\theta^j; x, \tau)\, G_{n-j}(\theta^{n-j}; x, \tau)\, d\tau, \qquad n = 2,3, \ldots \tag{2.14}$$

Setting $\theta^n = 0$, we obtain the transition probabilities $p_{1n}(t)$ of the linear birth-and-death process:

$$p_{1n}(t) = (b/d)^{n-1} p_{11}(t) [p_{10}(t)]^{n-1}, \qquad n \in N^*. \tag{2.15}$$

2.2. Population processes in Euclidean space

2.2.1. Homogeneous spatial models

2.2.1.1. We shall start with a brief presentation of the generalized homogeneous spatial Poisson process $\xi(S)$ where S denotes a bounded region of the plane or space. Suppose, by analogy with the one-dimensional case, that

(i) The random variable $\xi(S)$ assumes only nonnegative integer values and $0 < \mathsf{P}(\xi(S) = 0) < 1$ if $A(S) > 0$, where $A(S)$ denotes the area or volume of S;

(ii) The probability distribution of $\xi(S)$ depends on S only through the value of $A(S)$ with the further property that if $A(S) \to 0$ then $\mathsf{P}(\xi(S) \geqslant 1) \to 0$;

(iii) If $S_1, \ldots, S_m$, $m \geqslant 1$, are disjoint regions, then $\xi(S_1), \ldots, \xi(S_m)$ are mutually independent random variables, and

$$\xi(S_1 \cup \ldots \cup S_m) = \xi(S_1) + \ldots + \xi(S_m);$$

(iv) One has

$$\lim_{A(S)\to 0} \frac{\mathsf{P}(\xi(S) \geqslant 1)}{\mathsf{P}(\xi(S) = 1)} = 1.$$

Under these conditions we have (see KARLIN (1966, Ch. 12))

$$\mathsf{P}(\xi(S) = k) = \mathrm{e}^{-\lambda A(S)} \frac{[\lambda A(S)]^k}{k!}, \qquad k \in N$$

where λ is a positive constant (the so-called intensity parameter).

2.2.1.2. The familiar example of a homogeneous plane Poisson process is that of a plant spreading in two-dimensional space, which may, however, be easily replaced by other similar examples. In Chapters 12 and 13 of *Origin of species*, DARWIN gave some examples of accidental spreading, e.g., fish carried great distances away by cyclones.

In its ecological context, the problem of random dispersal and migration was studied for the first time by PEARSON and BLAKEMAN (1906). One should also recall the work of SKELLAM (1951) who discusses the spread of oaks in Great Britain in the post-glacial period, and that of Cox and SMITH (1957) on the distribution of *Tribolium confusum* in a container. THOMPSON (1955) considered a clustering model where a Poisson "parent" process had a family of "offspring" following each parent, except that offspring will then be distributed spatially round any parent (see also BARTLETT (1964 b)). KARLIN (1966, Ch. 12.6) deduces for the above-mentioned type of processes the probability generating function of the number of progeny in a region R produced by all the parents located in a region S.

See also MORGAN and WELSH's (1965) remarkable paper on a Poisson process on a square lattice.

2.2.1.3. CHAPMAN (1967) maintains, however, that the spread of the population can be regarded not as a function of the distribution of all surviving progeny in the kth generation, but of the *distance* of the most distant of the progeny from its parent. This led to extreme value theory.

We shall recall in the same context the problem of the spatial distribution to rth nearest neighbour in a population of randomly distributed objects (THOMPSON (1956)). The probability model is the following (HOLGATE (1965 a)): Suppose first that a single sample object (point) is chosen and consider the random distance $\delta_{11}(=\delta)$ to the nearest object, where the spatial distribution of objects is supposed to be a realization of a Poisson point process with density λ. It is a function on the space of all possible realizations of the process, and if the sample point is chosen independently of the realization, δ has the probability density function given by SKELLAM (1951):

$$f(d) = 2\pi\lambda d \exp(-\pi\lambda d^2).$$

It is convenient to transform to a new variable $\eta = \pi\delta^2$, which consequently has the probability density function $g(y) = \lambda \exp(-\lambda y)$. See also HOLGATE (1965 b) and ROBERTS (1969).

2.2.1.4. We conclude this paragraph with an interesting interpretation of age-dependent branching processes with an arbitrary state space (MODE and BIRCHER (1970)). A population of B-objects occupies some space S (a habitat). The process starts with an initial object of age 0 at some point $x \in S$ in the habitat. The lifespan of a B-object as well as the number of offspring it produces at the end of its life depend on its location in S. At the end of their lifespans each of these offspring in turn produce further offspring who scatter independently in S and so the process continues as long as there are live individuals in the population (see also Subparagraph 2.5.3.3.).

One can also imagine a class of age-dependent branching processes in which B-objects contribute offspring to the population at random points during its lifespan. Such processes have already been considered by CRUMP and MODE (1968, 1969) and independently by JAGERS (1969 a) for the case of one type, and a detailed account of this class of stochastic processes for the case of a finite number of types may be found in MODE's (1971) book.

2.2.2. Birth, death and migration processes in R^2 and R^3

2.2.2.1. BAILEY (1968 a) considered populations which are distributed over the points of a line, or over the nodes of a two- or three-dimensional lattice, in such a way that each point or node may support a colony (that is an aggregate of B-objects, parents and offspring) subject to stochastic birth, death and migratory exchange with nearest neighbours. The biologic prototypes of these models might be the development of a tumor, the growth of viral plaques or the population spread in the theory of urban development.

Let us first suppose that individual colonies are distributed over Z^2 and the size of the subpopulations at $(i, j) \in Z^2$ and at time t is given by the random variable $\xi_{ij}(t)$. At time $t = 0$ we can take, quite generally,

$$\xi_{ij}(0) = a_{ij}.$$

Each subpopulation is subject to a linear birth-and-death process, with the birth and death parameters b and d, respectively, and with emigration rates of $1/4\nu$, to each of the four nearest neighbours. More precisely, the ijth colony will increase by one unit in the time interval $(t, t + \Delta t)$ due to a new birth with probability $b\xi_{ij}(t)\Delta t$. Similarly, one unit will be lost due to death with probability $d\xi_{ij}(t)\Delta t$. We must consider all such transitions for $(i, j) \in Z^2$. Next, in specifying the migration from any one colony to an adjacent one it will be sufficient to examine first the migration from the ijth colony to the four nearest neighbours, and then to let (i, j) vary over Z^2. Migrations to the ijth colony from neighbouring colonies are then automatically taken care of. In the time interval $(t, t + \Delta t)$, the ijth colony will decrease by one unit due to migration to the $(i, j + 1)$th colony, say, with probability $1/4\nu$ $\xi_{ij}(t)\Delta t$. The same probability is to be assigned to migrations to the $(i, j - 1)$th, $(i + 1, j)$th and $(i - 1, j)$th colony. Assume also to first order in Δt, only one transition of any kind can occur in the time interval $(t, t + \Delta t)$.

If we consider the (formal) moment-generating function

$$M(\ldots, \theta_{ij}, \ldots; t) = \mathsf{E}\left(\exp \sum_{i,j \in Z} \theta_{ij}\, \xi_{ij}(t)\right),$$

by writing

$$M(\ldots, \theta_{ij}, \ldots; t + \Delta t) - M(\ldots, \theta_{ij}, \ldots; t) =$$

$$= \mathsf{E}\left(\exp \sum_{i,j \in Z} \theta_{ij}\, \xi_{ij}(t) \left\{\exp \sum_{i,j \in Z} \theta_{ij} [\xi_{ij}(t + \Delta t) - \xi_{ij}(t)] - 1\right\}\right),$$

it is not difficult to deduce the partial differential equation:

$$\frac{\partial M}{\partial t} = \sum_{i,j \in Z} \left[b(e^{\theta_{ij}} - 1) + d(e^{-\theta_{ij}} - 1) + \frac{1}{4}\nu(e^{-\theta_{ij} + \theta_{i,j+1}} - 1) + \right.$$

$$+ \frac{1}{4}\nu(e^{-\theta_{ij} + \theta_{i,j-1}} - 1) + \frac{1}{4}\nu(e^{-\theta_{ij} + \theta_{i+1,j_1}} - 1) +$$

$$\left. + \frac{1}{4}\nu(e^{-\theta_{ij} + \theta_{i-1,j}} - 1) \right] \frac{\partial M}{\partial \theta_{ij}}, \tag{2.16}$$

with the initial condition

$$M(\ldots, \theta_{ij}, \ldots; 0) = \exp \sum_{i,j \in Z} a_{ij}\, \theta_{ij}. \tag{2.17}$$

Setting

$$\mathsf{E}\, \xi_{ij}(t) = m_{ij}(t),$$

(2.16) leads to the doubly infinite set of simultaneous ordinary differential equations

$$\frac{dm_{ij}}{dt} = (b - d - \nu)\, m_{ij} + \frac{1}{4}\, \nu\, (m_{i,j+1} + m_{i,j-1} + m_{i+1,j} + m_{i-1,j}),$$

$$m_{ij}(0) = a_{ij}, \qquad i, j \in Z. \tag{2.18}$$

These equations can be solved explicitly by using the Laplace transform

$$\mathscr{L}(\lambda) = \int_0^\infty e^{-\lambda t} L(t)\, dt, \;\; \mathrm{Re}\lambda > 0.$$

Applying the transformation to equations (2.18) we deduce

$$(\lambda - b + d - \nu)\, m_{ij} =$$

$$= a_{ij} + \frac{1}{4}\, \nu\, (m_{i,j+1} + m_{i,j-1} + m_{i+1,j} + m_{i-1,j}). \tag{2.19}$$

We now introduce the generating function for mean values and the corresponding transforms

$$G(x, y; t) \sum_{i,j \in Z} m_{ij}\, x^i y^j,$$

$$\mathscr{G}(x, y; \lambda) = \sum_{i,j \in Z} m_{ij}\, x^i y^j.$$

We shall assume the existence of these Laurent expansions over some appropriately chosen domains of x and y, for complex x, y. Multiplying each side of (2.19) by $x^i y^j$ and summing over Z^2, we have

$$(\lambda - b + d + \nu)\, \mathscr{G} = \sum_{i,j \in Z} a_{ij}\, x^i y^j + \frac{1}{4}\, \nu \left(x + \frac{1}{x} + y + \frac{1}{y} \right) \mathscr{G},$$

which gives

$$\mathcal{G}(x, y; \lambda) = \left[\lambda - b + d + \nu - \frac{1}{4}\nu\left(x + \frac{1}{x} + y + \frac{1}{y}\right)\right]^{-1} \times$$

$$\times \sum_{i,j \in Z} a_{ij}\, x^i\, y^j.$$

The reverse is

$$G(x, y; t) = [\sum_{i,j \in Z} a_{ij}\, x^i\, y^j] \times$$

$$\times \exp\left(\left[b - d - \nu + \frac{1}{4}\nu\left(x + \frac{1}{x} + y + \frac{1}{y}\right)\right] t\right). \tag{2.20}$$

Using the general expansion in terms of modified Bessel functions of the first kind given by

$$\exp\left(\frac{1}{2} z (w + w^{-1})\right) = \sum_{k \in Z} w^k I_k(z),$$

we can transcribe (2.20) as

$$G(x, y; t) =$$

$$= e^{(b-d-\nu)t} \sum_{i,j \in Z} a_{ij}\, x^i y^j \sum_{u \in Z} x^u I_u\left(\frac{1}{2}\nu t\right) \sum_{v \in Z} y^v I_v\left(\frac{1}{2}\nu t\right). \tag{2.21}$$

Picking out the coefficient of $x^i y^j$ in the above expression, we deduce

$$m_{ij}(t) = e^{(b-d-\nu)t} \sum_{u,v \in Z} a_{uv}\, I_{i-u}\left(\frac{1}{2}\nu t\right) I_{j-v}\left(\frac{1}{2}\nu t\right). \tag{2.22}$$

If we admit that we initially had a single centre (node) of development so that $a_{00} = a$ and all other $a_{ij} = 0$, we will have

$$m_{ij}(t) = a\, e^{(b-d-\nu)t}\, I_i\left(\frac{1}{2}\nu t\right) I_j\left(\frac{1}{2}\nu t\right),$$

which becomes in the special case $b = d$

$$m_{ij}(t) = a\,\mathrm{e}^{-\nu t}\, I_i\left(\frac{1}{2}\nu t\right) I_j\left(\frac{1}{2}\nu t\right).$$

It is now possible to view the behaviour of the process in terms of damped waves spreading out from the origin in a two-dimensional way, although an exact description of the phenomenon is somewhat complicated.

2.2.2.2. The problem of the development in R^3 does not raise great difficulties, at least fundamentally. This time, each node has six nearest neighbours, so that emigration rates equal 1/6 ν. The random variables $\xi_{ijk}(t)$ will represent the size of the subpopulations at $(i, j, k) \in Z^3$ and at time t. Equation (2.18) would then be replaced by

$$\frac{\mathrm{d}m_{ijk}}{\mathrm{d}t} = (b - d - \nu)\, m_{ijk} + \frac{1}{6}\,\nu\,(m_{i,j,k+1} + m_{i,j,k-1} +$$

$$+ m_{i,j+1,k} + m_{i,j-1,k} + m_{i+1,j,k} + m_{i-1,j,k}),$$

with initial conditions $m_{ijk}(0) = a_{ijk}$, $i, j, k \in Z$. The analogue of equation (2.21) will be

$$G(x, y; t) = \mathrm{e}^{(b-d-\nu)t} \sum_{i,j,k \in Z} a_{ijk}\, x\, y^j z^k \sum_{u \in Z} x^u I_u\left(\frac{1}{3}\nu t\right) \times$$

$$\times \sum_{v \in Z} y^v I_v\left(\frac{1}{3}\nu t\right) \sum_{w \in Z} z^w\, I_w\left(\frac{1}{3}\nu t\right),$$

and the average colony size at point (i, j, k) at time t is

$$m_{ijk}(t) = \mathrm{e}^{(b-d-\nu)t} \sum_{u,v,w \in Z} a_{uvw}\, I_{i-u}\left(\frac{1}{3}\nu t\right) I_{j-v}\left(\frac{1}{3}\nu t\right) I_{k-w}\left(\frac{1}{3}\nu t\right).$$

2.3. Intrinsic processes

2.3.1. Multiple-phase processes

2.3.1.1. We have shown in Subparagraph 1.2.5.1 that a fundamental time distance for some B-objects is their division time τ, which measures the distance between the moment of the birth of a B-object and the moment of its subdivision into new B-objects. Let G be the generation time distribution, defined as $\mathsf{P}(\tau \leqslant t) = G(t)$. Its density G' is the probability-theoretic counterpart of the observed relative frequency of the life-lengths of B-objects. Explicit forms of G will yield a series of mathematical models of growth and will form an expression of the intrinsic nature of the biological process.

In fact, the processes studied up to now involve exponentially distributed random variables, in that the individual times spent in a particular state are exponentially distributed. But as is known from the paper of KELLY and RAHN (1932), the division time τ has no exponential distribution for various types of bacteria. In general, τ has an average value of 20 to 30 minutes, with limits between 9 minutes for *B. megatherium* and 18 hours for *B. tuberculosis*. The velocity of the division process is a characteristic of most bacteria but also of the embryonic cells of higher organisms. In the particular case of the growth of *B. aerogenes* (cultivated at 30°C), the average value of τ was about 30 minutes, the individual division times are scattered about this mean value with a coefficient of variation of the order of 20%, and the empirical distribution curve has a single peak in the neighbourhood of the mean. The distribution of τ is therefore not exponential and the growth process is not Markovian.

The alternative of a deterministic process where τ has a constant value $1/b$ is always possible in the case of large-sized populations; as mentioned by KENDALL (1953), it is only when we are concerned with the growth of small colonies or of colonies which have grown from a small number of common ancestors that the probability effects are large enough to be noticeable.

We shall consider processes involving nonexponential distributions, i.e., non-Markovian processes, assuming that during the course of its division a bacterium passes through k phases. We point out that this statement does not mean that τ is divided into k intervals (or phases), because these phases may be either *simultaneous* or *successive*. Thus the assumption considers k independent subprocesses, a fact which caused us to call the stochastic population processes involving this assumption (or other assumptions about some subprocesses) as *intrinsic processes*.

2.3.1.2. Experimental researches carried out in the last 10 to 12 years, have brought to light the fact that, at least for bacteria, the generation times are not entirely independent (as suggested by RAHN, for example); on the contrary, there are connections between the generation times of cells for an intermediate (some three generations) period. The dependence of the generation times of daughter cells upon those of their mothers is usually not revealed in correlation tests, but this dependence becomes evident when the sizes of cells at division are considered. The deterministic models of KOCH and SCHAECHTER (1962) are based upon this observation (see also TAKAHASHI, ISHIDA and KUROKAWA (1964)).

From the data of POWELL (1955, 1958) and POWELL and ERRINGTON (1963) it was found that the generation rate distribution is a Pearson Type III distribution, while the data of SCHAECHTER et al. (1962) showed this distribution to be normal; KUBITSCHEK (1961) found a distribution highly skewed toward longer generation times. These studies seem to lead to three different distributions of generation time for the same bacterial species. In order to obtain valid generation time, or rate, distributions, KUBITSCHEK (1967) proposes three guiding conditions: (a) the choice of a good criterion for the measuring of τ, (b) the observation of cells in balanced growth, (c) the calculation of distribution by avoiding data so selected that the distribution is biased.

In connection with the first condition, one must emphasize the fact that the terminal instant must be clearly distinguishable and have but small variability in the sequence of events composing the cell division cycle. For example, if the terminal instant is considered as that of bacterial cell wall separation, we shall speak of an *o*-fission, and if this instant is the cytoplasmic separation, it will be a *p*-fission (POWELL (1958)). As regards the second condition, we recall that one understands by balanced growth the type of growth in which every extensive cellular property (mass, number, composition) must increase at the same rate. Statistical fluctuations would occur for small numbers of cells, but these would be negligible if the culture were indefinitely large. We thus come to the problem of synchronized cell division. As we know, if in an exponentially growing culture, one can cause (by change in incubation temperature, food, starvation, selection, etc.) a fraction of the dividing cells to grow above values normally observed and the log growth curve to show a stepwise increase in the population number, we shall say in this case that the culture is *synchronous* or *synchronized* (SCHERBAUM (1962)). However, it has been observed (BURNETT-HALL and WAUGH (1967 a)) that there are two causes of departure from a state of synchrony: the stochastic fluctuations in the lifetimes of B-objects and the distribution of ages at the start of observation. But it is the second cause which contributes to the decay of synchrony with time. For this purpose we can without loss of generality treat the problem in terms of a population descendent from a single B-object which is of age zero at time $t = 0$. The mean growth of a population which is "perfectly synchronized" at time zero, in the sense that it consists of a large number of objects all of the same age (= zero) at the start, will follow at once. According to ENGELBERG (1961), any disturbance of a population away from the state of steady exponential growth induces some degree of synchrony. SANKOFF (1971) derives an index for synchrony for initially synchronized populations having Pearson Type III generation time distributions.

2.3.1.3. Let us now pass to the study of some intrinsic processes and consider first the case of simultaneous phases. Let us suppose that (a) a B-object contains k components (for example, kinetosomes, ribosomes, chromosomes, genes, plastides, etc.), k being a nonrandom positive integer, (b) these components divide simultaneously and independently during the time interval τ, (c) the lengths of simultaneous divisions are independent random variables, (d) the length τ_j of division subprocess j $(1 \leqslant j \leqslant k)$ has the probability density $\alpha_j\, e^{-\alpha_j t}$, where α_j is a constant which can be, hypothetically, common to all the k subprocesses, (e) at the end of all k subprocesses of division, the initial B-object is transformed into two new objects. Clearly, the division time τ will be

$$\tau = \max\ (\tau_1, \ldots, \tau_k).$$

It was shown by O. RAHN that in case $\alpha_j = \alpha,\ 1 \leqslant j \leqslant k$,

$$G(t) = \mathsf{P}(\tau \leqslant t) = \int_0^t k\,[1 - e^{-\alpha\tau}]^{k-1}\, e^{-\alpha\tau}\, d\tau,\ 0 < t < \infty, \tag{2.23}$$

which has been called (KENDALL (1952)) *Rahn's law*.

From (2.23) one deduces

$$\mathsf{E}\,\tau = \frac{1}{\alpha} \sum_{j=1}^{k} \frac{1}{j} \simeq \frac{\gamma + \log k}{\alpha},$$

where γ ($\gamma = 0.5572\ldots$) is the Euler constant (see FINNEY and MARTIN (1951)). The experimental data obtained by O. RAHN show that the parameter k takes values between 22 and 1010.

One can arrive at the distribution (2.23), as observed by KENDALL (1952), by noting that the variables

$$U = \frac{u_1}{1} + \frac{u_2}{2} + \ldots + \frac{u_k}{k},$$

$$V = \max\,(u_1, \ldots, u_k),$$

have the same distribution when the independent variables $u_1, \ldots, u_k$ have the common probability density e^{-u} $(0 < u < \infty)$.

2.3.1.4. Secondly, consider the case of successive phases. Suppose that (a) an object goes through k successive phases during its life, where k is a nonrandom positive integer, (b) the lengths of those successive phases are independent random variables, (c) the length of time τ_j $(1 \leqslant j \leqslant k)$ which the object spends in a phase j has a negative exponential probability density $b_j e^{-b_j t}$ and (d) at the end of the kth phase the object is transformed into two new objects, each of which then begins phase 1. In this case the division time will be $\tau = \sum_{1}^{k} \tau_j$, which will have the same distribution as,

$$U = \frac{u_1}{b_1} + \frac{u_1}{b_2} + \ldots + \frac{u_k}{b_k},$$

where the variables $u_1, \ldots, u_k$ are independent, each having the probability density e^{-u} $(0 < u < \infty)$. Thus, in case $b_j = j\alpha$, $1 \leqslant j \leqslant k$, we get again the Rahn distribution (2.23).

If $b_1 = b_2 = \ldots = b_k$ $(= kb)$, the τ-distribution will (apart from a scale factor) be that of a χ^2-variable having $2k$ degrees of freedom,

$$G(t) = \int_0^t \frac{b^k k^k}{\Gamma(k)} \tau^{k-1} e^{-bk\tau} d\tau. \tag{2.24}$$

The distribution (2.24) is called *Erlangian*: it has a unique mode equal to $\frac{1 - 1/k}{b}$, $E\tau = 1/b$ and the coefficient of variation of τ is equal to $1/\sqrt{k}$. The analysis of the experimental results shows that k takes values between 12 and 47 (KENDALL (1952)). For *B. aerogenes* one has $k = 20$ and $b = 40$, when time is measured in hours. The biological significance of the k hypothetical phases was initially given by L. MARTIN (1948—1949), who allowed the successive replication of bacterial k genes (or gene groups). Our present knowledge encourages this assumption, since one is aware on the one hand of the replications forks of circular bacterial chromosomes and the semiconservative DNA replication, and, on the other hand, of the importance of replicones and replicators. It is thus possible to consider k as the number of replicating operations switched on by the genetic replicators. This assumption suggests the detailed analysis of the four well-known phases of cellular multiplication: G_1, S_1, G_2 and M (see Subparagraph 1.2.5.4).

2.3.1.5. Let us note an important fact arising in connection with the case of successive phases, for $b_1 = b_2 = \ldots = b_k = kb$. Let ξ_i be the number of B-objects in the ith phase $(1 \leqslant i \leqslant k)$. If we consider the k-dimensional process $\boldsymbol{\xi} = (\xi_k)_{1 \leqslant k \leqslant n}$, it is easily seen that it is Markovian. For details see BAILEY (1964 a, p. 132) and KENDALL (1948 b).

2.3.1.6. Let us think of a B-object, viz. a bacterium or an embryo cell, as a "customer" who arrives at a "server", say, the coordination center of DNA replication. According to such an interpretation, we can associate with the process of successive phases a queueing system with service in phases. The arrivals are say, Poisson with mean λ; with the nth customer there is associated a service time distributed according to (2.24). In the usual notation, we have a $M/E_k/1$ queue. Let us notice that the service time can be thought of as consisting of k successive phases, their lengths being independent random variables with the negative exponential density $\frac{k}{b} \exp\left(-\frac{k}{b}\tau\right)$. If we consider the fact that the arrival of each customer introduces k phases into the system, then denoting by $Q_1(t)$ the number of phases in the system at time t, it can be shown that $(Q_1(t))_{t\geqslant 0}$ is a Markov jump process (PRABHU (1965 b)).

We shall now mention some alternatives for building the queueing system. If we assume that the B-customers arrive at regular intervals of time, we will have a $D/E_k/1$ system. It is interesting to note that such a system was studied by BAILEY (1952, 1955), for the purpose of studying the consultation time (i.e., service time) distribution of patients in a hospital (see also PIKE (1963)).

A supplementary assumption can be made even on the phase service, either admitting a random number of phases (LUCHAK (1958)) or by leaving to the customer the choice of the number of phases (GUPTA (1965)).

2.3.1.7. GOODMAN (1967 b) gave a generalization of Kendall's multiphase birth process, changing, however, the fundamental assumption to generation succession. L. A. GOODMAN assumes that the generation of descendents may occur before the closing of the last phase, an assumption allowing application of the model also to multicellular organisms. For a B-object existing in the xth phase (stage) of life ($0 \leqslant x \leqslant k$) at time t the following events are possible with the corresponding probabilities:

1°. $a(x)b_{i1}(x)\,\Delta t$, the probability that the B-object will move into the $(x+1)$th phase and will also have produced i new individuals, $i \in N$, which will be in the 0-th phase during the subsequent small time-interval $(t, t+\Delta t)$, with $b_{i1}(k) = 0$;

2°. $a(x)b_{i2}(x)\,\Delta t$, the probability that the B-object will die during the time interval $(t, t+\Delta t)$ but will produce i new individuals, $i \in N$, who will be in the 0-th phase during this time interval;

3°. $a(x)b_{i3}(x)\Delta t$, the probability that the B-object will remain in the xth phase during the time interval $(t, t+\Delta t)$ but will produce i new individuals, $i \in N^*$, who will be in the 0-th phase during this time interval.

For a B-object in the xth phase at time t the probability is $a(x)\,\Delta t$ that one of the three events considered above will occur; and $b_{ij}(x)$, $j = 1, 2, 3$, is the conditional probability of the corresponding event, given that one of the three listed events has occurred. Let

$$\sum_{j=1}^{3} b_{ij}(x) = b_i(x), \quad i \in N^*,$$

$$b(x) = \sum_{i \in N^*} i b_i(x),$$

$$b_{0j}(x) + b_{1j}(x) + \ldots = c_j(x), \quad j = 1, 2$$

$$b_{13}(x) + b_{23}(x) + \ldots = c_3(x),$$

$$\sum_{j=1}^{3} c_j(x) = 1,$$

and assume that $a(x) \neq 0$ and $c_3(x) \neq 1$, for $0 \leqslant x \leqslant k$.

Taking into account the fundamental assumption of generation succession, we shall consider that the first generation of the B-object in the xth phase at time $t = t_0$ is defined when one of the events $1^\circ - 3^\circ$ occurs for the first time after time $t = t_0$, and this generation consists of either

a) the i new objects, $i \in N$, in the 0-th phase that the object produced at that time and one "new" object in the $(x + 1)$th phase, if event 1° occurred; or

b) the i new objects, $i \in N$, in the 0-th phase that the object produced at that time, if event 2° occurred; or

c) the i new objects, $i \in N^*$, in the 0-th phase that the object produced and one "new" object in the xth phase, if event 3° occurred.

Consequently, the nth generation of the B-object in the xth phase at time $t = t_0$ is defined as consisting of the new objects in the first generation of the new objects in his $(n - 1)$th generation.

Let $Z_0^{(x)}$ denote the number of B-objects in the xth phase at $t = 0$. Considering the population generated by these objects, let $Z_n^{(x)}$ denote the number of B-objects in the nth generation which are in the xth phase, let

$$Z_n = Z_n^{(0)} + \ldots + Z_n^{(k)},$$

and let

$$\mathbf{Z}_n = \{Z_n^{(0)}, \ldots, Z_n^{(k)}\}.$$

The sequence $\mathbf{Z}_0, \mathbf{Z}_1, \ldots$ is a $(k+1)$-type Galton-Watson chain (see I; 1.2.2.7). Let $p_n(x)$ denote the probability that $Z_n = 0$ when $Z_0^{(x)} = 1$, $Z_0^{(i)} = 0$, $i \neq x$, and let $p(x) = \lim_{n \to \infty} p_n(x)$ denote the probability that $Z_n \to 0$.

Goodman's paper deals with the determination of the extinction probabilities for this process, since these probabilities are also the extinction probabilities for our multiphase birth-and-death process. Thus we shall have

$$p_{n+1}(x) = \sum_{i \in N} [b_{i2}(x) + b_{i1}(x)\, p_n(x+1) + b_{i3}(x)\, p_n(x)]\, [p_n(0)]^i,$$

where we define $b_{03}(x) = 0$, $0 \leqslant x \leqslant k$. Since the right side of this equation is a continuous function of $p_n(x)$ and $p_n(x+1)$ for $0 \leqslant p_n(x) \leqslant \leqslant 1$, we have

$$p(x) = f_2(x) + f_1(x)\, p(x+1) + f_3(x)\, p(x), \tag{2.25}$$

where

$$f_j(x) = \sum_{i \in N} b_{ij}(x)\, [p(0)]^i, \quad j = 1, 2, 3. \tag{2.26}$$

Since $f_3(x) < 1$, we can rewrite (2.26) as

$$p(x) = f_2^*(x) + f_1^*(x)\, p(x+1), \tag{2.27}$$

where

$$f_j^*(x) = \frac{f_j(x)}{1 - f_3(x)}, \quad j = 1, 2. \tag{2.28}$$

The solution of the system (2.27) is

$$p(x) = \sum_{i=x}^{k} \Phi_x^*(i), \quad x \leqslant i \leqslant k, \tag{2.29}$$

where

$$\Phi_x^*(i) = f_2^*(i) \prod_{j=x}^{i-1} f_i^*(j),$$

with the convention $\prod_{j=x}^{x-1} = 1$. Note that in (2.29), $p(x)$ appears as the probability that a B-object in the xth phase at $t = 0$ will die while it is in the ith phase ($x \leqslant i \leqslant k$) and that all the new objects it produced from $t = 0$ until its death will also eventually die. From (2.29) we have an explicit expression for $p(x)$, for $x > 0$, in terms of $p(0)$, where $p(0)$ satisfies the equation

$$p(0) = \sum_{i=0}^{k} \Phi_0^*(i) = g[p(0)]. \tag{2.30}$$

We find that if $p(0) = 1$, then $p(x) = 1$, $1 \leqslant x \leqslant k$, and if $p(0) < 1$ then $p(x) < 1$, except when $b_0(y) = 1$, $x \leqslant y \leqslant k$, in which case $p(x) = 1$. If $p(x) = 1$ for a given value of x, then $p(y) = 1$, $(x+1) \leqslant y \leqslant k$.

To solve (2.30), we shall first introduce a new probability, q_j, the probability that a B-object in the 0-th stage at $t = 0$ will produce a total of j new objects during its lifetime. We shall thus rewrite (2.30) as

$$p(0) = \sum_{j \in N} q_j [p(0)]^j = g[p(0)], \tag{2.31}$$

the smallest nonnegative solution of which is $p(0)$. Let us assume without any loss of generality that $c_2(i) \neq 1$, $0 \leqslant i \leqslant k$. In this case $q_1 = 1$ iff $c_3(i) = 0$, and there is a single value i_0 $(0 \leqslant i_0 \leqslant k)$ such that

$$\begin{aligned} b_1(i) &= 1 && \text{for} \quad i = i_0, \\ b_{12}(i) + b_{01}(i) &= 1 && \text{for} \quad i < i_0, \\ b_{02}(i) + b_{01}(i) &= 1 && \text{for} \quad i > i_0. \end{aligned}$$

If $q_1 = 1$, then the smallest nonnegative solution of (2.31) is $p(0) = 0$. If $q_1 \neq 1$, but $q_0 + q_1 = 1$, then $g[p(0)]$ is a linear function of $p(0)$ and (2.31) is satisfied only for $p(0) = 1$. If $q_0 + q_1 < 1$, then $g[p(0)]$ is a strictly convex function of $p(0)$, and the smallest solution will be

$$p(0) = \left. \begin{array}{l} 1, \text{ if } R \leqslant 1 \\ < 1, \text{ otherwise} \end{array} \right\}$$

where $R = \sum_{j \in N} j q_j$; R is the expected number of new objects produced during its lifetime by an object in the 0-th phase at $t = 0$. It can be interpreted as the reproductive value of a B-object, often called *the Malthusian rate*. Consequently, $p(0) = 0$ iff $q_0 = 0$, which will be the case iff there is at least one value of i, say $i_0 (0 \leqslant i_0 \leqslant k)$, such that $b_{01}(i_0) = 0$ and $b_{02}(j) = 0$ for $j \leqslant i_0$. In addition, we find that by taking any trial value $p_1 (0 \leqslant p_1 \leqslant 1)$, and then calculating

$$p_{n+1} = g[p_n], \quad n \in N^*$$

from (2.30), the limit of the sequence p_n will be the desired solution of (2.30), or, equivalently, of (2.31), if $q_1 \neq 1$; and if $q_1 = 1$ then $p_n = p_1$, which would indicate that $p(0) = 0$.

2.3.1.8. The general model presented by L. A. GOODMAN includes as special cases the Kendall multiphase birth process as well as the linear birth-and-death process.

For the first process, $b_{01}(x) = 1$, $0 \leqslant x \leqslant k - 1$, and $b_{22}(k) = 1$, so that obviously $q_0 = 0$ and $p(0) = 0$.

For the second process, $k = 0$, $b_{02}(0) = d/(b + d)$, $b_{22}(0)=b/(b+d)$, $R = 2b/(b + d)$, $p(0) = \dfrac{d + b[p(0)]^2}{b + d}$, so that we obtain the well-known result that $p(0) = d/b$ if $d \leqslant b$ and $p(0) = 1$ otherwise.

2.3.2. The life cycle process

2.3.2.1. The so-called life cycle model, independent of the multiphase birth-and-death model and differing from a model of biologic rhythms is justified by certain natural phenomena and by the relations of the growing organism to its biotic and abiotic environments. Organisms can couple certain types of light responses to their biological clock. It is possible, for example, for organisms to respond in a qualitatively different way to light, depending on whether the light acts during phases of the physiological cycle that normally coincide with the hours of light or during phases that normally coincide with the hours of darkness (BÜNNING (1967)). However, there is an annual fluctuation in responsiveness to the day-length signals with respect to the growth of gonads in lizards and birds. That is, a specific day length may stimulate the growth of gonads quite easily in cetain months, but only very little, or not at all, in other months. Let us also recall the periods of pause (*diapause*) in the developing of growth processes in insects. Larval diapause appears in the butterfly larva when it reaches the age when it can transform itself into a chrysalis; in autumn anopheles females which have sucked blood, the ovaries do not develop (imaginal diapause) in contrast to the phenomenon observed in summer females. It is therefore clear that life cycle models must be based upon the assumption that the periods spent in each stage of development, excluding the possibility of death, are independent random variables with a characteristic form of probability distribution. Let us also assume that at every moment of its life the organism is liable to suffer death, either as a result of the action of predators, of accidents or for other reasons. The main features which must be taken into account are — following READ and ASHFORD (1968) — the distributions of the times of birth, of the periods spent in each stage of development and of mortality in the various stages.

2.3.2.2. However, before dealing with such a model, we shall present a phaseless model in which it is assumed that the total lifetime of a B-object is a random variable which may extend over several time units (interpreted as "seasons"), the death rate being age-dependent, but reproduction continuing in an age-dependent fashion throughout life (WHITTLE (1964)).

Let r_m the probability that a B-object which is alive at the end of i seasons (of "age i") will produce m offspring during its $(i + 1)$th season, having itself the probability p_i of surviving it. The offspring will be considered as being of age 1 at the end of $(i + 1)$th season. These probabilities are independent of variables relating to other individuals.

Let

$$G(z) = \sum_{m \in N} r_m z^m$$

be the probability generating function of progeny/individual/season, and let

$$H(z) = q_1 z + p_1 q_2 z^2 + p_1 p_2 q_3 z^3 + \dots \tag{2.32}$$

be the probability generating function of lifetime, where $q_i = 1 - p_i$. It follows that a complete description of the situation is given by $\varphi_k(z_1, z_2, \dots)$, the joint probability generating function of the numbers of individuals of ages 1, 2, ..., alive at the beginning of the kth season. This function will obey the relations

$$\varphi_{k+1}(z_1, z_2, \dots) = \varphi_k(\zeta_1, \zeta_2 \dots), \quad k \in N^* \tag{2.33}$$

where

$$\zeta_i = (p_i z_{i+1} + q_i)\, G(z_1).$$

P. WHITTLE considers an imbedded chain $(\xi_j, \theta_j)_{j \in N^*}$ where ξ_j is the total number of B-objects in the jth generation of descendants from an initial ancestor of age 1, and θ_j is the total lifetime of members of this generation, or, possibly, the total "service time" of this generation. Defining the corresponding probability generating functions as

$$X_j(z) = \mathsf{E}\, z^{\xi_j},$$

$$T_j(w) = \mathsf{E}\, w^{\theta_j},$$

we will deduce

$$X_{j+1}(z) = T_j[G(z)],$$

and

$$T_j(w) = X_j[H(w)],$$

so that

$$X_{j+1}(z) = X_j(H[G(z)]),$$
$$T_{j+1}(w) = T_j(G[H(w)]). \tag{2.34}$$

Equations (2.34) replace (2.33).

Let us consider the following events:

A_i: the population deriving from a single initial B-object of age i ultimately becomes extinct ($A = A_1$);

B: the population derived from the progeny of a single given B-object in a single given season ("age") of his life ultimately becomes extinct.

Let

$$a_i = P(A_i); a = a_1 = P(A),$$
$$b = P(B),$$

so that the above results yield

$$a = H(b),$$
$$b = G(a), \tag{2.35}$$

and, consequently

$$a = H[G(a)],$$
$$b = G[H(b)]. \tag{2.36}$$

For classical Galton-Watson chains (where the ideas of season and generation coincide), we have $A = B$, $H(z) = z$, so that $a = b$, and equations (2.35) and (2.36) all take the well-known form

$$a = G(a). \tag{2.37}$$

The pair of equations (2.35) or the individual equations (2.36) can be regarded as generalizations of (2.37) to the case when a generation is no longer identical with a season.

Let us now return to the imbedded chain $(\xi_j, \theta_j)_{j \in N^*}$. We see that $(\xi_j)_{j \in N^*}$ and $(\theta_j)_{j \in N^*}$ individually constitute Galton-Watson chains with progeny probability generating functions $H[G(z)]$ and, respectively $G[H(w)]$. We thus observe that A is identical with the ultimate extinction of $(\xi_j)_{j \in N^*}$. Likewise, we observe that B is identical with the ultimate extinction of $(\theta_j)_{j \in N^*}$. We mention that $(\xi_j, \theta_j)_{j \in N^*}$ is equivalent to an *alter-*

nating process, i.e., one that is simple except that the objects alternate between two types (e.g., juvenile/adult) in consecutive generations. In the case of the joint processes considered, instead of these two types we shall have objects and seasons of life. It is easily seen that either both a and b are unity, or both are less than unity — the second case being true iff the expected number of progeny in a lifetime $H'(1)\, G'(1)$ strictly exceeds unity.

To obtain information on the relative magnitudes of a and b, we must consider the fact that by "no births" one also understands "still-births" so that a lifetime $i = 0$ can never occur. Then there is no constant term in the Taylor expansion (2.32) of $H(z)$, so that $H(z) < z$ strictly in (0,1). Consequently, $a < b$ unless $a = b = 1$ or 0, and, by the same argument,

$$b \leqslant r_0 + (1 - r_0)\, a.$$

For completeness, let us note that it is possible to show that

$$a_i = (p_i\, a_{i+1} + q_i)\, G(a_1), \quad i \in N^*.$$

P. Whittle also describes the analogous model in continuous time

2.3.2.3. Let us now assume that the life cycle of a B-object is divided in distinct k-phases, which run through in succession until it attains maturity by reaching phase k or accidental death intervenes. Let us also assume that once phase k is attained, the object eventually dies in that phase, either from accidental or natural causes. The expression "eventually dies" leaves open the possibility of interpretating the cycles as periodical oscillations, which allows us, for instance, to model some biochemical processes (the tricarboxylic acid cycle, the pentose-phosphate cycle, etc.). As one knows, such processes are treated in the simplest way as a deterministic sinusoidal oscillation,

$$X_n = a \cos (n\, \omega_0).$$

The transformation of this process into a random one could be made by considering a linear combination of sinusoidal oscillations with random phases, plus an uncorrelated error term:

$$X_n = \sum_{i=1}^{p} a_i \cos (n\, \omega_{0i} + \nu_i) + \beta_n.$$

In this expression, the chain $(\beta_n)_{n \in N}$ consists of uncorrelated random variables, the random variables $(\nu_i)_{1 \leqslant i \leqslant p}$ are independently rectangularly distributed over $(0, 2\pi)$, and $a_1, \ldots, a_p$; $\omega_{01}, \ldots, \omega_{0p}$ are constants (see Cox and Miller (1965, p. 284 ff)). This process is called a *system with hidden periodicities*.

2.3.2.4. If phase k is a terminal phase and phase 1 is taken as the birth, the model considered by READ and ASHFORD (1968) also includes phase 0 and phase $k + 1$. Phase 0 defines the existence of a B-object immediately prior to birth: in the case of an insect population this would correspond to the hatching of eggs. The transition from phase k to the fictitious phase $k + 1$ will be regarded as equivalent to death from *natural* causes in phase k.

The model is built on the basis of the fundamental assumption that for a B-object which survives to reach any nonterminal phase i, the period ξ_i spent in that phase may be regarded as the minimum of two other intervals;

ζ_i, the interval elapsing before transition to the next $(i + 1)$th phase, assuming that death does not take place, and,

η_i, the interval elapsing until death in state i, assuming that transition to phase $i + 1$ does not take place.

Since ζ_i and η_i will usually be random variables, it follows that $\xi_i = \min(\zeta_i, \eta_i)$ will also be a random variable.

Let us now examine possible assumptions on the distributions of the two variables. Insofar as the deaths within a nonterminal phase are largely due to predators, accidents or other chance happenings which occur randomly in time, it is reasonable to assume that η_i will follow a negative exponential distribution with probability density function $d_i \exp(-d_i x)$, $x > 0$. In general, the mortality in different phases will not be the same and the parameter d_i will vary from one phase to another.

We will assume that ζ_i must also be a nonnegative random variable which will follow a continuous distribution with a certain probability density function $g_i(x)$, $x \geqslant 0$. The distribution of ζ_i will also vary from phase to phase although g_i is likely to have the same general form for all phases. The probability density function of ζ_i will usually have a single peak at some non-zero value of x and will fall away to zero for very small or very large values of the abscissa.

A simple model is given by

$$g_i(x) = \frac{\lambda_i^a x^{a-1}}{(a-1)!} e^{-\lambda_i x} \quad , \quad x > 0, \ 0 \leqslant i \leqslant k,$$

where a is a given natural number. The illustration of the model was made on the basis of a study of the grasshopper, *Corthippus parallelus*. In this model, $a = 2$ and $a = 3$.

On the assumption that ζ_i and η_i are independent, the probability density function of ξ_i will be

$$f_i(x) = d_i e^{-d_i x} G_i(x) + g_i(x) e^{-d_i x} \quad , \quad x \geqslant 0 \qquad (2.38)$$

where $G_i(x) = \mathrm{P}(\zeta_i > x)$ is the survivor function of ζ_i.

Let $p_i(t)$ be the probability that a randomly chosen B-object is alive and in state $i (0 \leqslant i \leqslant k)$ at time $t \geqslant 0$. This probability describes the structure of the studied population. In order to calculate $p_i(t)$, we define:

$$\psi_i(x) = \lim_{\Delta x \to 0} \left[\frac{1}{\Delta x} \mathrm{P} \left(\begin{array}{l|l} \text{a B-object passes from} & \text{it survives} \\ \text{phase } i \text{ to phase } i+1 & \text{to enter} \\ \text{in the interval } (x, x+\Delta x) & \text{phase } i+1 \end{array} \right) \right],$$

$$\varphi_i(x) = \lim_{\Delta x \to 0} \left[\frac{1}{\Delta x} \mathrm{P} \left(\begin{array}{l|l} \text{the period spent by a B-object} & \text{it survives} \\ \text{in phase } i \text{ lies in the interval} & \text{to enter} \\ (x, x+\Delta x)\text{, and the object} & \text{phase } i+1 \\ \text{then passes into phase } i+1 & \end{array} \right) \right].$$

From the definitions above it follows that

$$\psi_0 = \varphi_0, \tag{2.39}$$

and

$$p_0(t) = F_0(t) = \mathrm{P}(\xi_0 > t). \tag{2.40}$$

We shall also have

$$\psi_i(x) = \int_0^x \psi_{i-1}(y)\, \varphi_i(x-y)\mathrm{d}y, \tag{2.41}$$

and

$$p_i(t) = \int_0^t \psi_{i-1}(y)\, F_i(t-y)\,\mathrm{d}y, \quad 1 \leqslant i \leqslant k \tag{2.42}$$

where $F_i(x) = \mathrm{P}(\xi_i > x)$ is the survivor function of ξ_i.

The proportion of organisms which transfer to state $i+1$ after spending a period of length x in phase i is

$$\frac{g_i(x)}{d_i\, G_i(x) + g_i(x)}.$$

It fol'ows then from (2.38) and from the definition of φ_i that

$$\varphi_i(x) = \exp(-d_i x)\, g_i(x).$$

By repeated application of (2.41), taking into account (2.39), we find that

$$\psi_s(x) = \int_0^x \int_0^{x-x_0} \cdots \int_0^{x-(x_0+\ldots+x_{s-2})} \exp\left(-d_s\left(x-[x_0+\ldots+x_{s-1}]\right)\right) \times$$
$$\times\, g_s\left(x-[x_0+\ldots+x_{s-1}]\right) \prod_{i=0}^{s-1} \mathrm{e}^{-d_i x} g_i(x)\,\mathrm{d}x, \tag{2.43}$$

Equations (2.42) and (2.43) then provide a formal solution for p_i.

2.3.3. The birth, death and marks transmission process

2.3.3.1. Let us assume that a B-object can be characterized by the presence of certain structures, which we will call "marks" in order to stress the fact that these are concrete indications of a single (or several) property. For example, the presence of some γ-globulin inclusions characterizes a certain type of leucocytes found not only in rheumatoid arthritis but also in a wide variety of diseases. These "marked" leucocytes, found in the peripheral blood as well as in the intra-articular fluids, have been called R. A. cells or ragocytes (see HOLLANDER et al. (1965)). Intracytoplasmic inclusions, as "inanimate marks" were recently found in a series of diseases among which we recall the Chediak-Higashi anomaly in man ("congenital gigantism of peroxidase granules") which has in children a malignant (lymphoma-like) character, or autoimmune haemolytic anemia of New Zealand Black (NZB) mice.

Some of the modern theories of immunology (e.g. G.J.V. NOSSAL and O. MÄKELÄ) are built upon the assumption that long-life lymphocytes (which live several months) form a heterogeneous population consisting of many cellular families containing the antigenic information that directs the production of a single type of antibody. According to this hypothesis, each cell is marked in a certain way, having a characteristic limited memory and reacting to a single type of antigen (or some related types). This memory is transmittable and is lost when the owner cell disappears. We may thus consider memory-carrying structures as marks, stressing the quantitative aspect of some problems of immunology. In the tissues of NZB mice with autoimmune haemolytic anemia and glomerulonephritis there were observed leukemia virus-like particles, which led to a working hypothesis on a relationship between autoimmunity and lymphoreticular neoplasms in man (MELLORS et al. (1969)).

2.3.3.2. In the light of these considerations, we shall now present the model suggested by WILLIAMS (1969) concerning "inanimate marks". Let us assume a population of B-objects obeying a nonhomogeneous birth-and-death process and suppose also that each object is initially labelled with m marks, the value of m being chosen randomly and independently from a common distribution. We assume that when an object dies or is destroyed, so also are the marks it bears, and that when it divides in two, each of the mark on the original object goes independently into two new objects with probability 1/2.

Let $N_m(t)$ be the expected number of objects bearing precisely m marks at time t, in terms of the given initial distribution, $N_m(0)$. We have

$$N_m(t + \Delta t) = N_m(t)\,[1 - (b(t) + d(t))\Delta t] +$$

$$+ 2b(t)\Delta t \sum_{k \geqslant m} 2^{-k} \binom{k}{m} N_k(t),$$

where the second series of terms obtains when an object bearing k marks splits, yielding the daughter pair $(m, k - m)$ or $(k - m, m)$, whence the factor 2. It should be noted that the factor 2 also obtains when 2 m marks go into a single daughter pair (m, m), since in this case the value m occurs twice. We thus deduce the equation

$$\frac{\partial}{\partial t} G(z, t) = -(b(t) + d(t))G(z, t) + 2b(t) \sum_{m \in N} z^m \sum_{k \geqslant m} 2^{-k} \binom{k}{m} N_k(t),$$

where $G(z, t)$ is the generating function

$$G(z, t) = \sum_{m \in N} N_m(t)\, z^m.$$

The double summation inverts to

$$\sum_{k \in N} 2^{-k} N_k(t) \sum_{m=0}^{k} \binom{k}{m} z^m = \sum_{k \in N} 2^{-k} (1 + z)^k N_k(t).$$

We are therefore led to the functional equation

$$(\partial/\partial t)\, G(z, t) = -(b(t) + d(t))\, G(z, t) + 2b(t)\; G(1/2 + 1/2z, t).$$

Let us now define

$$X_\nu(t) = \left[\frac{\partial^\nu}{\partial z^\nu} G(z, t)\right]_{z=1} =$$

$$= \sum_{m \geqslant \nu} m(m-1)\ldots(m-\nu-1)\, N_m(t), \quad \nu \in N^* \tag{2.44}$$

on the basis of which we get

$$G(z, t) = \sum_{\nu \in N} \frac{(z-1)^\nu}{\nu!} X_\nu(t),$$

so that

$$(d/dt) X_\nu(t) = -(b(t) + d(t)) X_\nu(t) + 2^{1-\nu} b(t) X_\nu(t),$$

with the solution

$$X_\nu(t) = X_\nu(0) \exp(-(1 - 2^{1-\nu}) B(t) - D(t)), \qquad (2.45)$$

where

$$B(t) \equiv \int_0^t b(u)\, du, \qquad D(t) \equiv \int_0^t d(u)\, du.$$

Hence one deduces the mean number of objects at time t

$$X_0(t) = \sum_{m \in N} N_m(t),$$

and the mean number of marks at time t,

$$X_1(t) = \sum_{m \in N^*} m N_m(t),$$

while from (2.45), we see that

$$\begin{aligned} X_0(t) &= X_0(0) \exp(B(t) - D(t)), \\ X_1(t) &= X_1(0) \exp(-D(t)). \end{aligned} \qquad (2.46)$$

The first equation (2.46) is the well-known result for the mean of a nonhomogeneous birth-and-death process (see I; 2.3.3.4.4); the subsequent equation reflects the fact that the non-replicating marks are subject *in toto* to a nonhomogeneous death process.

With the aid of (2.45), the expression of the generating function becomes

$$G(z, t) = \exp(-B(t) - D(t)) \sum_{\nu \in N} \frac{(z-1)^\nu}{\nu!} X_\nu(0) \exp(2^{1-\nu} B(t)). \quad (2.47)$$

But, from the first expression of $G(z, t)$

$$N_m(t) = \left[\frac{1}{m!} \frac{\partial^m}{\partial z^m} G(z, t) \right]_{z=0},$$

and, applying this to (2.47), we obtain

$$N_m(t) = \frac{1}{n!} \exp(-B(t) - D(t)) \sum_{\nu \in N} (-1)^\nu \frac{1}{\nu!} X_{\nu+m}(0) \exp(2^{1-\nu-m} B(t)), \tag{2.48}$$

the desired result, in conjunction with (2.44), which identifies the $X_m(0)$ as factorial moments of the known initial distribution.

2.3.3.3. In his paper T. WILLIAMS considers that a reasonable characterization of the process would be the mean proportion of cells bearing precisely m marks at time t, a number which, in the case of large-sized populations, is given to a good approximation by

$$\pi_m(t) = \frac{N_m(t)}{X_0(t)} = \frac{1}{m!} \exp(-2B(t)) \sum_{\nu \in N} (-1)^\nu \frac{1}{\nu!} \frac{X_{\nu+m}(0)}{X_0(0)} \times \tag{2.49}$$

$$\times \exp(2^{1-\nu-m} B(t)).$$

This mean proportion is entirely independent of the "death" component of the process. The most natural initial distribution of marks over cells is clearly the Poisson

$$\pi_m(0) = \frac{N_m(0)}{X_0(0)} = \frac{\lambda^m}{m!} e^{-\lambda},$$

by means of which (2.44) reduces to

$$\frac{X_\nu(0)}{X_0(0)} = \sum_{m \geq \nu} m(m-1) \ldots (m-\nu+1) \frac{\lambda^m}{m!} e^{-\lambda} = \lambda^\nu,$$

so that (2.49) becomes

$$\pi_m(t) = \frac{\lambda^m}{m!} \exp(-2B(t)) \sum_{\nu \in N} (-1)^\nu \frac{\lambda^\nu}{\nu!} \exp(2^{1-\nu-m} B(t)).$$

In terms of the function

$$s(x, y) \equiv \sum_{\nu \in N} \frac{x^\nu}{\nu!} \exp\left(\frac{y}{2^\nu}\right), \tag{2.50}$$

we get

$$\pi_m(t) = \frac{\lambda^m}{m!} \exp(-2B(t)) s(-\lambda, 2^{1-m} B(t)). \tag{2.51}$$

The function $s(x, y)$ possesses some interesting properties. Thus, we see at once that

$$\frac{\partial}{\partial x} s(x, y) = \sum_{\nu \in N} \frac{x^\nu}{\nu!} \exp \frac{y}{2^{\nu+1}} = s(x, y/2),$$

and similarly,

$$\frac{\partial}{\partial y} s(x, y) = s(x/2, y),$$

while the symmetric double series

$$s(x, y) = \sum_{\nu \in N} \frac{x^\nu}{\nu!} \sum_{d \in N} \frac{y^d}{2^{d\nu} d!}$$

that results from expanding the exponential in the initial expression (2.50) shows that

$$s(x, y) = s(y, x).$$

Since λ is merely the known mean initial number of marks per B-object, one can then invert (2.51), which is plainly monotonic, to obtain $B(t)$ for the given value of t, thus displaying B as a function of t. Next, $D(t)$ can be found from the growth curve (2.46), and finally, $b(t)$, and $d(t)$ may be computed by numerical differentiation. It can also be shown that $B(t)/\log 2$ is equal to the mean number of generations which have elapsed by time t.

2.3.4. Interdependent (self)-replicating process

2.3.4.1. In his paper on biological self-replicating systems, MOROWITZ (1967) has defined a self-replicating entity as one which takes components out of its environment and assembles two or more entities similar to the first such that at least two of them are again capable of initiating the assembly process. From this point of view, viruses represent an especially difficult case since the environment for virus assembly is very complex, consisting of the interior of a cell. In addition, virus replication is a composite process in which the cell plays an active part. This statement is justified by the simple enumeration of the fundamental stages of the virus replicating process: (1) virus attachment to the host cell wall, (2) entrance into the cell, (3) replication of the viral genome, (4) synthesis of viral proteins and assembly of the viral components, (5) virus release.

Obviously, a mathematical model of the replicating process that would consider all these stages cannot yet be built due to the complexity of each separate stage. For example, for stages (3) and (4), the utilization of a multiphase model is nonrealistic by itself, since we find here associated successive as well as simultaneous processes. It is sufficient to note that under the influence of the type 1 polyoma virus genome, there appear in the infected host cell (HeLa cell) approximately 14 new types of polypeptides, out of which only 4 play a structural role, the others changing the normal metabolism of the host cell (see SUMMERS et al. (1965)).

At the same time the synthesis of the viral genome occurs, i.e., of the specific nucleic acid molecules. The phase succession is demonstrated by the biosynthesis of "early" and "late" proteins. The synthesis of proteins associated with the infection of *E. coli* by virulent bacteriophage (or, simply, phage) occurs in a regulated temporal sequence. Early events include the cessation of host protein synthesis and the induction of enzymes concerned with the making of phage DNA, while the later events deal mostly with the synthesis of structural proteins and phage lysozyme.

As LURIA (1959) has pointed out, virus multiplication, as a biological process, belongs on the level of replication of subcellular elements, that is, on the level of cell growth rather than of cell multiplication.

We observe that in the case of this replicating process, the host cell must be considered as a new unit, the headquarters of the interaction between two systems denoted by the general terms *receptor (RR)* and *receptus (RS)*. The relations between *RR* and *RS* (in the case in question, between virus and host cell) may be interdependent or dependent. (For a general theory, see BIRTA's (1965) thesis on a formal approach to concepts of interaction.) Interdependence reactions are characterized by the synchronous replication of the fundamental components of both systems. The virus stays in the cell in a latent state, multiplying at a slow rate, while the host transmits the immature viral particles to its descendants, through its own division. The temperate (symbiotic) bacteriophage-lysogenic bacteria complex constitutes a representative example for this type of relation. In contrast to this, dependence relations lead either to the lysis of the host cell (cytocide effect) and the releasing of a very large number of infecting virions, or to a steady-state allowing the survival of the host cell. The cell response to a viral infection in the steady state is often a malignant transformation (for details, see DAWE's (1960) paper on cell sensitivity and specificity of response to polyoma virus).

2.3.4.2. We shall now present a simple stochastic model for the situation of bacteriophage-bacteria interdependence. The time interval T, from the infection up to the lysis of the bacterial cell can be divided into several stages, the most important one being the stage of *eclipse* of the virus, which corresponds to the phase in which the naked, coatless viral genome cannot be identified within the cell. The naked phage is called *vegetative* and the mature virus, *virion*.

As specified by GANI (1965 b), phage reproduction in a bacterium lends itself to formulation as a pure birth process during eclipse, followed by a birth-and-death process thereafter, until lysis occurs. In such a model (see STEINBERG and STAHL (1961), GANI (1962 a), OHLSEN (1963)), replication of the vegetative phage may be considered as a birth, while the final assembly of the DNA strand *(genome)* and protein coat *(capside)* into a mature phage is equivalent to death. But the replication process may also be formulated in terms of nonoverlapping generations of vegetative phage, with maturation starting after k generations and each phage mature independently with a probability 1/2 for each generation. Lysis is assumed to occur at the end of generation $(k + n)$.

Let v_i, $1 \leqslant i \leqslant k + n$, be the number of vegetative bacteriophages at the beginning of the ith generation (assuming $v_1 = 1$). Also let m_i, $1 \leqslant i \leqslant k + n$, be the number of bacteriophages maturing in the ith generation (obviously, $m_j = 0$ for $j \leqslant k$). At the beginning of the $(k + j + 1)$th generation there will be

$$v_{k+j+1} = 2^{k+j} - \sum_{i=1}^{j} 2^{j-i} m_{k+i} \quad , \quad 0 \leqslant j < n,$$

vegetative phages and m_{k+j} mature phages. The vectors (v_i, m_i), $k + 1 \leqslant i \leqslant k + n$, are Markov-dependent since $v_i = 2\,(v_{i-1} - m_{i-1})$ and

$$\mathrm{P}\,(m_i = m \mid v_{i-1}, m_{i-1}, v_{i-2}, m_{i-2}, \ldots) = \binom{v_i}{m}\left(\frac{1}{2}\right)^{v_i}.$$

KIMBALL (1965) has shown that the joint probability generating function G_i of the total numbers V_i and M_i of vegetative and mature phage, respectively, at the end of the ith generation satisfies the recursion formula

$$G_{i+1}(z_1, z_2) = G_i\,(1/4\,(z_1 + z_2)^2, z_2), \tag{2.52}$$

with $G_k(z_1, z_2) = z_1^{2k}$. Equation (2.52) may be used to obtain central moments as follows:

$$\mathsf{E}\; V_{k+n} = 2^k \quad ; \quad \mathsf{E} M_{k+n} = n2^k, \tag{2.53}$$

$$\mathsf{D}\; V_{k+n} = n2^{k-1} \quad ; \quad \mathsf{D}\; M_{k+n} = \frac{1}{3}\, n\,(2n^2 - 9n + 13)2^{k-2}, \tag{2.54}$$

$$\mathrm{Cov}\,[V_{k+n}, M_{k+n}] = n\,(n+3)\; 2^{k-2}. \tag{2.55}$$

2.3.4.3. We now present an interesting application of the fundamental ideas of renewal theory to the virion release process from the host cell (GANI and YEO (1962)). Let us assume that the number n of particles in a cell remains constant (the fixed size pool hypothesis). Upon elimination of a mature phage from the pool, a new vegetative phage immediately replaces it. We shall therefore consider a pool with ranked elements in the order of age, so that the ranked initial ages of the elements at time $T_0 = 0$ are

$$0 = x_1 < \ldots < x_n,$$

and where at each instant of replacement there is a probability p_i $(1 \leqslant i \leqslant n \;,\; 0 < p_i < 1, \;\sum_{i=1}^{n} p_i = 1)$ that the ith ranked element is replaced.

Let us suppose that such replacements are made at the times $T_1 = \tau_1$, $T_2 = \tau_1 + \tau_2, \ldots, T_m = \tau_1 + \ldots + \tau_m, \ldots$, where τ_m are mutually independent variables with a common distribution F. If replacement were age dependent, one might, for example, choose $p_{i+1} > p_i$ for all i, $1 \leqslant i \leqslant n-1$. For $x > 0$, let us denote by

$$F_i(x, T_j) = F_i(x, T_j | \, x_i, 0), \; j \in N$$

the age distribution function of the ith ranked virus particle immediately after the jth replacement (interpreted as a regeneration point), given that its age was x_i at $T_0 = 0$. In terms of the joint distribution function $L(x_{1n}, \ldots, x_{nn}; t)$ of ranked ages of the n objects and of the probability $p_{rn}(x_0\,; t)$ that r of the n virus particles have no greater age than x_0 at time t, it is clear that

$$F_i(x, T_j) = L(x, \ldots, x, x_{i+1} + T_j, \ldots, x_n + T_j; T_j) = \sum_{r=i}^{n} p_{rn}(x, T_j).$$

The Laplace transform of this age distribution function is

$$\mathcal{F}_i(\lambda, T_j) = \int_0^{\infty} \mathrm{e}^{-\lambda x}\, \mathrm{d}_x\, F_i(x, T_j) \quad , \quad \mathrm{Re}\,\lambda > 0.$$

By considering the process at times just after two consecutive replacements, it can be seen that

$$F_1(x, T_{j+1}) = F_1(x, T_j) = 1,$$

$$F_i(x, T_{j+1}) = (1 - q_{i-1}) F_i(x - \tau_{j+1}, T_j) + \tag{2.56}$$

$$+ q_{i-1} F_{i-1}(x - \tau_{j+1}, T_j) \quad , \quad 2 \leqslant i \leqslant n,$$

where $F_i(x, T_j) = 0$ for $x \leqslant 0$ for all i, j and $q_i = p_{i+1} + \dots + p_n$. Both $F_i(x, T_j)$ and $\mathcal{F}_i(\lambda, T_j)$, though clearly functions of $T_j, T_{j-1}, \dots,$ T_1, are written in the present shorter form for simplicity.

We shall now rewrite equations (2.56) in matrix form:

$$\mathbf{F}(x, T_{j+1}) = \mathbf{AF}(x - \tau_{j+1}, T_j) + \mathbf{F}_{j+1},$$

where $\mathbf{F}_{j+1}$ is the column vector with first component $F_1(x, T_{j+1}) - F_1(x - \tau_{j+1}, T_j)$ and all other components 0, while

$$\mathbf{A} = (a_{ij}) = \begin{pmatrix} 1 & & & \\ q_1 & 1 - q_1 & & \\ \cdot & \cdot & \cdot & \cdot \\ & & q_{n-1} & 1 - q_{n-1} \end{pmatrix}.$$

The equivalent matrix equation for the Laplace transforms is

$$\boldsymbol{\psi}(\lambda, T_{j+1}) = \mathbf{A}\, \boldsymbol{\psi}(\lambda, T_j)\, e^{-\lambda t_{j+1}} + \mathbf{L}_{j+1}(\lambda),$$

where $\mathbf{L}'_{j+1}(\lambda) = (1 - e^{-\lambda t_{j+1}}, 0, \dots, 0)$. Iterating the above equation, we finally obtain that at time T_m, just after the mth replacement,

$$\boldsymbol{\psi}(\lambda, T_m) = \sum_{i=0}^{m} e^{-\lambda(T_m - T_i)} \mathbf{A}^{m-i} \mathbf{L}_i(\lambda), \tag{2.57}$$

where $\mathbf{L}'_0(\lambda) = \boldsymbol{\psi}'(\lambda, 0) = (1, e^{-\lambda x_2}, \dots, e^{-\lambda x_n})$.

Now we write $\mathbf{A}$ in its canonical form, assuming that its roots are distinct:

$$\mathbf{A} = \mathbf{B}\, \boldsymbol{\Lambda}\, \mathbf{C},$$

where $\mathbf{\Lambda}$ is the diagonal matrix of the eigenvalues $1, 1 - q_1, \ldots, 1 - q_{n-1}$ of $\mathbf{A}$, and the matrices $\mathbf{B} = (b_{ij})$ and $\mathbf{C} = (c_{ij}) = \mathbf{B}^{-1}$ are lower triangular matrices with elements

$$b_{i+j,i} = \prod_{k=1}^{j} \frac{q_{i+j-k}}{q_{i+j-k} - q_{i-1}},$$

$$c_{i+j,i} = \prod_{k=1}^{j} \frac{q_{i+j-k}}{q_{i+j-k} - q_{i+j-1}}, \quad 1 \leqslant i \leqslant n, \quad j \geqslant 0,$$

$\prod_{k=1}^{0}$ being defined as 1. Then from (2.57) we obtain that the Laplace transform of the age distribution of the ith element just after T_m is given by

$$\mathcal{F}_i(\lambda, T_m) = \sum_{r=1}^{m} a_{i1}^{(m-r)} \exp\left(-\lambda(\tau_{r+1} + \ldots + \tau_m)\right)(1 - e^{-\lambda\tau_r}) +$$

$$+ \sum_{r=1}^{i} a_{ir}^{(m)} \exp\left(-\lambda(\tau_1 + \ldots + \tau_m) - \lambda x_r\right),$$

where the elements of $\mathbf{A}^m$ are

$$a_{11}^{(m)} = 1,$$

$$a_{ij}^{(m)} = \begin{cases} \sum_{k=j}^{i} b_{ik}\, c_{jk}\, (1 - q_{k-1})^m, & \text{when } i \geqslant j \text{ for } i > 1 \\ 0, & \text{when } i < j \text{ or } i > j, \quad i - j > m. \end{cases}$$

From the above expressions one obtains the following results:

— the age distribution $f_{im}(x)$ of the ith ranked element at the mth replacement (regardless of the time taken for these m changes), which can be written in the form

$$f_{im}(x) = \int_{\tau_1=0}^{\infty} \cdots \int_{\tau_m=0}^{\infty} F_i(x, T_m)\, \mathrm{d}F(\tau_1) \ldots \mathrm{d}F(\tau_m) =$$

$$= \sum_{r=0}^{m-1} a_{i1}^{(r)} [F^{(r)}(x) - F^{(r+1)}(x)] + \sum_{r=1}^{i} a_{ir}^{(m)} F^{(m)}(x - x_r),$$

where $F^{(r)}$ is the rth convolution of F;

— the age distribution $d_t f_{im}(x, t)$ of the ith ranked element if the mth replacement occurs in $(t, t + \Delta t)$ is found to be given by its Laplace transform $d_t \ell_{im}(\lambda, t)$ of the form

$$d_t \ell_{im}(\lambda, t) = \int_{\tau_1=0}^{t} \int_{\tau_2=0}^{t-T_1} \cdots \int_{\tau_{m-1}}^{t-T_{m-2}} \mathcal{F}_i(\lambda, T_m) \, dF(\tau_1) \ldots dF(t - T_{m-1}) =$$

$$= \sum_{r=1}^{m} a_{i1}^{(m-r)} \int_{u=0-}^{t} e^{-\lambda(t-u)} \{d_t F^{(m-r)}(t-u) \, dF^{(r)}(u) -$$

$$- d_t F^{(m-r+1)}(t-u) \, dF^{(r-1)}(u)\} + \sum_{r=1}^{i} + a_{ir}^{(m)} e^{-\lambda(t-x_r)} \, dF^{(m)}(t);$$

— the stationary age distribution $f_i(x)$ of the ith ranked element directly after replacement,

$$f_1(x) = 1$$

$$f_i(x) = \left(\prod_{r=1}^{i-1} q_r\right) \sum_{r=1}^{i-1} \left[\prod_{\substack{j=1 \\ j \neq r}}^{i-1} \frac{1 - q_r}{q_j - q_r}\right] \sum_{k \in N} (1 - q_r)^k F^{(k+i-1)}(x), \; 2 \leqslant i \leqslant n.$$

GANI (1965 a, b) has also given solutions to the fixed size model in continuous time in the general case where p_i are time-dependent as well as the distribution of interarrival time. DAVIS (1964) has studied the characteristic functional for the same model.

2.3.5. Point mutation processes

2.3.5.1. In this subparagraph we shall explain the point mutation process simply as an intrinsic process leading to the appearance of some discernible B-objects in a population initially formed by alteridem B-objects. The process takes place at the level of the genes and the mutation is one of the modalities of genotype modification. A gene has a high degree of specificity for a given character or phenotype; it is capable of precise replication, synchronously with cell division in the case of chromosomal genes; and it segregates and recombines as a unit in sexual reproduction. Finally, it is subject to change or mutation, and the resulting mutant gene again replicates precisely and continues to function with its changed specificity (TATUM (1964)). We must therefore insist on the definition of mutation as a hereditary change in a gene. The addition of the word "hereditary" is imperative, ZAMENHOF (1963) notes: changes in DNA can be induced by simply changing pH or temperature. Many such changes are transient and quite reversible and are not mutations. We add that in the case of point mutations these involve short segments of the DNA molecule, down to a single nucleotide. At the same time, these genetic events must be differentiated from the recombinations taking place at the level of chromosomes as well as of gene groups.

These modifications of gene structure, called "saltations" by DARWIN, were termed "mutations" by the Dutch botanist HUGO DE VRIES (1905, 1909). However, the term had been adopted since 1650, almost in its present meaning, by THOMAS BROWNE in his book *Pseudodoxia Epidemica — or Enquiries into Very many Received Tenents and commonly Presumed Truths* (see LOCK (1906), ZAMENHOF (1963)). Among the oldest examples of known mutations we shall recall the mutant of the plant *Chelidonium majus* (I. SPRENGER, Heidelberg, 1590), the Squarehead wheat in England of the dwarf sheep of Ancona (SETH WRIGHT, Massachusetts, 1791). Among the variations observed by DE VRIES on the plant *Oenothera Lamarckiana*, only one or two proved to be true mutations.

2.3.5.2. We study here the process of mutation without entering into an analysis of the evolutionist significances of the process. Within this frame the appearance and the development of mutants will be interpreted as an example of the growth of a heterogeneous population BARTLETT (1966, p. 126)). We shall thus consider a population formed of two types, subject to some kind of birth-and-death process, and study the occurrence and subsequent spread of the mutant type (see BAILEY (1964 a, p. 125), KARLIN (1966, p. 97)). We note that the mutation is a recurrent process and that the spontaneous mutation rates are controlled both by the environment and by the genetic constitution of the organism (FREESE and YOSHIDA (1965)).

For the purpose of building the stochastic model, we must first define the mutant as a B-object owning some modified attributes as compared to its parents or to an origin population taken as standard. The owning of some specific attributes has been long considered as proof that an object belongs to a certain population. This explains why in Euripide's Elektra a servant recognizes Orestes only because his hair is like Elektra's. It will therefore be necessary to establish a series of genetically controlled characteristics marking the difference between a standard object and the mutant object. In haploid cells this characterization is often possible. For example, the presence of the Z^+ gene assures the synthesis of β-galactosidase in *E. coli*; the mutation of this gene in the Z^- form will lead to the loss of the synthesis capacity of this enzyme. We have denoted here by a^+ the wild type gene and by a^- the mutant gene and we will call *forward mutations* events of type $a^+ \to a^-$ and *back* (reverse) *mutations* events of type $a^- \to a^+$.

In diploid cells the revelation of the phenomenon may present some difficulties, not only because of the fact that some mutations are difficult to discern but also because of the biologic peculiarities of the cell.

As stressed by MARKERT (1964), during cell division each daughter cell receives an identical set of chromosomes, hence an identical set of genes. Yet, these cells in embryos commonly differentiate into quite different cell types, as different as muscle, nerve, and pigment cells. The phenotypes of these cells are conspicuously different, although they have the same genotype. *Each cell characteristically manifests only a part of its genome.* The gene controlling the ability to synthetize the enzyme tyrosinase is an example (MARKERT (1958)). Animals with mutant, inactive forms of this gene cannot synthetize tyrosinase, therefore make no melanin pigment, and consequently become albinos. However, in a normal animal only the melanin-synthesizing cells show any evidence of possessing a gene for tyrosinase synthesis. The remaining cells are indistinguishable from the corresponding cells of an albino animal. This points out the fact that apparent gene function is dependent upon the state of differentiation of the cell and in turn, the state of differentiation is dependent at any time upon the integrated pattern of previous gene activity.

This important statement also finds its confirmation in studies of pharmacogenetics. For example, it was observed that drugs such as pamaquine, neosalvarsan, sulfanylamide, trinitrotoluene, etc. can induce an acute hemolytic anemia in some patients. The observed genetic defect, assigned to the inactivity of a gene controlling the synthesis of red cell glucose-6-phosphate dehydrogenase (a sex-linked trait), was discovered only in the older circulating erythrocyte (KIRKMAN (1968)). Among genes present in differentiated cells we mention histocompatibility genes controlling the response to tissue transplants. Thus one knows the H-2 region on chromosome 9 in the diploid cells of mice, a region subdivided in several loci $H\text{-}2^a, \ldots, H\text{-}2^s$. For example, a tumour arising in an A × SW hybrid mouse (genotype $H\text{-}2^a$ $H\text{-}2^s$) is unable to grow in the parental A-line (genotype $\mathrm{H\text{-}2}^a$ $\mathrm{H\text{-}2}^a$) unless a mutation occurred from $H\text{-}2^s$ to $H\text{-}2^a$ in one of the tumour cells (KLEIN and KLEIN (1956)).

The study of mutations in viruses and bacteria led to the denoting of those genetic *markers* (genetically controlled traits) phenotypically expressing the constitution of standard lines and of their mutants. For example, on attenuated (mutant) lines of poliomyelitis virus we shall call the lack of pathogenity in monkeys through intracerebral inoculation an *N character*, the growth inhibition in the presence of guanidine a *g character*, the reduced capacity of multiplying on monkey renal cells a *MS character*, etc. For bacteria and somatic cells, although chromosomal and certain other morphological markers could find accessory use in genetic studies, good selective characteristics such as complete nutritional dependence versus independence, or high-level resistance to a drug, virus, or antiserum versus sensitivity, are obviously the markers of choice (see SZYBALSKI and SZYBALSKA (1962)). It is clear that by a type $a^+ \to a^-$ mutation we will generally understand the modification of a single marker.

2.3.5.3. The second important question is that of knowing which model of *normal* growth must be chosen. According to the assumption made on the distribution of the division time, τ, KENDALL (1952, 1953) has defined three types of growth processes:

– growth processes of type D, associated with a deterministic model of the fission process in which τ is supposed to have a fixed value $1/b$;

– growth processes of type M, associated with a "Markovian" negative exponential distribution of τ;

– growth processes of type G, associating a perfectly general form with the division time distribution (as in multiple-phase processes).

We shall thus be able to denote the type of a mutation model by indicating the type of growth for the standard population and for the mutant one. Since in some cases there is a temporal disagreement between the modification of the genotype and its phenotypic expression ("phenotypic delay"), we shall denote by + or 0 whether or not the effect of this delay has been taken into account. Thus, the symbol $D/D/0$ will indicate the deterministic type of growth for both population and the ignoring of the phenotypic delay (LURIA and DELBRÜCK (1943), LEA and COULSON (1949), ARMITAGE (1952)).

A model involving purely deterministic growth would still retain a stochastic element due to the random occurrence of mutations. Supposing that the normal population grows purely deterministically, we should then have reduced a two-dimensional stochastic problem to a one-dimensional stochastic problem. As observed by BAILEY (1964 a), it is not unreasonable, at least as a first approximation, to assume that while both wild and mutant components are subject to birth with the same rate b, no death is involved.

BARTLETT (1951; see ARMITAGE (1952)) has built a model according to the scheme $M/M/0$; if we suppose that the mutant population grows according to a general law we get a $D/G/0$ model (HARRIS (1951)). The most important types of stochastic models of mutation can be found in the above quoted papers, particularly for bacteria mutations (see also KENDALL (1952, 1953)). For virus mutations, see STEINBERG and STAHL (1961), KIMBALL (1965), GANI and YEO (1965), and for somatic cell mutations see also FIRESCU and TAUTU (1967 b). In the following we shall study the growth of two subpopulations subject to mutations within the interconnected birth-and-death processes as described by PURI (1968 b).

2.3.5.4. The processes we shall consider are in fact two interconnected birth-and-death processes. Let $\xi_1(t)$ be the number of objects in wild state S_1 at time t, with $\xi_1(0) = 1$, and let $\xi_2(t)$ be the number of objects in mutant state S_2 at time t, with $\xi_2(0) = 0$. We shall now introduce the following probabilities:

1°. $b_i\Delta t + o(\Delta t)$, the probability that an object in state S_i $(i = 1,2)$ multiplies by binary fission in the short time interval $(t, t + \Delta t)$;

2°. $d_i \Delta t + o(\Delta t)$, the probability that an object in state $S_i (i = 1,2)$ dies in this state in the short time interval $(t, t + \Delta t)$;

3°. $\alpha \Delta t + o(\Delta t)$, the probability that an object in state S_1 transfers to state S_2 by a forward mutation, in the short time interval $(t, t + \Delta t)$;

4°. $\beta \Delta t + o(\Delta t)$, the probability that an object in state S_2 transfers to state S_1 by a reverse mutation, in the short time interval $(t, t + \Delta t)$;

5°. $1 - (b_1 + d_1 + \alpha) \Delta t + o(\Delta t)$ [or $1 - (b_2 + d_2 + \beta) \Delta t + o(\Delta t)$], the probability that an object in state S_1 [or S_2] undergoes no transition in the short time interval $(t, t + \Delta t)$;

6°. $o(\Delta t)$, the probability of having more than one mutation in the system during the short time interval $(t, t + \Delta t)$.

Furthermore, it is assumed that all the events that might occur to a B-object in $(t, t + \Delta t)$ are independent of the events occurring to other B-objects and of the events that occurred to this object in the past. Consider the probabilities

$$P_{x_1, x_2}(t) = \mathsf{P}(\xi_1(t) = x_1, \ \xi_2(t) = x_2 \mid \xi_1(0) = 1, \ \ \xi_2(0) = 0),$$

and let $G(z_1, z_2; t)$ denote their probability generating function, $|z_1|$, $|z_2| \leqslant 1$. Following the standard argument, the Kolmogorov forward differential equation for G is

$$\begin{aligned} \frac{\partial G}{\partial t} &+ [(1 - z_1)(b_1 z_1 - d_1) + \alpha (z_1 - z_2)] \frac{\partial G}{\partial z_1} + \\ &+ [(1 - z_2)(b_2 z_2 - d_2) + \beta (z_2 - z_1)] \frac{\partial G}{\partial z_2} = 0, \end{aligned} \tag{2.58}$$

with $G(z_1, z_2, 0) = z_1$. In its present generality, equation (2.58) seems intractable and we shall consider, for the moment, the case when all the rates are positive constants.

Besides $\xi_1(t)$ and $\xi_2(t)$, we also consider the random variables $\eta_1(t)$ and $\eta_2(t)$ which denote the numbers of transitions during $(0, t)$ from $S_1 \to S_2$ and from $S_2 \to S_1$ respectively. For $i, j = 1, 2$ let

$$g_i(z_1, z_2, \nu_1, \nu_2; t) = \mathsf{E}(z_1^{\xi_1(t)} z_2^{\xi_2(t)} \nu_1^{\eta_1(t)} \nu_2^{\eta_2(t)} \mid \xi_i(0) = 1, \ \ \xi_j(0) = 0, \ i \neq j).$$

By considering the first change wh'ch occurs in the period $(0, t)$, it is easy to establish that the probability generating functions g_1 and g_2 satisfy the integral equations

$$\begin{aligned} g_1(z_1, z_2, \nu_1, \nu_2; t) &= z_1 e^{-A_1 t} + \int_0^t e^{-A_1 \tau} [d_1 + b_1 g_1^2(z_1, z_2, \nu_1, \nu_2; t - \tau) + \\ &+ \alpha \nu_1 g_2(z_1, z_2, \nu_1, \nu_2; t - \tau)] \, d\tau, \\ g_2(z_1, z_2, \nu_1, \nu_2; t) &= z_2 e^{-A_2 t} + \int_0^t e^{-A_2 \tau} [d_2 + b_2 g_2^2(z_1, z_2, \nu_1, \nu_2; t - \tau) + \\ &+ \beta \nu_2 g_1(z_1, z_2, \nu_1, \nu_2; t - \tau)] \, d\tau, \end{aligned}$$

where $A_1 = b_1 + d_1 + \alpha$ and $A_2 = b_2 + d_2 + \beta$. Making a change of variable from τ to θ by putting $t - \tau = \theta$ in the integral of g_1, multiplying both sides of the integral equation by $e^{A_1 t}$ and finally by differentiating both sides with respect to t, we obtain the Kolmogorov backward equation for g_1:

$$\frac{dg_1}{dt} = b_1 g_1^2 - A_1 g_1 + \alpha v_1 g_2 + d_1. \tag{2.59}$$

In a similar way we obtain the corresponding equation for g_2:

$$\frac{dg_2}{dt} = b_2 g_2^2 - A_2 g_2 + \beta v_2 g_1 + d_2, \tag{2.60}$$

where both equations are subject to the initial condition $g_i(z_1, z_2, v_1, v_2; 0) = z_i$, $i = 1,2$, while the variables z_1, z_2, v_1 and v_2 are kept fixed.

2.3.5.5. Let us now consider the case in which the two birth-and-death processes are left disconnected ($\beta = 0$) and the remaining rates are positive constants. This is the case in which we assume that no B-object passes from the mutant state S_2 to the wild one, S_1.

Here it is dif cult to obtain an exact expression for the probability generating function g_1 from (2.59). The expression for g_2 of equation (2.60) is known for the present case:

$$g_2(z_2; t) = \frac{(b_2 z_2 - d_2) + d_2(1 - z_2)\, e^{(b_2 - d_2)t}}{(b_2 z_2 - d_2) + b_2(1 - z_2)\, e^{(b_2 - d_2)t}}.$$

The method that will yield an approximate solution for g_1 is based on the introduction of a sequence of stochastic processes $(\xi_1^{(n)}(t), \xi_2^{(n)}(t), \eta_1^{(n)}(t))_{t \geqslant 0}$ with $\xi_1^{(n)}(0) = 1$, $\xi_2^{(n)}(0) = \eta_1^{(n)}(0) = 0$, $n \in N$, such that the sequence $(f^{(n)}(z_1, z_2, v_1; t))_{n \in N}$ of their probability generating functions tends to $g_1(z_1, z_2, v_1; t)$ as $n \to \infty$. Here, since $\beta = 0, \eta_2(t) \equiv 0$. For convenience we define for $n = 0, f^{(0)}(z_1, z_2, v_1; t) \equiv 1$ for $t > 0$. This would imply that starting with $\xi_1^{(0)}(0) = 1$, the object is considered as dead right from the start.

Let us define the process $(\xi_1^{(1)}(t), \xi_2^{(1)}(t), \eta_1^{(1)}(t))_{t \geqslant 0}$ as follows: for a starting object in state S_1 we have the following three possibilities: of dying with rate d_1, of mutating in state S_2 with rate α (where it undergoes a linear birth-and-death process with parameters b_2 and d_2, respectively), or, finally, of giving birth with rate b_1 (with the property that as soon as the birth takes place it follows from then on the process corresponding to $n = 0$, whereas its progeny follows the process with $n = 1$). The interpretation of the nth process will be similar: when the B-object considered gives birth to an offspring, this follows the nth process, while the parental B-object follows the $(n-1)$th process. We then get the integral equation for $f^{(n)}$, $n \in N^*$ as

$$f^{(n)}(t) = z_1 e^{-A_1 t} + \int_0^t e^{-A_1(t-\tau)} [d_1 + b_1 f^{(n-1)}(\tau) f^{(n)}(\tau) + \alpha \nu_1 g_2(\tau)] \, d\tau,$$

which yields the differential equation

$$\frac{df^{(n)}}{dt} = b_1 f^{(n-1)} f^{(n)} - A_1 f^{(n)} + d_1 + \alpha \nu_1 g_2, \tag{2.61}$$

with $f^{(n)}(z_1, z_2, \nu_1; 0) = z_1$. Here the arguments z_1, z_2, and ν_1 of $f^{(n)}$, $f^{(n-1)}$, and g_2 are suppressed for convenience. Because of the recursive character of (2.61), it can be solved for $f^{(n)}$ recursively yielding for $n = 1,2, \ldots$

$$f^{(n)}(t) = z_1 \exp\left(-A_1 t + b_1 \int_0^t f^{(n-1)}(\tau) \, d\tau\right) +$$

$$+ \int_0^t \exp\left(-A_1(t-\tau) + b_1 \int_\tau^t f^{(n-1)}(u) \, du\right) \times$$

$$\times (d_1 + \alpha \nu_1 g_2(\tau)) \, d\tau.$$

P. S. PURI proved that $(f^{(n)}(z_1, z_2, \nu_1; t))_{n \in N}$ converges uniformly to $g_1(z_1, z_2, \nu_1; t)$ for $0 \leqslant z_1 \leqslant 1$, $0 \leqslant z_2 \leqslant 1$, $0 \leqslant \nu_1 \leqslant 1$, $t \in [0, T]$, (T finite but arbitrary), and that for every fixed (z_1, z_2, ν_1) with $0 \leqslant z_1$, $z_2 < 1$, $0 \leqslant \nu_1 \leqslant 1$, the limits $q \equiv \lim_{t \to \infty} g_2(z_2, t)$, $p_n \equiv \lim_{t \to \infty} f^{(n)}(z_1, z_2, \nu_1; t)$, $p \equiv \lim_{t \to \infty} g_1(z_1, z_2, \nu_1; t)$, $n \in N$, all exist. The following relations hold:

$$p_n = \frac{\alpha \nu_1 q + d_1}{A_1 - b_1 p_{n-1}}, \quad n \in N^*,$$

$$p = \lim_{n \to \infty} p_n.$$

Solving this relation for $p = p(\nu_1, q)$ we have

$$\lim_{t \to \infty} q_1(z_1, z_2, \nu_1; t) \equiv p(\nu_1, q) = \frac{1}{2b_1}(A_1 - [A_1^2 - 4b_1(d_1 + \alpha \nu_1 q)]^{1/2}).$$

Setting $p_{00} = \lim_{t \to \infty} \mathsf{P}(\xi_1(t) = \xi_2(t) = 0)$, the probability of ultimate extinction of both populations, from the above relation we obtain:

$$p_{00} = p(1, q) = \begin{cases} \dfrac{1}{2b_1}(A_1 - [A_1^2 - 4b_1(d_1 + \alpha q)]^{1/2}), & d_2 < b_2 \\ \dfrac{d_1 + \alpha}{b_1}, & d_2 \geqslant b_2,\ d_1 + \alpha < b_1 \\ 1, & d_2 \geqslant b_2,\ d_1 + \alpha \geqslant b_1. \end{cases}$$

When $p_{00} < 1$, with probability $1 - p_{00}$, $(\xi_1(t), \xi_2(t)) \to (0, \infty)$ or (∞, ∞) as $t \to \infty$, according as $d_1 + \alpha \geqslant b_1$ and $d_2 < b_2$ or $d_1 + \alpha < b_1$ and $d_2 < b_2$, respectively.

If we ignore the random variable $\xi_2(t)$, we can obtain the probability generating function $g_1(z_1, \nu_1; t)$ of $\xi_1(t)$ and $\eta_1(t)$ explicitly. Replacing g_2 by unity in (2.59), we have

$$\frac{\mathrm{d}g_1}{\mathrm{d}t} = b_1 g_1^2 - A_1 g_1 + \alpha\nu_1 + d_1.$$

This can be solved easily subject to the initial condition $g_1(z_1, \nu_1; 0) = z_1$ yielding

$$g_1(z_1, \nu_1; t) = W_2(\nu_1) + \frac{W_1(\nu_1) - W_2(\nu_1)}{1 - \dfrac{z_1 - W_1(\nu_1)}{z_1 - W_2(\nu_1)} \exp(b_1(W_1(\nu_1) - W_2(\nu_1))t)},$$

where $W_1(\nu_1)$ and $W_2(\nu_1)$ are, with positive and negative signs respectively, $\dfrac{1}{2b_1}(A_1 \pm [A_1^2 - 4b_1(\alpha\nu_1 + d_1)]^{1/2})$. Since it is a nondecreasing function of t, $\eta_1(t) \uparrow \eta$ a.s. as $t \to \infty$, where the probability generating function of η is given by

$$\lim_{t \to \infty} g_1(1, \nu_1; t) = W_2(\nu_1) = p(\nu_1, 1).$$

From this we obtain

$$\mathsf{P}(\eta = 0) = \frac{1}{2b_1}(A_1 - [A_1^2 - 4b_1 d_1]^{1/2}),$$

and for $k \in N^*$,

$$\mathsf{P}(\eta = k) = \frac{1}{b_1}\left[\frac{(2k-2)!}{k!(k-1)!} \cdot \frac{(\alpha b_1)^k}{(A_1^2 - 4b_1 d_1)^{k-1/2}}\right].$$

These probabilities add up to 1 only when $d_1 + \alpha \geqslant b_1$. When $d_1 + \alpha < b_1$, $W_2(1) < 1$, so that $\mathsf{P}(\eta = \infty) = 1 - W_2(1)$.

2.3.5.6. FIRESCU and TAUTU (1967 b) have tried to stress the genetic framework within which mutation models must be analysed by building two type $D/D/0$ models, one for haploid populations and the other for diploid populations. The importance of chromosomal make up was thus emphasized, particularly in the case of provoked mutations. The model for diploid populations starts from the experimental research carried out by PUCK and MARCUS (1956) on HeLa cell cultures. According to these authors, the proportion of survivors in a population of X-ray irradiated cells follows an empirical law of the form $[1-(1-\mathrm{e}^{-c/c_0})^n]^m$, where c is the administered X-ray dose, c_0 is the standard 37 per cent lethal dose (LD_{37}), n is the somatic number of B-objects ($n = 1$ for haploid cells, $n = 2$ for diploid cells), while m is the number of loci damaged by X-ray particles (see also HUG and KELLERER (1966)).

The results obtained by FIRESCU and TAUTU (1967 b) point out the fact that the growth process of the two populations (wild and mutant) is formally distinguished from the proper mutational process. In the case of haploid populations, the following assumptions are made: (i) the mutations occur in both directions, at rates α and β, respectively, (ii) the two rates are linearly dependent on the size of the respective populations, (iii) both populations, wild and mutant, are governed by a deterministic exponential law, with positive characteristic rates k_1 and k_2, respectively. By using classical procedures, one obtains the following differential equations for the means $M_{10}(t)$ and $M_{01}(t)$ of the random variables $\xi_1(t)$ = the number of wild haploid cells at time t, and $\xi_2(t)$ = the number of mutants at time t:

$$\frac{\mathrm{d}M_{10}(t)}{\mathrm{d}t} = -\alpha M_{10}(t) + \beta M_{01}(t) + k_1\, x_0 \mathrm{e}^{k_1 t},$$

$$\frac{\mathrm{d}M_{01}(t)}{\mathrm{d}t} = \alpha M_{10}(t) - \beta M_{01}(t) + k_2\, y_0\, \mathrm{e}^{k_2 t},$$

with the initial condition $\xi_1(0) = x_0$, and $\xi_2(0) = y_0$. Clearly,

$$\frac{\mathrm{d}M_{10}(t)}{\mathrm{d}t} + \frac{\mathrm{d}M_{01}(t)}{\mathrm{d}t} = k_1 x_0 \mathrm{e}^{k_1 t} + k_2 y_0 \mathrm{e}^{k_2 t}.$$

2.3.5.7. Karlin and McGregor (1967 b) have assumed that a mutation model which describes the fluctuation of the number of mutant lines over time consists, in fact, of two processes: (1) the stochastic process of formation of new allele (mutant) populations, (2) the stochastic process underlying the growth pattern of a particular mutant population.

Let $\eta(t)$ be the number of mutant lines coming into existence due to spontaneous mutations during the time interval $[0, t]$ On account of the assumption that a gene can have several alleles, we understand by "mutant lines" the subpopulations which appear after mutations within wild population. Thus, it is known that in *Drosophila melanogaster*, at locus w (the locus of the gene determining the white colour of the eye), there can be more than 20 alleles (denoting, for instance, by w^i the allele determining the white-ivory phenotype, by w^p the allele for white-pearl, by w^{bf} the allele for the white-buff colour, etc.). There are three alleles in man occupying the locus corresponding to serologic (group) differences of the blood: I^A, I^B, I^O. S. Karlin and J. McGregor state that the usual models involving the effects of mutation rates alone fail to account for these polymorphisms. For example, for populations of size 10^3, it would be necessary to ascribe a mutation rate as high as 10^{-3}, which is completely inconsistent with known mutation rates, which are much smaller. On the other hand, if these calculations are made by taking into account the biochemical structure of a gene, the results will likewise be questionable. If a typical gene is 10^3 nucleotides and there are 3×10^9 nucleotide pairs, then a mutation rate as low as 10^{-6} per gene yields three mutations per gamete per generation, seemingly too high to be credible, as Crow (1968) claims. The statement "one mutation – one phenotype modification" must be submitted to serious criticism. Thus there is the possibility of very fine adjustments to the environment by gene replacement that are not reflected in any overt change in the phenotype. Likewise, the possibility of a great deal of isoallelic variation among "normal" alleles has been made a reality by the finding of numerous polymorphisms for isozymes and other proteins (Crow (1968)). Since environmental conditions constantly change, the study of stationary or equilibrium mutation models is unable to account for the large number of alleles. The first model built by Karlin and McGregor (1967 b) thus treats a nonstationary situation with varying total population size. It describes the fluctuation of the number of mutant lines over time under the fundamental assumption that it consists of two processes: (1) the stochastic process of formation of new allele (mutant) populations, (2) the stochastic process which underlies the growth pattern of a particular mutant population.

The process of mutant lines formation $(\eta(t))_{t\geq 0}$ is called *the input process*. Here are three examples of $(\eta(t))$ of special interest:

(a) The number of mutant lines $\eta(t)$ is a homogeneous Poisson process with parameter λ or, more generally, $\lambda(t)$. The dependence of the parameter of the process on the time variable reflects the possibility of changing environmental conditions.

(b) The number of mutant lines $\eta(t)$ is a renewal process, that is, the times between the successive starts of new mutant lines are independent positive random variables with common distribution function $F(\tau)$, $0 < \tau < \infty$. The times of the creation of new mutant lines (in short, events) are thus taken to occur at $\tau_1, \tau_1 + \tau_2, \ldots, \tau_1 + \ldots + \tau_n, \ldots$, where $\tau_1, \tau_2, \ldots$, are independent positive random variables with distribution function F. In case F is a degenerate distribution corresponding to the value one, then a new mutant line comes into existence each unit of time.

(c) The number of mutants $\eta(t)$ is a general increasing point process of which a special example would be a Yule-Furry process, that is, the times of the creation of new mutant lines coincide with the times of events of a Yule-Furry process.

The above formulation postulates that at each event of $\eta(t)$ a single new mutant line is formed. This could obviously be generalized such that the number of mutant lines coming into existence at a given moment may be larger than one, in fact, possibly a random process. In this case the input process of example (b) becomes a compound renewal process.

The definition of the process and the added examples are intended to stress the important fact that the lifetime distribution of the population of a given allele may be a significant factor in accounting for large numbers of different alleles represented. Specifically, if the life of a specific allele in the population is very long, then even with very small mutation (and/or migration) rates the number of different alleles in existence becomes large if a long time has elapsed.

The nature of the life of a specific mutant population is the second process considered. The simplest assumption is that the fluctuation of a specific mutant population follows the laws of a Markov process $(\gamma(t))_{t\geqslant 0}$. If at time $t = 0$, a given mutant type is represented i times, then at time t its size is j with probability $p_{ij}(t)$. The state 0 is assumed to be absorbing, which means that the particular mutant form in question becomes extinct once state 0 is entered. One usually assumes that absorption into state 0 is a certain event. The process $(\gamma(t))_{t\geqslant 0}$ might, for example, be:

– A linear birth-and-death process for which $d \geqslant b$; or

– A continuous time branching process for which $f'(1) \leqslant 1$ (see I; Theorem 2.3.25).

Let us denote by $\xi_k(t)$ the number of mutant populations at time t consisting of exactly k members, that is the number of mutant populations in state k at time t. The distribution function of this random variable will depend on the nature of the two component processes.

Thus, if the input process is a nonhomogeneous Poisson with parameter $\lambda(t)$, then it is not difficult to show that the random processes $(\xi_k(t))_{t\geq 0}$, $k \in N^*$, are independent nonhomogeneous Poisson processes with joint generating function

$$\Phi(t; z_1, z_2, \ldots) = \sum_{i \in N^*} \sum_{x_1, x_2, \ldots} z_1^{x_1} z_2^{x_2} \ldots \mathsf{P}(\xi_i(t) = x_i, i \in N^*) =$$

$$= \exp\left(\sum_{k \in N^*} (z_k - 1) \int_0^t p_{1k}(t - \tau) \lambda(\tau) \, d\tau \right).$$

Further

$$\mathsf{E}\, \xi_k(t) = \int_0^t p_{1k}(t - \tau) \lambda(\tau) \, d\tau.$$

If $\lambda(t) = \lambda$ is constant, the distribution of $\xi_k = \lim_{t \to \infty} \xi_k(t)$, (convergence in law), is Poisson with parameter

$$\mathsf{E}\xi_k = \lambda \int_0^\infty p_{1k}(\tau) \, d\tau,$$

provided the integral exists, which is certainly the case if 0 is an absorbing state. The asymptotic expected number of alleles will be

$$m = \lambda \int_0^\infty \sum_{k \in N^*} p_{1k}(\tau) \, d\tau = \lambda \int_0^\infty [1 - F(\tau)] \, d\tau, \tag{2.62}$$

where

$$F(\tau) = 1 - \sum_{k \in N^*} p_{1k}(\tau)$$

is the distribution function of the time to extinction for a particular mutant line generated by a single initial parent. Notice that m is finite iff the distribution function $F(t)$ has finite mean. It is possible for $\mathsf{E}\xi_k$ to be finite while m is infinite.

Relation (2.62) is also valid when the input process is a renewal process, provided we interpret λ as the reciprocal of the mean interarrival time.

2.3.5.8. Consider now that the input process is a homogeneous Poisson one with parameter λ and the growth process $(\gamma(t))_{t\geq 0}$ is such that each mutant type ultimately becomes lost with probability 1 but the expected time to extinction is ∞. More specifically, we shall assume that the probability distribution $F(t)$ of the time until extinction (starting with a single initial parent) has the asymptotic growth behaviour

$$1 - F(t) = \sum_{k \in N^*} p_{1k}(t) \sim \frac{c}{t^a}, \tag{2.63}$$

as $t \to \infty$, where $0 < a \leqslant 1$, and c is a constant. For example, if $(\gamma(t))_{t \geqslant 0}$ is a linear birth-and-death process with both birth and death parameters equaling b, then by I; 2.3.4.7.5,

$$1 - F(t) \sim \frac{1}{bt}.$$

as $t \to \infty$. In this case we can determine the limiting growth behaviour of $\xi_k(t)$ and $\xi(t) = \sum_{k \in N^*} \xi_k(t)$, which is the total number of existing mutant types at time t, as $t \to \infty$. Explicitly, we get that $\xi_k(t)$ is Poisson distributed with mean

$$\lambda \int_0^t \frac{(b\tau)^{k-1}}{(1 + b\tau)^{k+1}} \, d\tau = \frac{\lambda}{b} \int_0^{bt/(1+bt)} u^{k-1} \, du = \frac{\lambda}{bk} \left(\frac{bt}{1 + bt} \right)^k.$$

Hence,

$$\lim_{t \to \infty} \mathsf{E}\, \xi_k(t) = \frac{\lambda}{bk}.$$

Thus, the number of populations with k members is of mean size λ/bk. Hence the total number of existing mutant types $\xi(t)$ has an asymptotic normal distribution with mean $\frac{\lambda}{b} \log (1 + bt)$ and variance $\frac{\lambda}{b} \log (1 + bt)$. A more general result is

Theorem 2.1. *Under* (2.63), $\xi(t)$ *has an asymptotic normal distribution (as* $t \to \infty$*) with mean and variance*

$$\mathsf{E}\, \xi(t) \sim \frac{\lambda c t^{1-a}}{1 - a}, \quad \mathsf{D}\, \xi(t) \sim \frac{\lambda c t^{1-a}}{1 - a},$$

when $0 < a < 1$, *and*

$$\mathsf{E}\, \xi(t) \sim \lambda c \log t, \quad \mathsf{D}\, \xi(t) \sim \lambda c \log t,$$

when $a = 1$.

Proof. From (2.62) we see that $\xi(t)$ is Poisson distributed with mean and variance given by

$$\lambda \int_0^t [1 - F(\tau)] \, d\tau \sim \begin{cases} \dfrac{\lambda c t^{1-a}}{1-a}, & 0 < a < 1 \\ \lambda c \log t, & a = 1 \end{cases}$$

for $t \to \infty$. The asymptotic normality follows immediately since the mean parameter tends to ∞ as $t \to \infty$. ◇

2.3.5.9. We observe that under the conditions in Subparagraph 2.3.5.7 if $m < \infty$, the number of different mutant lines maintained in the population achieves (as $t \to \infty$) a stable state (in the stochastic sense) which is Poisson distributed. The limiting total population size is random. On the other hand, subject to the hypothesis of Theorem 2.1, the expected number of alleles grows to infinity owing to the fact that the average lifetime of each allelic type is infinite even though each individual allele is ultimately lost with probability one. This asserts, in particular, that even if the rate of formation of new alleles, that is, the mutation rate, is exceedingly small, the number of existing alleles following an elapse of a sufficient duration of time is large and becomes infinite unless deterrents imply the nonapplicability of the model. In other words, the property that each specific mutant population possesses a long lifetime may be a significant factor in accounting for the large number of alleles observed in nature. See also SINGER (1970, 1971).

2.4. Stochastic demographic models

2.4.1. The discrete time model

2.4.1.1. The majority of demographic models is deterministic. These models exist in two forms: those using a discrete time variable and a discrete age scale, and those using a continuous time variable and a continuous age scale. The starting point of a mathematical analysis is either in discrete form as a set of linear first-order difference equations or in continuous form as an integral equation. The solution of the integral equation may proceed either by elementary calculus or by applying the Laplace transform to a sum of exponentials weighted by constants. The discrete first-order set of difference equations may be expressed directly as a matrix or made to depend on a higher order linear difference equation (see KEYFITZ (1967 a)).

The classical deterministic model of SHARPE and LOTKA (1911) is a continuous one, followed by models of the same type of LOTKA (1939) and RHODES (1940). Discrete deterministic models were suggested by BERNARDELLI (1941), LEWIS (1942) and LESLIE (1945, 1948). The latter are more accessible to actuarial practice and are preferable when the age-specific birth and death rates are to be given on the basis of empirical data. The important instrument of these models – known since 1895 by demographers such as E. CANNAN – is *population projection*, that is the calculation of the expected number of persons, age by age, at points in time subsequent to a census or other starting point, given a set of birth and death rates (KEYFITZ (1967 a, b)). Its matrix form was suggested by H. BERNARDELLI as well as by E. G. LEWIS and studied in detail by P. H. LESLIE (see other developments by KEYFITZ (1967 b), KEYFITZ and MURPHY (1967)). In a paper attempting to reconcile the two classes of models, KEYFITZ (1967 a) has demonstrated that the matrix gives the same result as the integral equation if the age intervals are made small enough and, with the five-year age intervals common in practice, gives a different finite approximation to the same result.

Taking up the same problem, in the same year and the same periodical where Keyfitz's paper was issued, L. A. GOODMAN found a new formula pertaining to the continuous model which in turn helped to clarify the relations between the theory of population growth and concepts pertaining to the "reproduction value" of a population – which were introduced by FISHER (1930) for the continuous model and by LESLIE (1948) for the discrete model. GOODMAN (1967 a) has also shown that Lotka's classical theory is limited in a number of respects: (1) it is concerned primarily with the growth of only one of the sexes (females) in the population, and it ignores for the most part the other sex; (2) it assumes that the birth and death rates per female depend only upon the age of the female, and it ignores for the most part other factors such as parity or nuptiality; (3) it assumes that instantaneous birth and death rates can be defined at each infinitesimal instant in the life of females, and it ignores the fact that in practice these rates are usually calculated only for five-year or one-year age-intervals; (4) the presentation of this theory does not include a correct statement of the conditions under which the main results of this theory are valid.

The stochastic analogue of Lotka's deterministic model was set up by KENDALL (1949), and the stochastic analogue of the discrete deterministic model by POLLARD (1966). However, it is interesting to remark that there are also deterministic analogues of some of the stochastic models. Thus, the discrete model when birth and death rates depend upon several factors (GOODMAN (1969)) is a special case of the deterministic analogue of multi-type Galton-Watson chains. Likewise, the model for birth, death and migration processes of spatially distributed populations (USHER and WILLIAMSON (1970)) is the deterministic analogue of BAILEY's (1968 a) model, presented under Paragraph 2.2.2.

We omit examples of finite Markov chains; an extensive study for demographic micromodels is that of FEICHTINGER (1971, Ch. 2); see also CHIANG (1968 Ch. 2). The term "micromodel" has been used by

HYRENIUS (1965). An applied account of ergodic chains (KEMENY and SNELL (1960, Ch. 5) in studying stable (and Malthusian) populations was given by SYKES (1969 b). See in this respect also FEICHTINGER (1971).

2.4.1.2. Pollard's discrete stochastic model starts from the following considerations:

1°. The female population (briefly, the F-population) is only measured at discrete intervals of time, $n = 0, 1, 2, \ldots$

2°. This population is divided into $(k+1)$ age-groups, usually denoted by 0-, 1-, ..., k-.

3°. The number of females (or, generally, F-individuals) in the age-group j- $(0 \leqslant j \leqslant k)$ at time n is a random variable $\gamma_n(j)$, γ being the first letter of the Greek word *gyne* (=female). We shall set $\mathsf{E}\gamma_n(j) = M_{j,n}$ and $\mathsf{D}\gamma_n(j) = D_{j,n}$.

4°. If an F-individual in the age-group j- at time $n-1$ gives birth to an F-individual at time n, the number of females in the age-group 0- at time n whose mothers were aged j at the time of the birth is a random variable $\gamma_n^{(j)}(0)$. Obviously,

$$\gamma_n(0) = \sum_{j=0}^{k} \gamma_n^{(j)}(0).$$

5°. The probability that an F-individual from age-group j- at time n, will survive to be in age-group $(j+1)$- after a unit interval of time is a fixed probability p_j, which for $j < k$ is strictly positive, $p_k = 0$. They are assumed independent. Hence $\gamma_{n+1}(j+1)$ is a binomial variable $B(\gamma_n(j), p_j)$, conditional on $\gamma_n(j)$. We shall write $q_j = 1 - p_j$.

6°. The probability that an F-individual in age-group j- at time n will give birth during the unit time interval $(n, n+1)$ to a single daughter, and that this daughter will be alive in age-group 0-at time $n+1$, is a fixed probability b_j. They are assumed independent. Hence $\gamma_{n+1}^{(j)}(0)$ is a binomial random variable $B(\gamma_n(j), b_j)$ conditional on $\gamma_n(j)$. We shall write $d_j = 1 - b_j$.

7°. The birth and the death processes are assumed independent.

8°. Changes in the male population structure are assumed to be consistent with the assumption of constant fertility measures $\{b_j\}$.

9°. Multiple births are ignored.

Before presenting the demographic model itself, it is necessary to know the following fundamental result: Let ξ and η be discrete random variables taking positive integral values. Let ξ_1' and ξ_2' be random variables with binomial distributions $B(\xi, p_1)$ and $B(\xi, p_2)$, respectively, conditional on ξ. Similarly, let η' be a random variable with binomial distribution $B(\eta, p_3)$ conditional on η. Then,

$$\left.\begin{aligned} \mathsf{E}\xi_1' &= p_1 \mathsf{E}\xi, \\ \mathsf{D}\xi_1' &= p_1^2 \mathsf{D}\xi + p_1 q_1 \mathsf{E}\xi, \end{aligned}\right\} \tag{2.64}$$

$$\left.\begin{aligned} \operatorname{Cov}[\xi_1', \xi_2'] &= p_1 p_2 \mathsf{D}\xi, \\ \operatorname{Cov}[\xi_1', \eta'] &= p_1 p_3 \operatorname{Cov}[\xi, \eta], \end{aligned}\right\} \tag{2.65}$$

where $q_1 = 1 - p_1$. Furthermore, if ξ is a binomial random variable, then ξ_1' is also a binomial random variable. This result was given by POLLARD (1966) and in a more general form by KENDALL (1949).

Considering again our model we can therefore write

$$\begin{aligned}
\mathsf{E}\,\Upsilon_{n+1}(0) &= M_{0,\,n+1} = \sum_{j=0}^{k} b_j\, M_{j,\,n}\,,\\
\mathsf{E}\,\Upsilon_{n+1}(1) &= M_{1,\,n+1} = p_0\, M_{0,\,n},\\
\mathsf{E}\,\Upsilon_{n+1}(2) &= M_{2,\,n+1} = p_1\, M_{1,\,n},\\
&\cdots\cdots\cdots\cdots\cdots\\
\mathsf{E}\,\Upsilon_{n+1}(k) &= M_{k,\,n+1} = p_{k-1}\, M_{k-1,\,n}.
\end{aligned} \tag{2.66}$$

Using relations (2.64) and (2.65) we obtain

$$\mathsf{D}\,\Upsilon_{n+1}(j+1) = D_{j+1,\,n+1} = p_j^2\, D_{j,\,n} + p_j\, q_j\, M_{j,\,n},\ j \geqslant 0 \tag{2.67}$$

$$\operatorname{Cov}[\Upsilon_{n+1}(j+1),\ \Upsilon_{n+1}(h+1)] = p_j\, p_h \operatorname{Cov}[\Upsilon_n(j), \Upsilon_n(h)],$$
$$j, h \geqslant 0, j \neq h \tag{2.68}$$

$$\operatorname{Cov}[\Upsilon_{n+1}^{(j)}(0), \Upsilon_{n+1}(h+1)] = b_j\, p_h\ \operatorname{Cov}[\Upsilon_n(j), \Upsilon_n(h)], j \neq h$$

$$\operatorname{Cov}[\Upsilon_{n+1}^{(j)}(0),\ \Upsilon_{n+1}^{(h)}(0)] = b_j\, b_h\ \operatorname{Cov}[\Upsilon_n(j), \Upsilon_n(h)], j \neq h$$

$$\mathsf{D}\Upsilon_{n+1}^{(j)}(0) = b_j^2\, D_{j,\,n}^2 + b_j\, d_j\, M_{j,\,n},\ j \geqslant 0.$$

Now by definition

$$\Upsilon_{n+1}(0) = \sum_{j=0}^{k} \Upsilon_{n+1}^{(j)}(0).$$

Therefore,

$$\mathsf{D}\Upsilon_{n+1}(0) = \sum_{j=0}^{k} \mathsf{D}\Upsilon_{n+1}^{(j)}(0) + \sum_{j \neq h}\sum \operatorname{Cov}[\Upsilon_{n+1}^{(j)}(0), \Upsilon_{n+1}^{(h)}(0)] =$$

$$= \sum_{j=0}^{k} (b_j^2\, D_{j,\,n} + b_j\, d_j\, M_{j,\,n}) + \sum_{j \neq h}\sum b_j\, b_h \operatorname{Cov}[\Upsilon_n(j), \Upsilon_n(h)]. \tag{2.69}$$

according to the relations above. In the same way we obtain

$$\operatorname{Cov}\left[\sum_{j=0}^{k} \gamma_{n+1}^{(j)}(0), \gamma_{n+1}(k+1)\right] =$$

$$= \operatorname{Cov}[\gamma_{n+1}^{(j)}(0), \gamma_{n+1}(h+1)] + \sum_{j \neq h} \operatorname{Cov}[\gamma_{n+1}^{(j)}(0), \gamma_{n+1}(h+1)] =$$

$$= b_h p_h D_{h,n} + \sum_{j \neq h} b_j p_h \operatorname{Cov}[\gamma_n(j), \gamma_n(h)].$$

Thus,

$$\operatorname{Cov}[\gamma_{n+1}(0), \gamma_{n+1}(h+1)] = \sum_{j=0}^{k} b_j p_h \operatorname{Cov}[\gamma_n(j), \gamma_n(h)]. \quad (2.70)$$

Equations (2.66) — (2.70) completely define the recurrence relations for the means, variances and covariances. Furthermore, they are linear recurrence relations, and may be written in the matrix form as

$$\begin{pmatrix} \mathbf{M}_{n+1} \\ \mathbf{V}_{n+1} \end{pmatrix} = \begin{pmatrix} \mathbf{A} & \mathbf{0} \\ \mathbf{B} & \mathbf{A} \times \mathbf{A} \end{pmatrix} \begin{pmatrix} \mathbf{M}_n \\ \mathbf{V}_n \end{pmatrix}, \quad (2.71)$$

where the vector $\mathbf{V}_n$ has as its elements the variances and covariances $D_{ij,n}$ and these are listed in dictionary order according to the subscripts i and j (for $i \neq j$, $D_{ij,n}$ and $D_{ji,n}$ are both listed). Here $\mathbf{A}$ is the $(k+1) \times \times (k+1)$ Leslie matrix, which is defined as

$$\mathbf{A} = \begin{pmatrix} b_0 & b_1 & \ldots & b_{k-1} & b_k \\ p_0 & 0 & \ldots & 0 & 0 \\ 0 & p_1 & \ldots & 0 & 0 \\ \cdot & \cdot & & \cdot & \cdot \\ \cdot & \cdot & & \cdot & \cdot \\ \cdot & \cdot & & \cdot & \cdot \\ 0 & 0 & \ldots & p_{k-1} & 0 \end{pmatrix}.$$

Here $\times$ denotes the direct product of matrices.

It is of interest to note that by successive application of recurrence relation (2.71), we can readily obtain

$$\begin{pmatrix} \mathbf{M}_n \\ \mathbf{V}_n \end{pmatrix} = \begin{pmatrix} \mathbf{A} & \mathbf{0} \\ \mathbf{B} & \mathbf{A} \times \mathbf{A} \end{pmatrix}^n \begin{pmatrix} \mathbf{M}_0 \\ \mathbf{V}_0 \end{pmatrix} = \begin{pmatrix} \mathbf{M}_n \\ (\mathbf{A} \times \mathbf{A})^n \mathbf{V}_0 + \sum_{i=1}^{n} (\mathbf{A} \times \mathbf{A})^{n-i} \mathbf{B} \mathbf{M}_{i-1} \end{pmatrix}.$$

That is, we have the following relation for the variances and covariances

$$\mathbf{V}_n = (\mathbf{A} \times \mathbf{A})^n \mathbf{V}_0 + \sum_{i=1}^{n} (\mathbf{A} \times \mathbf{A})^{n-i} \mathbf{B}\mathbf{M}_{i-1}.$$

Let us now write $\mathbf{V}_{[n]}$ for the vector of non-central quadratic product-moments at time n. Obviously,

$$\mathbf{V}_{[n]} = \mathbf{V}_n + (\mathbf{M}_n \times \mathbf{M}_n) = \mathbf{V}_n + (\mathbf{A} \times \mathbf{A})^n (\mathbf{M}_0 \times \mathbf{M}_0).$$

Hence we have

$$\mathbf{V}_{[n]} = (\mathbf{A} \times \mathbf{A})^n \mathbf{V}_{[0]} + \sum_{i=1}^{n} (\mathbf{A} \times \mathbf{A})^{n-i} \mathbf{B}\mathbf{M}_{i-1}. \tag{2.72}$$

This equation is of the type encountered in I; 1.2.2.7.2. In this way we arrive at one of the most important theoretical results: *whenever a Leslie deterministic method can be applied, a multitype Galton-Watson model can be used instead*, and the expectation and quadratic moments easily calculated by a recurrence equation similar in form to the corresponding deterministic iterative equation.

It is possible to generalize these results to deal with the higher-order moments about the origin in any multitype Galton-Watson chain. This interesting approach is to be found in Pollard (1966).

2.4.1.3. In the biology of population the expression "additional mortality" has been known for some time, meaning mortality due to disease, lack of food, predatory action or insufficient number of trophic niches. A simple example in the case of human populations is given by the change of meteorological conditions: a well-known fact is that a severe winter raises mortality rates — especially in the old and very young; conversely, a mild winter will mean that the mortality rates experienced are lower than usual. Thus, there may be occasions when it is reasonable to consider the mortality probabilities as random variables and therefore to generalize the usual multitype Galton-Watson chains by assuming that the conditional branching probabilities themselves are random variables.

Branching process calculations performed with fixed conditional probabilities and large populations usually lead to variances considerably smaller than those encountered in practical relations. Of the two major sources of variability, that is, statistical fluctuations due to the finite size of the population and fluctuations in the conditional probabilities themselves, the latter is often greater. The moment recurrence relation (2.71) for the general multitype Galton-Watson chains need in this case only minor modification to allow for the possibility of random probabilities. The relation was given by Pollard (1968 b) in a short note.

2.4.2. A model related to human populations

2.4.2.1. In the case of discrete-time stochastic population models, an efficient way to derive population parameters is to use probability generating functions, considering the fact that a probability distribution is totally defined by its generating function. According to the assumptions of the model (for example, age and time dependence), these functions receive special properties. Thus, the probability generating function of a sum of random variables is given by constructing the joint probability generating function and equating the parameters of the variables to be summed. For example, if $G(z_0, z_1, \ldots)$ is the joint probability generating function of $\gamma_0, \gamma_1, \ldots$, the probability generating function of the sum $\gamma_0 + \gamma_1 + \ldots$ is $G(z, z, \ldots)$.

It may likewise be shown that if independently of each other each unit of a random variable γ with probability generating function $G(z)$ generates units according to the probability generating function $H(z)$, then the probability generating function of the resultant distribution is given by $G[H(z)]$. Generalizing, one can show that if independently of each other each unit of the ith variable of a random vector $\boldsymbol{\gamma}$ with probability generating function $G(\mathbf{z})$ generates other variables with probability generating function $H_i(\mathbf{z})$ then the probability generating function of the resultant distribution is $G[\mathbf{H}(\mathbf{z})]$ where $\mathbf{H}(\mathbf{z}) = (H_i(\mathbf{z}))$.

By using these properties, THOMAS (1969) has studied a model which assumes age- and time-dependent birth and death distributions. The particular assumptions used for this are:

1°. The F-population only is considered, and it is assumed that the effect of a possibly atypical sex-ratio is negligible;

2°. The probability of an F-individual of age j at time n, surviving to age $j + 1$, at time $n + 1$, depends only on j and n;

3°. The distribution of individuals born by any F-individual of age j at time n depends only on j and n.

Live births only are considered in order to improve the accuracy of the model. This affirmation draws attention to possible implications and difficulties. To obtain negligible errors, one must first admit that all conditions remain satisfactorily constant for the interval between two moments $(n, n + 1)$. The population exists only at unit intervals of time, any change being a discontinuous jump from one time period to the next. Thus one must consider the position of each event versus the beginning or the end of the time interval. An F-individual's age will become zero the next time interval after birth. Thus, on the average (assuming a uniform distribution of births through the time interval), ages will be half a time period less than commonly understood. Birth and death rates must be adjusted accordingly. As a result of this interpretation, a birth is not counted if a child does not survive until the following time period; the child fails to reach age zero. In the same sense, V. J. THOMAS notes what he called the major difficulty of the system:

the fact that an individual born shortly after the beginning of a time interval lives for almost one interval before entering the system, whereas another, born at the end of the interval, enters the system almost immediately. This in turn leads to a certain fuzziness in applying age-dependent birth and death rates to individuals, and should (one presumes) have an appropriate variance component arising from the distribution of births through the interval. One may postulate a population-dependent time taken as the mean point in the interval of the "birth days" of all those individuals alive at some point in the interval. This is equivalent to having a different time point for each time interval and to transferring the fuzziness from the population to the time scales.

2.4.2.2. Let us now pass to the mathematical expression of the model with the help of generating functions. Let $\gamma_n(j)$ be the number of F-individuals of age j at time n, with probability generating function $G_n(z;j)$. Using $\boldsymbol{\gamma}_n$ to represent the random vector $(\gamma_n(0), \gamma_n(1), \ldots)$, let the corresponding probability generating function be $G_n(\mathbf{z})$, $\mathbf{z} = (z_0, z_1, z_2, \ldots)$. Evidently, $G_n(z;j) = G_n(\mathbf{z}')$, where $z_j' = z$ and $z_i' = 1$, $i \neq j$.

Consider now an F-individual of age j at time n. At time $n+1$ assuming it survives, it gives rise to two sub-populations: itself at age $j+1$, and any children it may have in the interval n to $n+1$, of age 0. Let the joint probability generating function be $H_n(z_0, z_{j+1}; j)$ and denote the vector $(H_n(z_0, z_1; 0), H_n(z_0, z_2; 1), \ldots)$ by $\mathbf{H}_n(\mathbf{z})$, $\mathbf{z} = (z_0, z_1, z_2, \ldots)$.

On the basis of the properties of probability generating functions, as shown above, z_i in G_n is replaced by $H_n(z_0, z_{i+1}; i)$ to give G_{n+1}. Hence the following recursive relationship holds:

$$G_{n+1}(\mathbf{z}) = G_n[\mathbf{H}_n(\mathbf{z})]. \tag{2.73}$$

This is the basic expression, successive differentiations of which lead to a set of recursive equations for the population parameters.

2.4.2.3. Here the basic expression (2.73) will be found again in the case of the introduction of some supplementary assumptions, such as immigration, dependence of birth rates on marriage, on time since marriage or on time since last child.

Thus, let $\iota_n(j)$ be the number of immigrants of age j between times n and $n+1$. Denote by R_n the probability generating function of the random vector $(\iota_n(0), \iota_n(1), \ldots)$ and suppose it to be independent of all other probability generating functions. Consequently, we get

$$G_{n+1}(\mathbf{z}) = G_n[\mathbf{H}_n(\mathbf{z})]\, R_n(\mathbf{z}). \tag{2.74}$$

According to some demographers, for nuptial births, time since marriage is a better parameter by which to vary the birth rate than age. A simplified model can be built on the basis of the following assumptions:

1°. Ex-nuptial births are negligible.

2°. Once married, always married.

3°. To simplify the notation, birth, marriage and death rates are assumed independent of time.

Within this framework, let w_{ju} be the parameter related to the variable $\gamma_n(j, u)$, the number of women of age j, married for u time units, at time n, and let $\mathbf{w} = (w_{ik})$. Use z_j as before, for the single populations,

$$S(z_{j+1}, w_{(j+1),0}; j)$$

for the probability generating function for the single populations, and

$$Z_u(z_0, w_{(j+1),(u+1)}; j)$$

for the married population. Putting now

$$\mathbf{Z} = (Z_k(z_0, w_{(i+1),(k+1)}; i))$$

and

$$\mathbf{S} = (S(z_{i+1}, w_{(i+1),0}; i)),$$

we deduce

$$G_{n+1}(\mathbf{z}, \mathbf{w}) = G_n(\mathbf{S}, \mathbf{Z}).$$

The adaptation of the above approach to the case in which the birth rate depends on the time since the last child is straightforward. In this case we shall denote by $\mathbf{z}$ the parameters for childless women of various ages, while specifying that w_{ju} will now represent the parameter for an F-individual of age j whose last child was born u time units ago. As above, $\mathbf{w} = (w_{ik})$. Childless women have a probability generating function $S(z_0, z_{j+1}, w_{(j+1),0}; j)$ and women with children a probability generating function $C_u(z_0, w_{(j+1),(u+1)}, w_{(j+1),0}; j)$. With $\mathbf{S} = (S(z_0, z_{j+1}, w_{(j+1),0}; j))$ and $\mathbf{C} = (C_u)$, the basic recursive relation becomes

$$G_{n+1}(\mathbf{z}, \mathbf{w}) = G_n[\mathbf{S}, \mathbf{C}]. \tag{2.75}$$

2.4.2.4. The reproduction rate of a population is usually measured by the expected number of female children that would be born to a hypothetical F-individual who lived her entire child-bearing period of life under current conditions. This rate is gross if it is given that she does not die before reaching an age such that probability of any further children is zero. If the probability of her dying is considered, the rate is net.

Using Thomas' elementary model, the gross reproduction is the mean of the distribution with probability generating function $H(z, 1; 0)$ $H(z, 1; 1) \dots H(z, 1; h)$, where the birth-and-death probability generating function is $H(z_0, z_{j+1}; j)$ and h is sufficiently large that $H(z, 1; h+j) = 1$ for all $j > 0$.

Of more interest is the net reproduction rate. It is the mean of $H[z, H(z, \dots H(z, 1; h) \dots); 0]$, which is the probability generating function of the number of children the hypothetical woman would bear. Put

$$K(z; i) = H[z, H(z, \dots H(z, 1; h) \dots; i+1)].$$

Hence

$$K(z; i) = H(z, K(z; i+1); i),$$

and $K(z; i)$ is the probability generating function of the number of children produced by an F-individual once she is of age i, and thereafter. From the relation above, one deduces by the usual calculations, the means and variances of the number of births resulting from one F-individual during its lifetime, under current birth and death rates, and, likewise, the net reproduction rate.

2.4.3 Population growth of the sexes

2.4.3.1. As already shown at the beginning of this paragraph, usual demographic practice considers the population growth of only one sex, usually female, or, as we have denoted, the F-population growth. In the spirit of what we have presented up to now, one may deduce that a discrete parameter model of two-sex population growth may be also regarded as a special case of multitype Galton-Watson chains and can be studied by means of known methods. However, since some of the two-sex models are not positive regular (see I; 1.2.2.7.2), one cannot directly apply the more usual asymptotic results which depend upon positive regularity (see I; 1.2.2.7.5).

In the literature there are few two-sex stochastic models of population growth: we refer to the works of KENDALL (1949), GOODMAN (1953), JOSHI (1954) and LAMENS (1957). The models presented start from two simplifying assumptions, considering that (a) the birth and death rates are equal for both subpopulations, the F-subpopulation and the M-subpopulation, (b) the birth and death rates per person are constants that are independent of the age of the person, and also independent of other relevant variables. Another auxiliary assumption was that one sex is marriage-dominant, i.e., the joint distribution of the numbers of girls and boys born at time t depends only upon the number of the dominant sex in the population in the immediately preceding time-period. This hypothesis would be appropriate for a society in which births are associated with the mother, or father, independent of the number of possible fathers, or mothers who are available. Various societies will fall somewhere between the two extremes of female or male marriage-dominance (GOODMAN (1958)).

2.4.3.2. The model presented here belongs to L. A. GOODMAN and represents a model in which neither females nor males are marriage-dominant, i.e., where the joint distribution of the numbers of boys and girls born at time n depends upon both the numbers of females and males in the various age-intervals in the immediately preceding time period. The model is discrete-time. Denoting the successive age--intervals by 0-, 1-, ..., k-, and the corresponding successive time points by $n = 0, 1, 2, \ldots$, we shall now consider the general case of this stochastic model defined by the following assumptions:

1°. For an F-individual in the j-age-interval $(0 \leqslant j \leqslant k)$ at time n, the probability is $\gamma_{vw1}(j)$ that it will survive to the $(j+1)$-age-interval at time $n+1$, and will also produce v daughters and w sons $(v, w \in N)$ who will be in the 0-age-interval at time $n+1$, with $\gamma_{vw1}(k) = 0$;

2°. For an F-individual in the j-age-interval $(0 \leqslant j \leqslant k)$ at time n, the probability is $\gamma_{vw2}(j)$ that it will die during the time period from n to $n+1$, but will produce v daughters and w sons $(v, w \in N)$ who will be in the 0-age-interval at time $n+1$;

3°. For an M-individual in the j-age-interval $(0 \leqslant j \leqslant k)$ at time n, the probability is $\alpha_{vw1}(j)$ that it will survive to be in the $(j+1)$-age-interval at time $n+1$, and will also produce v daughters and w sons $(v, w \in N)$ who will be in the 0-age-interval at time $n+1$, with $\alpha_{vw1}(k) = 0$;

4°. For an M-individual in the j-age-interval $(0 \leqslant j \leqslant k)$ at time n, the probability is $\alpha_{vw2}(j)$ that it will die during the time-period from n to $n+1$, but will produce v daughters and w sons $(v, w \in N)$ who will be in the 0-age-interval at time $n+1$;

5°. The probabilities defined above may vary with the age j, but they are independent of time n;

6°. The subpopulations generated by two coexisting F-individuals develop independently of one another.

For an F-individual in the j-age-interval at time n, $\gamma_{v.1}(j)$ and $\gamma_{.w1}(j)$ are the marginal probabilities that it will survive to be in the $(j+1)$-age-interval at time $n+1$ and will also produce v daughters and w sons respectively, who will be in the 0-age-interval at time $n+1$. In a similar way, we shall denote by $\gamma_{v..}(j)$ and $\gamma_{.w.}(j)$ the marginal probabilities that an F-individual in the j-age-interval at time n will produce v daughters or w sons who will be in the 0-age-interval at time $n+1$, without any specification of the fate of the mother. Therefore, we shall have

$$\gamma_{v\cdot s}(j) = \sum_{w \in N} \gamma_{vws}(j), \qquad \alpha_{v\cdot s}(j) = \sum_{w \in N} \alpha_{vws}(j),$$

$$\gamma_{\cdot ws}(j) = \sum_{v \in N} \gamma_{vws}(j), \qquad \alpha_{\cdot ws}(j) = \sum_{v \in N} \alpha_{vws}(j), \quad s = 1,2,$$

$$\gamma_{v\cdot\cdot}(j) = \sum_{s=1}^{2} \gamma_{v\cdot s}(j), \qquad \alpha_{v\cdot\cdot}(j) = \sum_{s=1}^{2} \alpha_{v\cdot s}(j),$$

$$\gamma_{\cdot w\cdot}(j) = \sum_{s=1}^{2} \gamma_{\cdot ws}(j), \qquad \alpha_{\cdot w\cdot}(j) = \sum_{s=1}^{2} \alpha_{\cdot ws}(j),$$

$$a(j) = \sum_{v \in N} v\gamma_{v\cdot\cdot}(j), \qquad a^{+}(j) = \sum_{v \in N} v\alpha_{v\cdot\cdot}(j),$$

$$b(j) = \sum_{w \in N} w\gamma_{\cdot w\cdot}(j), \qquad b^{+}(j) = \sum_{w \in N} w\alpha_{\cdot w\cdot}(j),$$

$$\nu(j) = \sum_{v \in N} \gamma_{v\cdot 1}(j) = \sum_{w \in N} \gamma_{\cdot w1}(j),$$

$$\nu^{+}(j) = \sum_{v \in N} \alpha_{v\cdot 1}(j) = \sum_{w \in N} \alpha_{\cdot w1}(j),$$

$$\mu(j) = \sum_{v \in N} \gamma_{v\cdot 2}(j) = \sum_{w \in N} \gamma_{\cdot w2}(j),$$

$$\mu^{+}(j) = \sum_{v \in N} \alpha_{v\cdot 2}(j) = \sum_{w \in N} \alpha_{\cdot w2}(j).$$

With these definitions, $\mu^{+}(j) = 1 - \nu^{+}(j)$. Here $\nu(j)$ is the marginal probability that an F-individual in the j-age-interval at time n will survive to the $(j + 1)$-age-interval at time $n + 1$, and $\mu(j)$ is the marginal probability that it will die during the time-period from n to $n + 1$. Also, $a(j)$ and $b(j)$ are the average number of daughters and sons, respectively, born by any F-individual in the j-age-interval at time n who will be in the 0-age-interval at time $n + 1$.

Let $\bar{a}(j)$ and $\bar{b}(j)$ denote the values of $a(j)$ and $b(j)$, respectively, when there is female marriage-dominance. Let $\bar{a}^{+}(j)$ and $\bar{b}^{+}(j)$ denote the values of $a^{+}(j)$ and $b^{+}(j)$, respectively, when there is male marriage-dominance.

When neither females nor males are marriage-dominant, we might assume that

$$a(j) = \bar{a}(j)\,\Delta, \qquad a^+(j) = [1-\Delta]\,\bar{a}^+(j),$$

$$b(j) = \bar{b}(j)\,\Delta, \qquad b^+(j) = \bar{b}^+(j)\,[1-\Delta],$$

where Δ is a proportion of births associated with the mother, as has been done by GOODMAN (1953) and KEYFITZ and MURPHY (1967). We can consider the more general situation where a proportion Δ_w of the births of sons is associated with the mother, a proportion $1-\Delta_w$ of the births of sons is associated with the father, a proportion Δ_v of the births of daughters is associated with the father, and a proportion $1-\Delta_v$ of the births of daughters is associated with the mother. In POLLARD's (1948) deterministic model $\Delta_v = \Delta_w = 1$, whereas in GOODMAN's (1967 a) and KEYFITZ and MURPHY's (1967) papers $\Delta_w = 1-\Delta_v$. We get

$$a(j) = \bar{a}(j)\,[1-\Delta_v], \qquad a^+(j) = \bar{a}^+(j)\,\Delta_v,$$

$$b(j) = \bar{b}(j)\,\Delta_w, \qquad b^+(j) = \bar{b}^+(j)\,[1-\Delta_w].$$

Let $f_{j,n}$ and $m_{j,n}$ denote the number of F-and M-individuals, respectively, in the j-age-interval at time n. Let $\mathbf{f}_n$ and $\mathbf{m}_n$ denote the $(k+1)$-dimensional column vectors $(f_{0,n}, f_{1,n}, \ldots, f_{k,n})'$ and $(m_{0,n}, m_{1,n}, \ldots, m_{k,n})'$, respectively, and let $\mathbf{z}_n$ denote the $2(k+1)$-dimensional column vector $(f_{0,n}, \ldots, f_{k,n}, m_{0,n}, \ldots, m_{k,n})' = (z_{in})'_{1\leqslant i\leqslant 2(k+1)}$. The sequence $\mathbf{z}_0, \mathbf{z}_1, \ldots,$ is a multitype Galton-Watson chain.

The first moments are defined by

$$c_{hg} = \mathsf{E}[z_{g1} \mid \mathbf{z}_0 = \mathbf{e}_h], \quad 1 \leqslant h, g \leqslant 2(k+1),$$

where $\mathbf{e}_h$ is a $2(k+1)$-dimensional column vector whose hth component is 1 and whose other components are 0. The matrix $\mathbf{C} = (c_{hg})$ can be expressed as

$$\mathbf{C} = \begin{pmatrix} \mathbf{C}_{00} & \mathbf{C}_{01} \\ \mathbf{C}_{10} & \mathbf{C}_{11} \end{pmatrix},$$

in which $\mathbf{C}_{rs} = (c_{rsil})_{0\leqslant i,\, l\leqslant k}$ is a $(k+1)\times(k+1)$ matrix for $r, s = 0, 1$ with

$$c_{00il} = \mathsf{E}[f_{l1} \mid \mathbf{z}_0 = \mathbf{e}_{i+1}],$$

$$c_{01il} = \mathsf{E}[m_{l1} \mid \mathbf{z}_0 = \mathbf{e}_{i+1}],$$

$$c_{10il} = \mathsf{E}[f_{l1} \mid \mathbf{z}_0 = \mathbf{e}_{i+k+1}],$$

$$c_{11il} = \mathsf{E}[m_{l1} \mid \mathbf{z}_0 = \mathbf{e}_{i+k+1}].$$

Thus,

$$c_{00i0} = a(i),$$

$$c_{01i0} = b(i),$$

$$c_{0,0,i,i+1} = \nu(i),$$

and $c_{rsil} = 0$ otherwise. The submatrices $\mathbf{C}_{00}$ and $\mathbf{C}_{01}$ are defined by the above entries, while the submatrices $\mathbf{C}_{10}$ and $\mathbf{C}_{11}$ have the following entries:

$$c_{rsil} = \begin{cases} a^+(i), \text{ for } r = 1,\ s = 0,\ l = 0 \\ b^+(i), \text{ for } r = s = 1,\ l = 0 \\ \nu^+(i), \text{ for } r = s = 1,\ l = i + 1 \end{cases}$$

with $c_{11li} = c_{10il} = 0$ otherwise.

From the usual results for multitype Galton-Watson chains, we see that the expected value of $\mathbf{z}_n$, given $\mathbf{z}_0$ is

$$\mathsf{E}\mathbf{z}_n = \mathbf{C}^n\mathbf{z}_0. \tag{2.76}$$

In addition to the exact calculation of means, variances and covariances, GOODMAN (1968) has developed asymptotic methods which can be applied under rather general conditions.

2.4.4. The reproductive process

2.4.4.1. Mathematical methods on the reproductive performance or on the human fertility are rather recent: as far as we know, they began in 1942 with C. GINI's paper presented at the International Mathematics Congress in Toronto.

These mathematical formulations are intended to develop demographic research and to support medical research regarding the biologic factors of fertility and child health. This is why the modelling of the fertility process, together with problems of family planning and contraceptive practices have taken on particular importance. It is a commonplace that in modern societies natural selection by deaths has been replaced by the social selection of births. The "vital revolution" or the demographic transition was defined by KIRK (1968) as the transition from wastefully high birth and death rates to the much more efficient lower birth and death rates. This opinion was taken over by the Polish demographer ROSSET (1964) who differentiated between wasteful and economical fertility. Economical fertility means that the children born survive, whereas wasteful fertility means that a significant number of born children dies.

According to NEEL and CHAGNON (1968), there are three principal demographic stages in relatively recent human history and prehistory. The first and earliest stage is a period of "intermediate" *effective* fertility and early mortality. With the advent of agriculture (and the ascendancy of religions emphasizing fruitfulness) a period of higher fertility with a corresponding increase in pre-reproductive mortality was introduced. In the modern industrial period there is a strong trend toward relatively low fertility but even lower mortality.

Mathematical models of fecundity may be extended to the whole class of mammals and research work carried out on 200 pairs (matings) of laboratory mice has recently been published (SHEPS, DOOLITTLE, and NEW (1969)). There are obviously a number of peculiarities regarding the human species in which one discerns a differential fertility. Demographic changes in four areas affect the opportunity for natural selection through differential fertility: (i) mating and marriage, (ii) childlessness, (iii) number of offspring, and (iv) age of childbearing and mean length of generation (KIRK (1968)). Let us start by remarking that in the year 1800 the average American woman passing through her reproductive life had seven children; today she has three or less. Today marriage is earlier, also intervals between marriage and first and subsequent births have been reduced so that births are increasingly concentrated in the early childbearing years. A similar situation was signalled in Poland (see ROSSET's paper and THEISS' comments (1964)): the cycle of reproduction begins and ends sooner; according to THEISS, Poland belongs to the few European countries in which the first, second and third births together are sufficient to secure the full replacement of generations.

The concentration of childbearing in young ages at low parities (viz. secundi-, tertiparity) has direct genetic influence since a number of genetic disorders are correlated with age of mother, birth order, and number of children per family.

Although the social, economic and psychological factors that affect birth rates can exert their effects only through modifying the biological determinants of the process, we shall not present here the biological factors which have been treated in detail by PERRIN and SHEPS (1964) and SHEPS (1967).

2.4.4.2. The most simple stochastic model proposed belongs to CHIANG (1971) and consists of three states: state 1 = fecundable state; state 2 = pregnant or infecundable state; state 3 = death. A female is said to be in the fecundable state 1 if she is able of being impregnated; she is in the infecundable state 2 if she is either pregnant or in the post-partum sterile period. A transition $1 \to 2$ means conception, whereas a transition $2 \to 1$ indicates that a female, after having given birth to infant or fetal death, has become fecundable again. Thus, in the present model, the frequency of a woman's pregnancies is represented by the number of her transitions from 1 to 2. When a woman dies, we say that she enters the absorbing state 3 from either 1 or 2 depending upon whether she was fecundable at the time of death or not.

A characteristic of the fertility process is reflected in the transition between the two transient states 1 and 2. During the female reproduction period a transition may take place at any time from fecundable state 1 to pregnant state 2 through conception. But transition $2 \to 1$ is quite restricted. Generally, there is a minimum length of time that a female must remain in the pregnant or infecundable state 2 before she can leave that state to become fecundable again. On the other hand, a female may not stay in the infecundable state 2 for more than a maximum period of time. This nonsusceptible variable period was also called "lost time" consisting of pregnancy (gestation) plus the interval after its termination before resumption of ovulation. This is why in this model two limits are proposed, τ_1 and τ_2, $\tau_2 > \tau_1$, such that a female can exit from 2 to 1 only if she has been in 2 for a duration of τ, for $\tau_1 < \tau < \tau_2$, and she cannot remain in 2 for a period greater than τ_2. Transitions from state 2 to the absorbing state 3, however, may take place anytime and are not subject to these restrictions.

Let a_0 be the initial age of a fertile female, such as age at puberty. Consider then a fertile female at age a_0. During the age interval (a_0, a) she may become pregnant a number of times, and at age a she may be either in state 1 or in infecundable state 2 or she may have died. Taking into account the number of transitions she makes during the interval (a_0, a) and the state that she is in at age a, C. L. Chiang introduces the following (multiple) transition probabilities:

1°. The probability $P_{1i}^{(m)}(a_0, a)$ that a female in state 1 at age a_0 will leave 1 to become pregnant m times in interval (a_0, a) and will be in state $i (i = 1,2)$ at age a.

2°. The probability $Q_{1i}^{(m)}(a_0, a)$ that a female in state 1 at age a_0, will leave 1 to become pregnant m times and die when she is in state $i (i = 1,2)$ prior to or at age a.

For $m = 1$, for example, $P_{11}^{(1)}(a_0, a)$ is the probability that a female will become pregnant once and will be fecundable again at age a. If a female is in state 2 at initial age a_0, the corresponding probabilities $P_{2i}^{(m)}(a_0, a)$ and $Q_{2i}^{(m)}(a_0, a)$ are similarly defined.

We shall also consider:

3°. The probability $P_{22}^{(0)}(t - a, t)$ that a female became pregnant at age $t - a$ will remain in state 2 till age t.

To derive formulas for the multiple transition probabilities, we introduce the intensity functions underlying this model related to the characteristics of transitions $1 \to 2$ and $2 \to 1$. For each real number $a > a_0$ and for the interval $(a, a + \Delta a)$ we let:

– the probability that a female of age a in state 1 will become pregnant in the interval $(a, a + \Delta a)$ is $p_{12}(a)\,\Delta a + o(\Delta a)$;

— the probability that a female of age a in state 1 will enter the death state 3 in the interval $(a, a + \Delta a)$ is $d_1(a)\,\Delta a + o(\Delta a)$;

— the probability that a female of age a in state 1 will remain in this state in the interval $(a, a + \Delta a)$ is $1 + p_{11}(a)\,\Delta a + o\,(\Delta a)$, where $p_{11}(a) = -\,[p_{12}(a) + d_1(a)]$;

— the probability that a female of age a who has been in state 2, for a duration τ will leave 2 for state 1 in the interval $(a, a + \Delta a)$ is $p_{21}(a; \tau)\,\Delta a + o\,(\Delta a)$;

— the probability that a female of age a who has been in state 2, for a duration τ will enter the death state 3 in the interval $(a, a + \Delta a)$ is $d_2(a; \tau)\,\Delta a + o(\Delta\, a)$;

— the probability that a female of age a who has been in state 2, for a duration τ will remain in state 2 in the interval $(a, a + \Delta a)$ is $1 + p_{22}(a; \tau)\,\Delta a + o\,(\Delta a)$, where $p_{22}(a; \tau) = -[p_{21}(a; \tau) + d_2(a; \tau)]$.

These intensity functions completely determine the transition probabilities and indeed the theoretical model of the fertility process itself. Using the intensity functions we see that for $m = 0$, the corresponding probabilities satisfy the differential equations:

$$\frac{\mathrm{d}}{\mathrm{d}a} P_{11}^{(0)}(a_0, a) = P_{11}^{(0)}(a_0, a)\, p_{11}(a),$$

$$\frac{\mathrm{d}}{\mathrm{d}a} P_{22}^{(0)}(a-\tau, a) = P_{22}^{(0)}(a-\tau, a)\, p_{22}(a, \tau),$$

and hence the solution

$$P_{11}^{(0)}(a_0, a) = \exp\left(\int_{a_0}^{a} p_{11}(\theta)\, \mathrm{d}\theta\right),$$

$$P_{22}^{(0)}(a-\tau, a) = \exp\left(\int_{a-\tau}^{a} p_{22}(\theta, \theta - (a-\tau))\, \mathrm{d}\theta\right). \tag{2.77}$$

For $m = 1$, we have the differential equation

$$\frac{\mathrm{d}}{\mathrm{d}a} P_{11}^{(1)}(a_0, a) = P_{11}^{(1)}(a_0, a)\, p_{11}(a) +$$

$$+ \int_{\tau_1}^{\tau_2} P_{11}^{(0)}(a_0, a-\tau)\, p_{12}(a-\tau) P_{22}^{(0)}(a-\tau, a)\, p_{21}(a, \tau)\, \mathrm{d}\tau,$$

and the solution

$$P_{11}^{(1)}(a_0, a) =$$
$$= \int_{a_0+\tau_1}^{a} \left[\int_{\tau_1}^{\tau_2} P_{11}^{(0)}(a_0, a_1 - \tau)\, p_{12}(a_1 - \tau)\, P_{22}^{(0)}(a_1 - \tau, a_1) \times \right. \tag{2.78}$$
$$\left. \times p_{21}(a_1, \tau)\, d\tau \right] P_{11}^{(0)}(a_1, a)\, da_1,$$

where $P_{11}^{(0)}(\cdot,\cdot)$ and $P_{22}^{(0)}(\cdot,\cdot)$ are given above.

The solution in (2.78) may be explained intuitively as follows: for a female in state 1 at age a_0 to be pregnant once and return to state 1 within the interval (a_0, a), she must

- stay in state 1 until age $(a_1 - \tau)$,
- become pregnant in the interval $(a_1 - \tau, a_1 - \tau + d\tau)$,
- remain in state 2 for a duration τ,
- return to fecundable state 1 in the interval $(a_1, a_1 + da_1)$,
- remain in state 1 from a_1 to a.

The whole process can occur for each possible value of τ, $\tau_1 \leqslant \tau \leqslant \tau_2$.

For $m > 1$ we have the formula

$$P_{11}^{(m)}(a_0, a) = \int_{a_0+m\tau_1}^{a} \left[\int_{\tau_1}^{\tau_2} P_{11}^{(m-1)}(a_0, a_1 - \tau)\, p_{12}(a_1 - \tau) \times \right. \tag{2.79}$$
$$\left. \times P_{22}^{(0)}(a_1 - \tau, a_1)\, p_{21}(a_1, \tau)\, d\tau \right] P_{11}^{(0)}(a_1, a)\, da_1.$$

Formula (2.79) presents a recursive relationship between $P_{11}^{(m)}(a_0, a)$ and $P_{11}^{(m-1)}(a_0, a)$; therefore, the explicit formula can be successively derived starting with $m = 2$.

The multiple transition probability $P_{12}^{(m)}(a_0, a)$ may be derived directly, or it may be obtained through its relation with $P_{11}^{(m)}(a_0, a)$. This relation is

$$P_{12}^{(m)}(a_0, a) = \int_0^{\tau_2} P_{11}^{(m-1)}(a_0, a - \tau)\, p_{12}(a - \tau)\, P_{22}^{(0)}(a - \tau, a)\, d\tau. \tag{2.80}$$

The multiple transition probabilities leading to death can also be obtained by using similar relations, namely

$$Q_{11}^{(m)}(a_0, a) = \int_{a_0+m\tau_1}^{a} P_{11}^{(m)}(a_0, a)\, d_1(a)\, da,$$
$$Q_{12}^{(m)}(a_0, a) = \int_{a-\tau_2}^{a} P_{12}^{(m)}(a_0, a)\, d_2(a, a_0 - a)\, da. \tag{2.81}$$

Expressions (2.80) and (2.81) provide the complete solution for the multiple transition probabilities. These probabilities describe the fertility process in terms of the number of pregnancies and the survival state of a woman at each of the ages j. The model presented (see also CHIANG (1968)) is a special case of an illness-death process originally suggested by FIX and NEYMAN (1951).

2.4.4.3. Let us now consider a stochastic model due to PERRIN and SHEPS (1964) (see also PERRIN (1967), SHEPS (1967), SHEPS and PERRIN (1966)), which takes into account the whole reproductive history of a woman. At the beginning of a sexual union she is not pregnant and is susceptible to conception. This state will be called state 0. After a random period of time, a woman may enter pregnancy (state 1), from which she will enter the postpartum nonsusceptible state following a surviving live birth (state 2), a still-birth (state 3) or an early fetal death (state 4). After a stay of variable duration in any of these states, she again becomes susceptible to a passage to state 0, and so on. The model may be supplemented with new states in the nonsusceptible period (e.g., live birth followed by early death, induced early fetal death, etc.) and by denoting the fact that when resuming the cycle each successive visit to a state may be different from earlier ones.

The length of stay in each state is viewed as a random variable, the time spent in the susceptible nonpregnant state 0 before passage to state 1 having the probability density function $f(t)$. The duration of stay in pregnancy (state 1) is assumed to depend on the outcome of pregnancy, $b_i(t)$, $i = 2,3,4$, being the (conditional) probability density function of the time spent in state 1, given that the next state visited is state i ($i = 2,3,4$), and the probability density function of the time spent in the postpartum state i being $g_i(t)$. Finally, $p_i (i = 2, 3, 4)$ is the probability that a given pregnancy ends with a transition to state i. These important components of the system are presented in Table 1 (see PERRIN (1967)):

Table 1

States of the reproductive process

State	Probability density function of length of stay in state:	Probability the next state will be				
		0	1	2	3	4
0	$f(t)$	0	1	0	0	0
1	$b_i(t)$, given $i = 2, 3, 4$ visited next	0	0	p_2	p_3	p_4
2	$g_2(t)$	1	0	0	0	0
3	$g_3(t)$	1	0	0	0	0
4	$g_4(t)$	1	0	0	0	0

The defined process is semi-Markov (see I; 2.3.6.1) with the specified states i, $0 \leqslant i \leqslant 4$. To study the process of fertility it is important to know the number of passages into a specified state, the fertility rate or pregnancy rate per unit of time, or the probability distribution of states after the process has continued for time t.

Let $\mathcal{T}_{ij}$ be the Laplace-Stieltjes transform of the probability density function of the first passage time from state i to state j, $0 \leqslant i, j \leqslant 4$. To derive the matrix $\mathbf{T} = (\mathcal{T}_{ij})_{0 \leqslant i,j \leqslant 4}$ we consider the matrix $\mathbf{\Phi} = (\mathcal{F}_{ij})_{0 \leqslant i,j \leqslant 4}$ consisting of Laplace-Stieltjes transforms of the appropriate transition distributions. For example, the probability that from pregnancy (state 1) the next passage will be to state 2 and will occur by time t is $p_2 \int_0^t b(x)\,\mathrm{d}x$. Consequently,

$$\mathcal{F}_{12}(\lambda) = p_2 \int_0^\infty \mathrm{e}^{-\lambda x} b_2(x)\,\mathrm{d}x = p_2 \mathscr{b}_2(\lambda), \tag{2.82}$$

where $\mathscr{b}_i(\lambda)$ is the Laplace-Stieltjes transform of $b_i(t)$. Denoting in a similar way the Laplace-Stieltjes transforms of $f(t)$ and $g_i(t)$ by $\mathscr{f}$ and $\mathscr{g}_i$, we shall have

$$\mathbf{\Phi} = \begin{pmatrix} 0 & \mathscr{f} & 0 & 0 & 0 \\ 0 & 0 & p_2\mathscr{b}_2 & p_3\mathscr{b}_3 & p_4\mathscr{b}_4 \\ \mathscr{g}_2 & 0 & 0 & 0 & 0 \\ \mathscr{g}_3 & 0 & 0 & 0 & 0 \\ \mathscr{g}_4 & 0 & 0 & 0 & 0 \end{pmatrix}. \tag{2.83}$$

The matrix $\mathbf{T}$ is given by

$$\mathbf{T} = \mathbf{\Phi}[\mathbf{I} - \mathbf{\Phi}]^{-1} [\{(\mathbf{I} - \mathbf{\Phi})^{-1}\}_{\mathrm{dg}}]^{-1}, \tag{2.84}$$

where $\mathbf{I}$ is the identity matrix, so that

$$\mathbf{T} = \begin{pmatrix} \mathscr{f} \sum_{i=2}^{4} p_i \mathscr{b}_i \mathscr{g}_i & \mathscr{f} & \frac{p_2 \mathscr{f} \mathscr{b}_2}{w_2} & \frac{p_3 \mathscr{f} \mathscr{b}_3}{w_3} & \frac{p_4 \mathscr{f} \mathscr{b}_4}{w_4} \\ \sum_{i=2}^{4} p_i \mathscr{b}_i \mathscr{g}_i & \mathscr{f} \sum_{i=2}^{4} p_i \mathscr{b}_i \mathscr{g}_i & \frac{p_2 \mathscr{b}_2}{w_2} & \frac{p_3 \mathscr{b}_3}{w_3} & \frac{p_4 \mathscr{b}_4}{w_4} \\ \mathscr{g}_2 & \mathscr{f}\mathscr{g}_2 & \frac{p_2 \mathscr{f} \mathscr{b}_2 \mathscr{g}_2}{w_2} & \frac{p_3 \mathscr{f} \mathscr{b}_3 \mathscr{g}_2}{w_3} & \frac{p_4 \mathscr{f} \mathscr{b}_4 \mathscr{g}_2}{w_4} \\ \mathscr{g}_3 & \mathscr{f}\mathscr{g}_3 & \frac{p_2 \mathscr{f} \mathscr{b}_2 \mathscr{g}_3}{w_2} & \frac{p_3 \mathscr{f} \mathscr{b}_3 \mathscr{g}_3}{w_3} & \frac{p_4 \mathscr{f} \mathscr{b}_4 \mathscr{g}_3}{w_4} \\ \mathscr{g}_4 & \mathscr{f}\mathscr{g}_4 & \frac{p_2 \mathscr{f} \mathscr{b}_2 \mathscr{g}_4}{w_2} & \frac{p_3 \mathscr{f} \mathscr{b}_3 \mathscr{g}_4}{w_3} & \frac{p_4 \mathscr{f} \mathscr{b}_4 \mathscr{g}_4}{w_4} \end{pmatrix} \tag{2.85}$$

with

$$w_i = -1 / \sum_{j \neq i} p_j b_j g_j, \quad 2 \leqslant i \leqslant 4.$$

The moments of the first passage times are now obtained as

$$\theta_{ij}^{(r)} = [-1]^r \mathcal{T}_{ij}^{(r)}(0),$$

where $\mathcal{T}_{ij}^{(r)}(0)$ is the rth order derivative of $\mathcal{T}_{ij}$ with respect to λ, evaluated at zero. In particular, the first two moments of the intervals between successive pregnancy outcomes $i(2 \leqslant i \leqslant 4)$ are

$$\theta_{ii}^{(1)} = \frac{1}{p_i} \sum_{j=2}^{4} p_j \alpha_j^{(1)},$$

$$\theta_{ii}^{(2)} = \frac{1}{p_i} \sum_{j=2}^{4} p_j \alpha_j^{(2)} + \frac{1}{p_i^2} \sum_{j=2}^{4} p_j \alpha_j^{(1)} \left(\sum_{h \neq i} p_h \alpha_h^{(1)} \right),$$

where

$$\alpha_i^{(r)} = \int_0^\infty \int_0^\infty \int_0^\infty (x + y + z)^r f(x)\, b_i(y)\, g_i(z)\, \mathrm{d}x \mathrm{d}y \mathrm{d}z.$$

We have also

$$\theta_{00}^{(r)} = \theta_{11}^{(r)} = \sum_{i=2}^{4} p_i \alpha_i^{(r)}, \quad r \in N^*.$$

As $\theta_{11}^{(1)}$ is the mean interval between live births, the asymptotic fertility rate φ is given by

$$\varphi \sim \frac{1}{\theta_{11}^{(1)}},$$

according to the renewal theory (see I; Theorem 2.3.54).

2.4.4.4. If it is assumed that the total "lost time", i.e., the nonsusceptible periods, for each outcome has a fixed integral value, and that fecundity is constant, the model considered may be reduced to a Markov chain. For details see SHEPS (1967, p. 121).

The results obtained have implications both for the analysis of data on the distributions of births to a group of women in a calendar period and for the planning and evaluation of family planning programs. In this direction family building models have been set up (see SHEPS, MENKEN, RADICK (1969)), also a model for studying short-term effects of change in reproductive behaviour (SHEPS and MENKEN (1971)) and a model for conception delays in a heterogeneous group (SHEPS (1967)).

2.5. Other growth models and derived processes

2.5.1. The cumulative process

2.5.1.1. Let us consider a population A with alteridem B-objects developing according to the rules of a birth-and-death process $(\xi(t))_{t\geqslant 0}$.

If we do not keep in view any criterion capable of differentiating living from dead B-objects (if the latter have not disintegrated), we can state that population A forms the generator of a population B including all the objects born in the time interval $(0, t)$. The growth of population B follows a random process $(\eta(t))_{t\geqslant 0}$. In accordance with the known scheme, when the size of population A increases by a unit, the size of population B presents an identical increase; however, when the size of population A decreases by a unit, the size of B remains unchanged. Therefore, the random variable $\eta(t)$ represents the number of all B-objects born up to the moment t, irrespective of the fact that these objects are alive or not, while the random variable $\xi(t)$ represents the size of the actual population alive at time t.

The situation described here may be observed on cell populations cultivated *in vitro*, where a simple count under the microscope cannot usually distinguish living from dead cells. For this purpose, discriminatory techniques are necessary, such as vital staining. Generalizing, we may consider that the cumulative process $(\xi(t), \eta(t))_{t\geqslant 0}$ (see e.g. KENDALL (1948 b)) represents a growth process for populations in which B-objects own pairs of qualities which can become discernible: living-dead, healthy-ill, free-fixed, etc. For example, in epidemic theory, it is often easier to observe with accuracy the total size of the epidemic, i.e., the total number of individuals who have ever been infected after the elapse of a sufficiently long period of time, than to record the actual numbers of infectious persons circulating in the population at any moment (BAILEY (1964 a, p. 121)).

Suppose that $(\xi(t))_{t\geqslant 0}$ is a Feller-Arley process with parameters b and d. Owing to the relations between the two random variables $\xi(t)$ and $\eta(t)$, the common generating function $G(z_1, z_2; t)$ satisfies the equation (see e.g. BAILEY (1964 a, p. 123)):

$$\frac{\partial G}{\partial t} - [bz_1^2 z_2^2 - (b + d) z_1 + d] \frac{\partial G}{\partial z_1} = 0,$$

under the initial condition $G(z_1, z_2; 0) = z_1 z_2$ (assuming that $\xi(0) = \eta(0) = 1$). Using this equation it is easily shown that

$$\mathsf{E}\,\eta(t) = \frac{1}{d-b}\,(d - b\mathrm{e}^{(b-d)t}),$$

$$\mathsf{D}\,\eta(t) = \frac{b(d+b)}{(d-b)^2}\,(1 - \mathrm{e}^{(b-d)t}) - \frac{4b^2\,dt\mathrm{e}^{(b-d)t}}{(d-b)^2} + $$

$$+ \frac{b^2(d+b)}{(d-b)^3}\,(1 - \mathrm{e}^{2(b-d)t}),$$

$$\mathrm{Cov}\,[\xi(t), \eta(t)] = \frac{b\mathrm{e}^{(b-d)t}}{d-b}\left(2\,dt - \frac{d+b}{d-b}\,(1 - \mathrm{e}^{(b-d)t})\right).$$

It follows that for $b < d$,

$$\lim_{t\to\infty} \mathsf{E}\,\eta(t) = \frac{d}{d-b},$$

$$\lim_{t\to\infty} \mathsf{D}\,\eta(t) = \frac{bd\,(b+d)}{(d-b)^3},$$

$$\lim_{t\to\infty} \mathrm{Cov}\,[\xi(t), \eta(t)] = 0.$$

This means that, as $t \to \infty$, the random variables $\xi(t)$ and $\eta(t)$ are non-correlated.

2.5.1.2. A generalization of this process has been given by PURI-(1968 a) who took as starting point a Feller-Arley process $(\xi(t))_{t\geqslant 0}$ with parameters b and d. Let $\zeta(t)$ be the number of deaths and $\eta(t)$ the total number of events (births and deaths combined) occurring during $(0, t)$ and consider the process $(\xi(t), \zeta(t), \eta(t))_{t\geqslant 0}$. Clearly, $\eta(t) = 2\zeta(t) + \xi(t) - \xi(0)$. Finally, let

$$\gamma(t) = \int_0^t \xi(\tau)\,\mathrm{d}\tau.$$

From P. S. PURI's paper we shall mention the limiting results concerning $(\eta(t))_{t\geqslant 0}$. As $t \to \infty$, $\eta(t)$ converges almost surely to a random variable η such that

$$\mathsf{P}(\eta = 2n+1) = \frac{(2n)!}{n!\,(n+1)!}\left(\frac{b}{b+d}\right)^n \left(\frac{d}{b+d}\right)^{n+1}, \quad n \in N. \qquad (2.86)$$

For $d \geqslant b$, η is finite a.s. If $d < b$, then $\mathsf{P}(\eta = \infty) = 1 - (d/b)$. Equation (2.86) indicates that in the long run, in order to have $\eta = 2n + 1$, besides the death of the parent organism starting at $t = 0$, there must occur exactly n births which in turn should die in due course, where $b/(b + d)$ and $d/(b + d)$, respectively, are the probabilities of an event being a birth or a death. Again since $\zeta(t) \equiv \frac{1}{2}[\eta(t) - \xi(t) + 1]$ and $\xi(t) \to 0$ as $t \to \infty$ with probability one if $d \geqslant b$, and with probability d/b if $d < b$, we conclude that given that $\xi(t) \to 0$ as $t \to \infty$, $\zeta(t) \to \to \frac{\eta + 1}{2}$. Furthermore, it can be easily seen that whenever $\xi(t) \to \infty$, as is the case when $b > d$, $\zeta(t)$ and $\eta(t)$ both tend to ∞ as $t \to \infty$. These considerations indicate that, with ζ denoting the stochastic limit of $\zeta(t)$, we have

$$\zeta = \frac{1}{2}(\eta + 1), \text{ a.s.} \tag{2.87}$$

Also, from (2.86) and (2.87), with $b > d$, we have the conditional probability

$$\mathsf{P}(\zeta = n + 1 \mid \zeta < \infty) = \frac{(2n)!}{n!\,(n + 1)!}\left(\frac{d}{b + d}\right)^n \left(\frac{b}{b + d}\right)^{n+1}, \quad n \in N.$$

PURI (1966) has also shown that the limiting distribution of $\gamma(t)$ as $t \to \infty$ is a weighted mean of certain χ^2 distributions.

2.5.2. The Prendiville (logistic) process

2.5.2.1. FELLER (1939) studied a so-called *binomial birth-and-death process* whose birth and death rates are given by

$$b_n \equiv \alpha(n_2 - n) \quad \text{for} \quad n_1 \leqslant n \leqslant n_2 \;, \quad \text{and 0 otherwise,}$$

$$d_n \equiv \beta(n - n_1) \quad \text{for} \quad n_1 \leqslant n \leqslant n_2 \;, \quad \text{and 0 otherwise,}$$

where $\alpha > 0$, $\beta > 0$, $n_1 \in N$ and $n_2 \in N^*$ are absolute constants. A detailed discussion of this process has already been given by KENDALL (1949) (For a special case see Subparagraph 3.2.4.3; notice also that the case $n_1 = 0$, $\alpha, \beta > 0$, $\alpha + \beta = 1$, is known as the *continuous time Ehrenfest model*.) At the same time, B.J. PRENDIVILLE proposed another type of finite state birth-and-death process with birth and death rates given by

$$b_n = \begin{cases} \alpha\left[\dfrac{n_2}{n} - 1\right] & , \quad 0 < n_1 \leqslant n \leqslant n_2 \\ 0 & , \quad \text{otherwise} \end{cases}$$

$$d_n = \begin{cases} \beta\left[1 - \dfrac{n_1}{n}\right] & , \quad 0 < n_1 \leqslant n \leqslant n_2 \\ 0 & , \quad \text{otherwise.} \end{cases}$$

Such a birth-and-death process $(\xi(t))_{t\geqslant 0}$ will be said to be a *Prendiville process.*

2.5.2.2. To analyse the process considered we shall follow TAKASHIMA (1956). Let

$$\mathsf{P}(\xi(t) = n) = p_n(t),$$

i.e., $p_n(t)$ is the probability that the population size $\xi(t)$ has the value n at time t. We shall have the differential equations (see I; 2.3.4.3.4)

$$\frac{\mathrm{d}}{\mathrm{d}t} p_{n_1}(t) = -[n_1 b_{n_1} + n_1 d_{n_1}]\, p_{n_1}(t) + (n_1 + 1)\, d_{n_1+1}\, p_{n_1+1}(t),$$

$$\frac{\mathrm{d}}{\mathrm{d}t} p_n(t) = -[nb_n + nd_n]\, p_n(t) + (n+1)\, d_{n+1}\, p_{n+1}(t) +$$

$$+ (n-1) b_{n-1}\, p_{n-1}(t) \quad , \quad n = n_1 + 1, \ldots, n_2 - 1$$

$$\frac{\mathrm{d}}{\mathrm{d}t} p_{n_2}(t) = -[n_2 b_{n_2} + n_2 d_{n_2}]\, p_{n_2}(t) + (n_2 - 1)\, b_{n_2-1}\, p_{n_2-1}(t).$$

Let $G(z, t)$ be the probability generating function of $\xi(t)$. We therefore get the following differential equation for G:

$$\frac{\partial G}{\partial t} = (1 - z)(\alpha z + \beta)\frac{\partial G}{\partial z} + (z - 1)\left(\alpha n_2 + \frac{\beta n_1}{z}\right) G. \quad (2.88)$$

Equation (2.88) can be completely solved. Let us set the following initial condition:

$$G(z, t) = z^{n_0} \quad , \quad n_1 \leqslant n_0 \leqslant n_2.$$

In other words, we assume that the population size at time $t = 0$ has the value n_0. The characteristic differential system of (2.88) is

$$\frac{\mathrm{d}t}{1} = \frac{\mathrm{d}z}{(z-1)(\alpha z + \beta)} = \frac{\mathrm{d}G}{(z-1)[\alpha n_1 + (\beta n_1/z)]\,G}.$$

From this system we obtain

$$\Phi\left(\mathrm{e}^{-(\alpha+\beta)t}\frac{z-1}{\alpha z + \beta}\right) = \frac{z^{n_1}}{(\alpha z + \beta)^{n_1 - n_2}} \cdot \frac{1}{G},$$

where Φ is an arbitrary function. From the above initial condition, we can determine the form of Φ, i.e.,

$$\Phi(u) = \frac{(1 + \beta u)^{n_1 - n_0}(1 - \alpha u)^{n_0 - n_2}}{(\alpha + \beta)^{n_1 - n_2}}.$$

Therefore,

$$G(z, t) = \frac{1}{(\alpha+\beta)^{n_2 - n_1}} z^{n_1} [(\alpha + \beta \mathrm{e}^{-(\alpha+\beta)t}) z + \tag{2.89}$$

$$+ \beta(1 - \mathrm{e}^{-(\alpha+\beta)t})]^{n_0 - n_1} [\alpha(1 - \mathrm{e}^{-(\alpha+\beta)t}) z + (\alpha \mathrm{e}^{-(\alpha+\beta)t} + \beta)]^{n_2 - n_0}.$$

From (2.89) one can obtain the moments

$$m(t) = \frac{1}{\alpha+\beta}[(\alpha n_2 + \beta n_1) - \{(\alpha n_2 + \beta n_1) - n_0(\alpha+\beta)\}\,\mathrm{e}^{-(\alpha+\beta)t}],$$

$$\sigma^2(t) = \left[\frac{n_0}{\alpha+\beta} - \frac{\alpha(\alpha n_2 + \beta n_1)}{(\alpha+\beta)^2}\right] + \left[n_0 - \frac{\alpha n_2 + \beta n_1 + 2\alpha n_0}{\alpha+\beta} + \right.$$

$$\left. + \frac{2\alpha(\alpha n_2 + \beta n_1)}{(\alpha+\beta)^2}\right] \mathrm{e}^{-(\alpha+\beta)t} + \left[-n_0 + \frac{\beta n_1 + 2\alpha n_0}{\alpha+\beta} - \right.$$

$$\left. - \frac{\alpha(\alpha n_2 + \beta n_1)}{(\alpha+\beta)^2}\right] \mathrm{e}^{-2(\alpha+\beta)t}.$$

Now using (2.89) we obtain

$$\lim_{t\to\infty} G(z, t) = G(z, \infty) = \frac{1}{(\alpha+\beta)^{n_2-n_1}} z^{n_1} (\alpha z + \beta)^{n_2-n_1} =$$

$$= \frac{1}{(\alpha+\beta)^{n_2-n_1}} \sum_{x=0}^{n_2-n_1} \binom{n_2-n_1}{x} \alpha^x \beta^{(n_2-n_1)-x} z^{x+n_1}.$$

The function $G(z, \infty)$ generates probabilities which represent the ultimate "stable" distribution of the values of the population size. The probability that the ultimate population size equals $n_1 + x$, will be

$$p_{n_1+x} = \frac{1}{(\alpha+\beta)^{n_2-n_1}} \binom{n_2-n_1}{x} \alpha^x \beta^{(n_2-n_1)-x} \quad , \quad 0 \leqslant x \leqslant n_2 - n_1.$$

The mean value of $G(z, \infty)$ is

$$\frac{\alpha n_2 + \beta n_1}{\alpha + \beta},$$

which equals the limiting value of the deterministic logistic model.

2.5.3. Some problems of survival and extinction

2.5.3.1. We present in this subparagraph some problems regarding the fate of populations with regard to their ultimate behaviour. These problems are generally known and have already been described so we shall limit ourselves to some papers.

Let us consider, for example, a bacterial population put into contact with any type of bactericide. In this case the problem of the bacterial extinction time is easily solved (see Epstein (1967)). Let ξ_0 be the number of bacteria in a sample just prior to the introduction, at time $t = 0$, of a bactericide. This random variable is supposed to be Poisson with mean λ, and the lifetime of a bacterium in the presence of a fixed dose of bactericide has a known distribution function $F(t)$, $t \geqslant 0$. As the probability that x bacteria are in the sample at $t = 0$ is given by

$$\mathsf{P}(\xi_0 = x) = \frac{\lambda^x \exp(-\lambda)}{x!},$$

and the probability that all x bacteria die by time t is $[F(t)]^x$ (assuming that bacterial extinction times are mutually independent), it follows that the distribution function of the time to extinction of the sample considered is

$$G(t) = \sum_{x \in N} \frac{\lambda^x \exp(-\lambda)}{x!} [F(t)]^x = \exp(-\lambda[1 - F(t)]) \ , \quad t > 0.$$

If bacterial lives are distributed with $F(t) = 1 - \exp(\alpha t)$, $G(t)$ becomes

$$G(t) = \exp(-\lambda \exp(-\alpha t)), t > 0. \tag{2.90}$$

The distribution (2.90) was called by E.J. GUMBEL the Type I distribution of largest values within the framework of extreme value theory. B. EPSTEIN has pointed out the fact that GART's (1965) model for the distribution of response times after inoculating a host with a solution containing microorganisms can also be considered in the same manner, the only difference being that response times are distributed as smallest values (see also KODLIN (1967)).

2.5.3.2. DALEY (1968) studied the probability of extinction of a modified bisexual Galton-Watson chain. Let there be γ_n females and α_n males in the nth generation, $n \in N$, and suppose the $(\gamma_{n+1}, \alpha_{n+1})$ offspring in the $(n+1)$th generation are derived from $Z_n = \zeta(\gamma_n, \alpha_n)$ mating units according to the following rule, where $\zeta(\cdot, \cdot)$ is an integer-valued function nondecreasing in both arguments. Independently of the generation number and independently of all other units, each mating has ω_1 female and ω_2 male offspring, where ω_1 and ω_2 are random variables with probability generating function

$$G(z_1, z_2) = \mathsf{E}\, z_1^{\omega_1} z_2^{\omega_2} , \tag{2.91}$$

so that for $|z_1| \leqslant 1$, $|z_2| \leqslant 1$,

$$\mathsf{E}[z_1^{\gamma_{n+1}} z_2^{\alpha_{n+1}} | Z_n = j] = [G(z_1, z_2)]^j. \tag{2.92}$$

D. J. DALEY considers two particular cases of $G(\cdot,\cdot)$:

(a) The total number of offspring $\omega = \omega_1 + \omega_2$ has the probability generating function $G(z) = \sum_{j \in N} \mathfrak{p}_j z^j$, and each offspring is female with probability ξ and male with probability $(1 - \xi)$, independently of the other offspring so that

$$G(z_1, z_2) = G(\xi z_1 + (1 - \xi) z_2). \tag{2.93}$$

(b) The number of female offspring ω_1 is independent of the number of male offspring ω_2 so that

$$G(z_1, z_2) = G_1(z_1) G_2(z_2),$$

where $G_1(\cdot)$ and $G_2(\cdot)$ are the probability generating functions of ω_1 and ω_2, respectively.

These two cases coincide when the offspring distribution is Poisson, i.e., for $G(z)=e^{-\lambda(1-z)}$. For the sake of simplicity one may assume that the first case (with probability generating function (2.93)) is always pertinent and to avoid triviality one can assume that $0 < \xi < 1$, that $\mathfrak{p}_0 + \mathfrak{p}_1 < 1$, and that $G(1) = \sum_{j \in N} \mathfrak{p}_j = 1$.

Moreover, suppose that $\zeta(x, y) = x \min(1, y)$, that is,

$$Z_n = \begin{cases} \Upsilon_n, & \alpha_n > 0 \\ 0, & \alpha_n = 0. \end{cases} \tag{2.94}$$

This can be interpreted as meaning that conditions of "complete promiscuity" prevail with males of infinitely great reproductive power. Combining (2.91) with (2.92) and (2.93) one finds that $(Z_n)_{n \in N}$ is a homogeneous Markov chain with transition probabilities p_{jk} given by the probability generating function

$$\mathsf{E}[z^{Z_{n+1}} | Z_n = j] \equiv \sum_{k \in N} p_{jk} z^k \equiv \sum_{k \in N} \mathsf{P}(Z_{n+1} = k \mid Z_n = j)\, z^k =$$
$$= [G(\xi z + 1 - \xi)]^j - [G(\xi z)]^j + [G(\xi)]^j.$$

To eliminate trivial cases we have to exclude the case $Z_0 = 1$ when $\mathfrak{p}_0 + \mathfrak{p}_1 + \mathfrak{p}_2 = 1$. Then we have

Theorem 2.2 $\lim_{n \to \infty} \mathsf{P}(Z_n = 0 \mid Z_0 > 0) = 1$, *iff* $\xi G'(1) \leqslant 1$.

Proof. See Daley (1968). ◇

If one considers the situation of "polygamous mating with perfect fidelity", assuming that each male can have up to w wives (w, a positive integer), the function $\zeta(\cdot, \cdot)$ will have the form $\zeta(x, y) = \min(x, wy)$. In this case we have

Theorem 2.3 $\lim_{n \to \infty} \mathsf{P}(Z_n = 0 | Z_0 > 0) = 1$, *iff* $\min(\xi G'(1), (1-\xi) w G'(1)) \leqslant 1$.

Proof. See Daley (1968). ◇

See also Bishir's (1962) paper on the (maximum) population size in a branching process; also Adke (1964) and Jagers (1969, 1970).

2.5.3.3. For other contexts in which extinction probabilities are studied the reader is referred to Conner (1964) and Goodman (1967 b).

2.5.3.4. Explicit probabilities of births, survivors, and death for the class of nonhomogeneous birth-and-death processes, for which $d(t)/b(t)$ does not depend on t, were found by Gani and Yeo (1965). See also Klonecki (1967).

3. Population dynamics processes

By dynamics processes we mean those random processes which can represent the behaviour of populations with alterexter B-objects as well as the situations created by the relations between these objects. Thus, interacting populations such as ecologic communities of differentiating groups fall within the framework of our considerations. The easiest and simplest approach of this type of process is in our opinion, the study of certain multidimensional jump processes, with which this section begins.

3.1. Some multidimensional Markov jump processes

3.1.1. Consael processes

3.1.1.1. We shall present as Consael processes three types of Markov jump processes studied by CONSAEL (1949). HAIGHT (1963, p. 66) discusses under the name of Consael process a bivariate mixed Poisson process, which can be regarded as a bridge between a bivariate Poisson distribution in which the two variables are dependent and a bivariate Poisson distribution in which the two variables are independent (see CONSAEL (1952)).

The population we consider consists of two types ("species") of B-objects, which we shall simply denote by 1 and 2. The joint feature of the populations studied in these three Consael processes is the fact that B-objects of type 1 generate (or become) B-objects of type 2, and reversely. Let $\xi_1(t)$ be the number of type 1 objects at time t and let $\xi_2(t)$ be the number of type 2 objects at time t. The modifications taking place within the whole population in the time interval $(t, t + \Delta t)$ are given, for the three types of bidimensional processes $(\xi_1(t), \xi_2(t))_{t\geqslant 0}$ in Table 2.

Table 2

Type of Consael process	Type of transition in the interval $(t, t+\Delta t)$		Transition probabilities
	$\xi_1(t) = x$	$\xi_2(t) = y$	
I	$x-1$	$y+1$	$xb(t)\,\Delta t + o(\Delta t)$
	$x+1$	$y-1$	$yb(t)\,\Delta t + o(\Delta t)$
	$x-1$	y	$xd(t)\,\Delta t + o(\Delta t)$
	x	$y-1$	$yd(t)\,\Delta t + o(\Delta t)$
	x	y	$1-(x+y)(b(t)+d(t))\Delta t+o(\Delta t)$
II	x	$y+1$	$xb(t)\,\Delta t + o(\Delta t)$
	$x+1$	y	$yb(t)\,\Delta t + o(\Delta t)$
	x	y	$1-(x+y)\,b(t)\,\Delta t + o(\Delta t)$
III	x	$y+1$	$xb(t)\,\Delta t + o(\Delta t)$
	$x+2$	$y-1$	$yb(t)\,\Delta t + o(\Delta t)$
	x	y	$1-(x+y)\,b(t)\,\Delta t + o(\Delta t)$

The particularities of Consael models are easy to interpret; for example, Process I can be called a model of intermigration and death, since one B-object of type 1 (2) is transformed into a B-object of type 2 (1) (i.e., the disappearance of a type 1 (2) object is simultaneous with the appearance of a type 2 (1) object and the intensity of this modification is proportional to the size of the type 1 (2) population). Process II involves only births: a B-object of type 1 (2) gives birth to an offspring of type 2 (1). In the case of Process III, a B-object of type 2 gives birth to two offspring of type 1 and disappears from the population, and a B-object of type 1 gives birth to a single B-object of type 2.

We shall describe below the Consael Process III with the initial conditions $\xi_1(0) = 0$ and $\xi_2(0) = 1$. Let

$$P(x, y; t) = \mathsf{P}(\xi_1(t) = x,\ \xi_2(t) = y \mid \xi_1(0) = 0,\ \xi_2(0) = 1).$$

The function considered satisfies the system of differential equations (see I; 2.3.3.2.1):

$$\frac{\mathrm{d}}{\mathrm{d}t} P(x, y; t) = b(t)xP(x, y-1; t) + b(t)(y+1)P(x-2, y+1; t) -$$

$$- b(t)(x+y)P(x, y; t), \qquad x \geqslant 2, y \geqslant 1,$$

$$\frac{\mathrm{d}}{\mathrm{d}t} P(x, 0; t) = b(t)P(x-2, 1; t) - b(t), xP(x, 0; t), \quad x \geqslant 2,$$

$$\frac{d}{dt} P(0, y; t) = -b(t)\, y P(0, y; t), \qquad y \geqslant 1,$$

$$\frac{d}{dt} P(1, y; t) = b(t) P(1, y-1; t) - b(t)(y+1) P(1, y; t), \qquad y \geqslant 1,$$

$$\frac{d}{dt} P(1, 0; t) = -\, b(t) P(1, 0; t),$$

$$\frac{d}{dt} P(0, 0; t) = 0, \tag{3.1}$$

with initial conditions

$$P(0, 1; 0) = 1, \qquad P(x, y; 0) = 0, \qquad x \neq 0, \qquad y \neq 1,$$

Let now

$$Q(n; t) = \mathsf{P}(\xi_1(t) + \xi_2(t) = n),$$

so that

$$Q(n; t) = \sum_{x+y=n} P(x, y; t) = \sum_{x=0}^{n} P(x, n-x; t).$$

It is easily seen that

$$\sum_{x+y=n} [xP(x, y-1; t) + (y+1)\, P(x-2, y+1; t)] =$$
$$= (n-1)\, Q(n-1; t).$$

Hence, the system (3.1) becomes

$$\frac{d}{dt} Q(n; t) = b(t)(n-1)\, Q(n-1; t) - b(t) n\, Q(n; t), \qquad n \geqslant 1,$$

$$\frac{d}{dt} Q(0; t) = 0,$$

with initial conditions $Q(1; 0) = 1$, $Q(n; 0) = 0$, $n \neq 1$.

We introduce the generating functions

$$G_1(z_1, z_2; t) = \sum_{x\in N} \sum_{y\in N} P(x, y; t)\, z_1^x z_2^y, \qquad |z_1| < 1, \quad |z_2| < 1,$$

$$G_2(z, t) = \sum_{n\in N} Q(n; t)\, z^n, \qquad |z| < 1.$$

Clearly, $G_2(z, t) = G_1(z, z; t)$, Taking (3.1) into account we observe that $G_1(z_1, z_2; t)$ satisfies the linear homogeneous partial differential equation

$$\frac{\partial G_1}{\partial t} = b(t)\,[z_1(z_2 - 1)]\,\frac{\partial G_1}{\partial z_1} + b(t)(z_1^2 - z_2)\,\frac{\partial G_1}{\partial z_2}, \tag{3.2}$$

with $G_1(z_1, z_2; 0) = z_2$ and $G_1(1, 1; t) = 1$.

Similarly, the probability generating function $G_2(z, t)$ satisfies the linear, homogeneous, partial differential equation

$$\frac{\partial G_2}{\partial t} = b(t)\,z\,(z - 1)\,\frac{\partial G_2}{\partial z},$$

with $G_2(z; 0) = z$ and $G_2(1, t) = 1$. By denoting

$$\alpha(t) = \int_0^t b(\tau)\,\mathrm{d}\tau,$$

and

$$f(t) = 1 - \mathrm{e}^{-\alpha(t)},$$

it is easily seen that

$$G_2(z, t) = \frac{(1 - f(t))\,z}{1 - f(t)z}\,. \tag{3.3}$$

To solve (3.2) we consider the equations

$$\frac{\mathrm{d}z_1}{\mathrm{d}t} = z_1(1 - z_2)\,b(t),$$

$$\frac{\mathrm{d}z_2}{\mathrm{d}t} = (z_2 - z_1^2)\,b(t),$$

or

$$\frac{d(z_1 - z_2)}{z_1 - z_2} = (1 + z_1)b(t)\,dt,$$
$$\frac{d(z_1 + z_2)}{z_1 + z_2} = (1 - z_1)b(t)\,dt. \tag{3.4}$$

We thus get

$$(z_1^2 - z_2^2)\,e^{-2\alpha(t)} = c_1 \text{ (constant)}.$$

The second equation of the system (3.4) thus becomes

$$\frac{d}{dt}(z_1 + z_2) = -\frac{b(t)}{2}c_1 e^{2\alpha} + b(t)(z_1 + z_2) - \frac{b(t)}{2}(z_1 + z_2)^2,$$

which is a Riccati equation admitting a particular solution

$$z_1 + z_2 = i\sqrt{c_1 e^{\alpha}}.$$

We thus deduce

$$\frac{2i\sqrt{c_1 \exp(\alpha - f\sqrt{z_2^2 - z_1^2})} - (z_1 + z_2 - i\sqrt{c_1 e^{\alpha}})(1 - \exp(-f\sqrt{z_2^2 - z_1^2}))}{2i\sqrt{c_1}(z_1 + z_2 - \sqrt{z_2^2 - z_1^2})} = c_2.$$

The general integral of equation (3.2) will be

$$G_1(z_1, z_2; t) = \Phi(c_1, c_2).$$

In conjunction with the initial conditions, denoting $w = \sqrt{z_2^2 - z_1^2}$, we obtain

$$G_1(z_1, z_2; t) = w(1 - f(t))\frac{z_2 \cosh wf(t) - w \sinh wf(t)}{w \cosh wf(t) - z_2 \sinh wf(t)}. \tag{3.5}$$

Formula (3.5) may be rewritten in the form

$$G_1(z_1, z_2; t) = [1 - f(t)]\frac{z_2 H_1(z_1, z_2; t) - H_3(z_1, z_2; t)}{H_1(z_1, z_2; t) - f(t)\,H_2(z_1, z_2; t)z_2}, \tag{3.6}$$

where

$$H_1(z_1, z_2; t) = \sum_{y \in N} \frac{1}{(2y)!} (z_2^2 - z_1^2)^y f^{2y}(t),$$

$$H_2(z_1, z_2; t) = \sum_{y \in N} \frac{1}{(2y+1)!} (z_2^2 - z_1^2)^y f^{2y}(t),$$

$$H_3(z_1, z_2; t) = \sum_{y \in N} \frac{1}{(2y+1)!} (z_2^2 - z_1^2)^{y+1} f^{2y+1}(t).$$

Note that for $z_1 = z_2 = z$ these relations are reduced to (3.3). The probabilities $P(x, y; t)$, $x, y \in N$, can be computed step by step. For example,

$$P(1, 0; t) = 0, \quad P(0, 1; t) = 1 - f(t),$$

$$P(2, 0; t) = [1 - f(t)]f(t),$$

$$P(1, 1; t) = 0, \quad P(0, 2; t) = 0,$$

and so on. We get $Q(n; t) = [1 - f(t)]f^{n-1}(t)$, and

$$P(2, y; t) = \frac{2^y}{(y+1)!} [1 - f(t)] f^{y+1}(t),$$

$$P(4, y; t) = \frac{2^{y+1}(2^{y+2} - (y+3))}{(y+3)!} [1 - f(t)] f^{y+3}(t), \qquad y \geqslant 1$$

and so on.

From (3.6) we deduce by the usual procedure,

$$\mathsf{E}\xi_1(t) = \frac{2}{3} e^{\alpha(t)} - \frac{2}{3} e^{-2\alpha(t)},$$

$$\mathsf{E}\xi_2(t) = \frac{1}{3} e^{\alpha(t)} + \frac{1}{3} e^{-2\alpha(t)},$$

$$\mathsf{D}\xi_1(t) = \frac{1}{45} e^{2\alpha(t)}(20 - 12e^{-\alpha(t)} + 80e^{-3\alpha(t)} - 60e^{-4\alpha(t)} - 28e^{-6\alpha(t)}),$$

$$\mathsf{D}\xi_2(t) = \frac{1}{45} e^{2\alpha(t)}(5 + 3e^{-\alpha(t)} - 40e^{-3\alpha(t)} + 60e^{-4\alpha(t)} - 28e^{-6\alpha(t)}).$$

From the results obtained we observe that if $\alpha(t)$ tends to a finite limit as $t \to \infty$, then the mean values of the two populations also tend to a finite limit. If $\alpha(t) \to \infty$, as $t \to \infty$, then the mean value of the total number of B-objects of both types increases indefinitely, as $t \to \infty$.

3.1.2. A generalized n-dimensional ($n \geqslant 2$) linear growth process

3.1.2.1. Now we shall present a two-dimensional linear growth process $\xi(t) = (\xi_1(t), \xi_2(t))_{t \geqslant 0}$, a generalization of the Consael Process I. Migration, mutation, differentiation or competition may be easily represented by means of this general model, which only poses the condition of linearity — that can in some cases be a first and good approximation for a more complicated process.

Let (x, y) be a two-dimensional vector of nonnegative integer components. If $\xi(t) = (x, y)$, the probability of an increase by one in the number of B-objects of one or the other of the two types occurring during the time interval $(t, t + \Delta t)$ is $b^{(1)}_{x,y}\Delta t + o(\Delta t)$ or $b^{(2)}_{x,y}\Delta t + o(\Delta t)$, respectively, where

$$b^{(1)}_{x,y} = c_{11}x + c_{12}y + c_{13},$$

$$b^{(2)}_{x,y} = c_{21}x + c_{22}y + c_{23},$$

where $c_{11}, \ldots, c_{23}$ are functions of t. Similarly, the probability of the disappearance of a B-object of type 1 or type 2 in the time interval $(t, t + \Delta t)$ is $d^{(1)}_{x,y}\Delta t + o(\Delta t)$ or $d^{(2)}_{x,y}\Delta t + o(\Delta t)$, respectively, where

$$d^{(1)}_{x,y} = c'_{11}x + c'_{12}y + c'_{13},$$

$$d^{(2)}_{x,y} = c'_{21}x + c'_{22}y + c'_{23},$$

where $c'_{11}, \ldots, c'_{23}$ are functions of t. We shall add the assumption that in the time interval $(t, t + \Delta t)$ no change has occurred in the composition of the population with probability $1 - [b^{(1)}_{x,y} + b^{(2)}_{x,y} + d^{(1)}_{x,y} + d^{(2)}_{x,y}]\Delta t + o(\Delta t)$.

3.1.2.2. We immediately observe that Consael Process II is a special case of the above linear model and one suggests, as a useful exercise, the construction of random analogues for some of the deterministic models of population kinetics, models set up by V. Volterra and V. A. Kostitzin (see Kostitzin (1937)).

In the cases

$$c_{11} = c_{12} = c_{13} = p \text{ (constant)} > 0,$$

$$c_{21} = c_{22} = c_{23} = q \text{ (constant)} > 0,$$

$$c'_{11} = c'_{22} = 1,$$

$$c'_{12} = c'_{13} = c'_{21} = c'_{23} = 0,$$

and

$$c_{11} = c_{13} = c_{22} = c_{23} = 1,$$

$$c_{12} = c_{21} = 0,$$

$$c'_{11} = c'_{12} = c'_{13} = p \text{ (constant)} > 0,$$

$$c'_{21} = c'_{22} = c'_{23} = q \text{ (constant)} > 0,$$

it is possible to obtain representation formulae for the transition probabilities of $\xi(t)$ in terms of classical orthogonal polynomials, as was done by KARLIN and McGREGOR (1957) for the one-dimensional case. The reader will find the details in MILCH (1968) as well as an extension to the $(n + 1)$ dimensional case, $n > 1$.

3.2. Immigration-emigration processes

3.2.1. The Kendall process

3.2.1.1. In Section 2 the populations were treated in the majority of the cases as isolated entities. Such situations are rather seldom encountered in nature, so that previous representations have largely covered experimental situations. Common sense shows us that in the case of an isolated population (in nature) of B-objects, these must have moved at least once to reach that place. We must therefore take into account the fact that one of the fundamental problems of biology is the understanding of the growth and change of populations of individuals, both in regard to their intrinsic properties and in relation to environmental influences (BAILEY (1967 b, p. 154)).

As claimed by biologists, a species is unequally distributed on a territory, being always represented by more or less discontinuous demes. Among these, there are spaces unfavourable to the respective species or occupied by other species representing stronger competitors and, generally, the geographical distribution of a species can have different representations — from marked spatial discontinuity of the demes up to the closest vicinities. Populations of sessile (or sedentary) organisms are exactly these populations growing through immigration. Spores, cysts or the active phases of these forms can be carried large distances (sometimes on the surface of or inside other animals); many species of microscopic planktonic organisms are scattered by ocean currents and are part of cosmopolitan or semicosmopolitan forms.

Let us now take an example from the highly motile animals, such as birds. In a local population of birds one distinguishes a sedentary nucleus (adult birds) as well as a group of young individuals coming from other demes. This mixture of population assures the maintenance of crossings between individuals of different ages and a vast exchange of gametes on a wide territory. In the case of other species the birth rate is supplemented or even replaced through immigration.

3.2.1.2. These biological considerations confirm a series of deductions obtained by the analysis of classical stochastic models. It is well-known that a Feller-Arley process is *always* unstable; even when its parameters b and d are equal, the population is unavoidably deemed to extinction. To achieve the stability of the process a "mechanism of recovery" (KENDALL (1952)) such as immigration must be added.

Let us now consider the Kendall process as it was defined in Subparagraph 2.1.2.7. It is a birth-and-death process $(\xi(t))_{t\geqslant 0}$ with birth and death rates $b_n = bn + \nu$, $d_n = dn$, $n \in N$; $b, d, \nu > 0$, where ν must be considered as the immigration rate, the probability that one immigrant enters into a local population in the time interval $(t, t + \Delta t)$ being $\nu\Delta t + o(\Delta t)$. The partial differential equation for the probability generating function $G(z, t)$ of $\xi(t)$, given the forward equations, is

$$\frac{\partial G}{\partial t} = (bz - d)(z - 1)\frac{\partial G}{\partial z} + \nu(z - 1)G(z, t), \tag{3.7}$$

with the initial condition $G(z, 0) = z^{n_0}$ when starting off with $\xi(0) = n_0$. Kendall's hypothesis is $n_0 = 0$. Equation (3.7) differs from the equation corresponding to the Feller-Arley process only in the addition of the last term on the right, which is clearly a Poisson process component. Standard computations yield

$$G(z, t) = \left(\frac{p}{1 - qz}\right)^{\nu/b} \left(\frac{1 + \left(\frac{q}{p} - \Lambda\right)(1 - z)}{1 + \frac{q}{q}(1 - z)}\right)^{n_0}, \tag{3.8}$$

where

$$\Lambda = e^{(b-d)t},$$

$$p = \begin{cases} \dfrac{b - d}{b\Lambda - d}, & b \neq d \\ (1 + bt)^{-1} & b = d \end{cases}$$

and $q = 1 - p$. Hence the mean of $\xi(t)$ equals

$$m(t) = \frac{\nu}{b} \cdot \frac{q}{p} + n_0\Lambda,$$

with the limiting properties as $t \to \infty$,

$$\left.\begin{aligned} m(t) &\sim \left(\frac{\nu}{b-d}+n_0\right)\Lambda, && b > d \\ &\sim \nu t, && b = d \\ &\sim \frac{\nu}{d-b}, && b < d \end{aligned}\right\} \tag{3.9}$$

Note also that in case $b < d$,

$$G(z, \infty) = \left(\frac{d - bz}{d - b}\right)^{-\nu/b}.$$

Hence, the extinction probability is less than one even for $b < d$.

3.2.1.3. It is interesting to know the behaviour of the Kendall process under the hypothesis that $n_0 = 0$ and the general effect of immigration is negligible, but sufficient to restart the process if it should happen to become extinct ($\nu \to 0$). We then rewrite the probability generating function as

$$G(z, t) = p^{\alpha}(1 - qz)^{-\alpha}, \tag{3.10}$$

where $\alpha = \nu/b$. If we let $\alpha \to 0$ in (3.10), we simply have a degenerate distribution with all probability concentrated at zero.

In some circumstances we are more interested in the distribution conditional on the absence of the zero class (BAILEY (1964 a, p. 100)). To this purpose, we remove the constant term of (3.10), factor out αp^{α}, and normalize the result. Thus we obtain a logarithmic distribution, for which the probability generating function is

$$G(z, t) = \frac{\log(1 - qz)}{\log(1 - q)}.$$

3.2.2. Immigration and emigration processes

3.2.2.1. Let us now consider a system of local populations (or demes), assuming that a B-object can migrate from one deme to another (for example, the behaviour of the young component of some bird populations). The nonlinear migration model set up by WHITTLE (1967) starts from the hypothesis that on any areal there are, at time t, m demes containing $\xi_1(t), \ldots, \xi_m(t)$ B-objects, respectively. Suppose now that $\boldsymbol{\xi} = (\xi_1(t), \ldots, \xi_m(t))_{t\geqslant 0}$ is a homogeneous Markov jump process. More specifically, migration of a single B-object from deme j to deme k takes place with intensity $\lambda_{jk}\nu_j(n_j)$, $1 \leqslant j, k \leqslant m$, $j \neq k$, if in deme j there were n_j B-objects. In the following, we shall agree that $\lambda_{jj} = 0$, $1 \leqslant j \leqslant m$.

The process is obviously reducible, because the total number of B-objects is conserved at each transition, so that

$$\sum_{j=1}^{m} \xi_j(t) = \mathfrak{n},$$

say, and no value of $\mathfrak{n}$ can be attained from another. However, we shall assume that there are no other reductions, so that once $\mathfrak{n}$ is specified, the process has a unique stationary distribution. Sufficient conditions for this are that all elements of some power of the matrix $(\lambda_{jk})_{1\leqslant j,\, k\leqslant m}$ are positive, and that $\nu_j(n) > 0$ if $n > 0$, $1 \leqslant j \leqslant m$, for these ensure that an object can never be trapped in any deme or subgroup of demes. We shall also require that $\nu_j(0) = 0$ for all j: there can be no emigration from an empty deme. Put

$$\mathbf{N} = \left\{\mathbf{n}\colon\ \mathbf{n} = (n_1, \ldots, n_m),\ n_i \geqslant 0,\ 1 \leqslant i \leqslant m,\ \sum_{i=1}^{m} n_i = \mathfrak{n}\right\}.$$

Let $(p_{\mathbf{n}})_{\mathbf{n}\in\mathbf{N}}$ be the stationary distribution of our process. Also let U_j be the "displacement" operator that increases the n_j argument of a function of $\mathbf{n} = (n_1, \ldots, n_m)$, $\mathbf{n} \in \mathbf{N}$, so that, for example:

$$U_1 f(n_1, \ldots, n_m) = f(n_1 + 1, n_2, \ldots, n_m).$$

The equations satisfied by the stationary distribution (see I; 2.3.4.3.4) can then be written as

$$\sum_{j=1}^{m} \sum_{k=1}^{m} \lambda_{jk}\nu_j(n_j + 1)\, U_j U_k^{-1} p_{\mathbf{n}} = p_{\mathbf{n}} \sum_{j=1}^{m} \sum_{k=1}^{m} \lambda_{jk}\nu_j(n_j), \qquad \mathbf{n} \in \mathbf{N}.$$

One can easily verify that

$$p_{\mathbf{n}} = c_{\mathfrak{n}} \prod_{j=1}^{m} \alpha_j(n_j), \qquad \mathbf{n} \in \mathbf{N},$$

where $c_{\mathfrak{n}}$ is a constant determined from the fact that

$$\sum_{\mathbf{n}\in\mathbf{N}} p_{\mathbf{n}} = 1,$$

and

$$\alpha_j(0) = 1,$$

$$\alpha_j(n) = \frac{\pi_j^n}{\nu_j(1)\,\nu_j(2)\ldots\nu_j(n)}, \qquad n \in N^*.$$

Here π_j, $1 \leqslant j \leqslant m$, is the solution of

$$\pi_j \sum_{k=1}^{m} \lambda_{jk} = \sum_{k=1}^{m} \pi_k \lambda_{kj}.$$

By the conditions imposed on (λ_{jk}) the vector (π_j) is determined uniquely, to within a multiplicative constant, whose arbitrariness is absorbed by $c_{\mathfrak{n}}$.

If the probability generating function of $(p_{\mathbf{n}})_{\mathbf{n} \in \mathbf{N}}$ is denoted by

$$G(\mathbf{z}) = \sum_{\mathbf{n} \in \mathbf{N}} p_{\mathbf{n}} \prod_{j=1}^{m} z_j^{n_j},$$

we see from the above that

$$G(\mathbf{z}) = \frac{\text{coefficient of } \theta^{\mathfrak{n}} \text{ in } \prod_{j=1}^{m} A_j(\theta z_j)}{\text{coefficient of } \theta^{\mathfrak{n}} \text{ in } \prod_{j=1}^{m} A_j(\theta)},$$

where

$$A_j(z) = 1 + \sum_{n \in N^*} \frac{(\pi_j z)^n}{\nu_j(1)\,\nu_j(2)\ldots\nu_j(n)}.$$

If $\mathfrak{n}$ is held finite then we need not concern ourselves with the convergence of these power series. However, we obtain a "natural" distribution of deme sizes by choosing

$$G(\mathbf{z}) = \prod_{j=1}^{m} \frac{A_j(\theta z_j)}{A_j(\theta)},$$

for some positive θ (which implies that the total population $\mathfrak{n}$ is now a random variable) and convergence of this expression for $|z_1| = \ldots = |z_m| = 1$ is necessary if the "natural" distribution is to be physically possible. Clearly, in this distribution the ξ_j, $1 \leqslant j \leqslant m$, are independent variables. For further considerations concerning an open system the reader is referred to WHITTLE (1968).

3.2.2.2. We note that if $\nu_j(n) = n$, for all j, then this is the classical linear migration model (BARTLETT (1949)), in which B-objects migrate independently between demes. If

$$\nu_j(n) = \begin{cases} 0, & n = 0 \\ 1, & n \in N^*, \end{cases}$$

then we have essentially a system of one-server queues with exponential service times, feeding into one another. In the jth queue, customers are served at rate $\sum_{k=1}^{m} \lambda_{jk}$, and on completion of service the customer enters the kth queue with probability $\lambda_{jk}\left[\sum_{k=1}^{m} \lambda_{jk}\right]^{-1}$. If

$$\nu_j(n) = \min(n, r),$$

for some integer r, then one has the same situation, but with the jth deme amounting to an r-server queue with exponential service times. A very interesting and unifying study concerning Markov population processes has been given by KINGMAN (1969).

3.2.2.3. It is interesting to add that Whittle's model contains the Ehrenfest model as a special case and includes also the so-called Fermi-Dirac case (see FELLER (1968, p. 41); LOÈVE (1963, p. 42)). We can achieve a Fermi-Dirac system by requiring that

$$\nu_j(n) = \begin{cases} 0, & n = 0 \\ 1, & n = 1 \\ +\infty, & n > 1 \end{cases}$$

so that a B-object can be expelled immediately from the deme as soon as the deme has more than one member. This makes the migration rate also dependent upon the number in the deme to which one is migrating. In this case, $A_j(z) = 1 + \pi_j z$.

3.2.2.4. We shall now present the multivariate emigration-immigration process, considered by RUBEN (1962). It is a vector stochastic process $\boldsymbol{\xi} = (\xi_1(t), \ldots, \xi_m(t))_{t \geqslant 0}$, with state space N^m and the parameters λ_{ij}, b_i and $d_i (1 \leqslant i, j \leqslant m, i \neq j)$. These parameters are defined as follows:

1°. The probability of a change from $(n_1, \ldots, n_m)$ to $(n_1, \ldots, n_i - 1, \ldots, n_j + 1, \ldots, n_m)$ in the time interval $(t, t + \Delta t)$ is $\lambda_{ij} n_i \Delta t + o(\Delta t)$;

2°. The probability of a change from $(n_1, \ldots, n_m)$ to $(n_1, \ldots, n_i + 1, \ldots, n_j - 1, \ldots, n_m)$ in the time interval $(t, t + \Delta t)$ is $\lambda_{ji} n_j \Delta t + o(\Delta t)$;

3°. The probability of a change from $(n_1, \ldots, n_m)$ to $(n_1, \ldots, n_i - 1, \ldots, n_m)$ in the time interval $(t, t + \Delta t)$ is $d_i n_i \Delta t + o(\Delta t)$;

4°. The probability of a change from $(n_1, \ldots, n_m)$ to $(n_1, \ldots, n_i + 1, \ldots, n_m)$ in the time interval $(t, t + \Delta t)$ is $b_i n_i \Delta t + o(\Delta t)$;

5°. The probability of no change from $(n_1, \ldots, n_m)$ is

$$1 - \sum_{i=1}^{m} \left(d_i + b_i + \sum_{j \neq i} \lambda_{ij} \right) n_i \Delta t + o(\Delta t),$$

in the time interval $(t, t + \Delta t)$.

It is assumed that the above probabilities are independent of the past realization of the process.

This immigration-emigration process is a natural analogue of the linear birth-and-death process in the sense that when $m = 1$, the former reduces to the latter. For $m > 1$, the rates may be regarded as the instantaneous mean interaction rates between the states of the process. The reader can find the mathematical treatment of the process in RUBEN (1962, 1963).

3.2.2.5. However, to be able to understand Ruben's model, a more detailed (and also wider) interpretation of the empirical process is necessary. Thus imagine that we are dealing with a finite set $\{1, \ldots, m\}$ of compartments that can be crossed *in both directions* by a B-object initially started from a generating compartment a. We shall say that this B-object "emigrates" from compartment i and "immigrates" into compartment j, when it moves in direction $i \to j$, at rate λ_{ij}. The reverse movement in direction $j \to i$ is also possible. When the B-object moves in the direction $i \to a$, we consider this a "recovery" of the object in the generating compartment. The term "compartment" (equivalent with that of deme under certain conditions: see Subparagraph 1.2.5.6) was used because of easily understandable intuitive considerations: we may consider the immigration-emigration process as a representation of the transitions of an animal group through various zones or ecological niches, i.e., the hunting zone, the raising zone, the reproduction zone, etc. The model may also be useful in paleo-biogeographic research; the existence of a generator compartment corresponds to the Darwinist hypothesis that a species has a single site of origin, and the reverse transitions through already visited compartments have a correspondent in the recurrence phenomenon. To be entirely correct in the present case, the model should admit different death rates for each compartment (niche, areal). Such a supplementary assumption would permit the representation of the asynchronous disappearance of some species (e.g. the *Chalicotherium*, a herbivorous mammal, disappeared from North America in the middle Myocene, while it survived in China, India and Africa until the Pleistocene).

3.2.2.6. Another interpretation of compartments is that each compartment represents a different state of the objects. In the first example, the problem of the modification of B-objects by passing from one compartment to another was not considered; in this example it will be supposed that by entering a certain compartment, the B-object obtains a distinctive property (or several properties). Let us thus consider compartment a as representing the "healthy state" of a population of B-objects and the other m compartments as the stages of any disease which can be traversed by a B-object in both senses. The model may represent the sequence of modifications undergone by a cell population submitted to the action of certain physical or chemical agents (e.g. X-rays, alkylating agents, etc.) assuming the existence of lesions of different degrees and the possibility of recovery. It may also be co sidered that under certain conditions the model can represent the intimate process of cell differentiation or dedifferentiation.

3.2.3. Intermigration and colonization

3.2.3.1. We return now to intermigration — already mentioned in connection with Consael processes — to also discuss some genetical implications. As is known, the frequencies of genes and genotypes remain the same from one generation to another (Hardy-Weinberg law), if there is no intervention of migration, mutation or selection processes (the so-called "directive forces"). The migration of some B-objects from one local population to another, or from a large population to a small one, contributes to the creation of a genotype equilibrium; intermigration gives birth to a complex population, with balanced polymorphism and a wider capacity for modification and adaptation. Geneticists call a *dispersion process* the process tending to dissipate the frequency of genes away from the values of a stable equilibrium. The developments of this process, i. e., (i) the differentiation of local populations, (ii) the reduction of genetic variation within these populations and (iii) the marked increase of the frequency of homozygotes (Wahlund's relation) may be opposed by the effect of directive forces. Local populations (with random mating) constitute the matter of the dispersion process and it has been calculated that, irrespective of the size of the local population, the immigration of a single individual at every second generation is sufficient to avoid an appreciable differentiation of this population (see Wright (1940), Falconer (1967)). If we refer to the set of genes considered, the population from which a B-object emigrates is, in fact, a "donor" while the population into which it immigrates is a "receptor". If the "receptor" is genetically strongly differentiated, the immigrant B-object is an alterexter B-object.

In the case of the Consael Type I Process, we need only the intermigration of B-objects successively belonging to one of the two separate demes as well as the death of these objects. The model is strictly demographic without any genetical implications. Using the definitions and notations in Table 2 and putting $\nu(t)$ instead of $b(t)$, supposing $\xi_1(0) = x_0$ and $\xi_2(0) = y_0$, and putting

$$P(x, y; t) = \mathsf{P}(\xi_1(t) = x, \xi_2(t) = y \mid \xi_1(0) = x_0, \xi_2(0) = y_0),$$

we shall have the system

$$\begin{aligned}
\frac{d}{dt} P(x, y; t) = {} & d(t)(x+1) P(x+1, y; t) + d(t)(y+1) P(x, y+1; t) - \\
& - (\nu(t) + d(t))(x+y) P(x, y; t) + \\
& + \nu(t)(x+1) P(x+1, y-1; t) + \\
& + \nu(t)(y+1) P(x-1, y+1; t), \qquad x, y \geqslant 1,
\end{aligned}$$

$$\begin{aligned}
\frac{d}{dt} P(x, 0; t) = {} & d(t)(x+1) P(x+1, 0; t) - x(\nu(t) + d(t)) P(x, 0; t) + \\
& + d(t) P(x, 1; t) + \nu(t) P(x-1, 1; t), \quad x \geqslant 1,
\end{aligned}$$

$$\begin{aligned}
\frac{d}{dt} P(0, y; t) = {} & d(t)(y+1) P(0, y+1; t) - y(\nu(t) + d(t)) P(0, y; t) + \\
& + d(t) P(1, y; t) + \nu(t) P(1, y-1; t), \quad y \geqslant 1,
\end{aligned}$$

$$\frac{d}{dt} P(0, 0; t) = d(t) P(1, 0; t) + d(t) P(0, 1; t), \tag{3.11}$$

under the initial conditions $P(x_0, y_0; 0) = 1$ and $P(x, y; 0) = 0$, $x \neq x_0$, $y \neq y_0$.

By using the same methods as in Subparagraph 3.1.1.1. (see CONSAEL (1949)), we arrive at the following expression for the probabilities:

$$\begin{aligned}
P(x, y; t) = {} & \frac{(1 - e^{-\alpha(t)})^{x_0 + y_0}}{(2(e^{\alpha(t)} - 1))^{x+y}} \times \\
& \times \sum_{\substack{m_1 + m_2 = x \\ n_1 + n_2 = y}} \frac{x_0!\, y_0!}{m_1!\, m_2!\, n_1!\, n_2!\, (x_0 - m_1 - n_1)!\, (y_0 - m_2 - n_2)!} \times \\
& \times (1 + e^{-2\beta(t)})^{m_1 + n_2} (1 - e^{-2\beta(t)})^{m_2 + n_1}, \quad x, y \geqslant 0,
\end{aligned} \tag{3.12}$$

where $\alpha(t) = \int_0^t d(\tau)\, d\tau$ and $\beta(t) = \int_0^t \nu(\tau)\, d\tau$.

The mean values will be given by the following relations

$$\mathsf{E}\xi_1(t) = \frac{1}{2}(x_0 + y_0)\,\mathrm{e}^{-\alpha(t)} + \frac{1}{2}(x_0 - y_0)\,\mathrm{e}^{-\alpha(t)-2\beta(t)},$$

$$\mathsf{E}\xi_2(t) = \frac{1}{2}(x_0 + y_0)\,\mathrm{e}^{-\alpha(t)} - \frac{1}{2}(x_0 - y_0)\,\mathrm{e}^{-\alpha(t)-2\beta(t)},$$

$$\mathsf{E}[\xi_1(t) + \xi_2(t)] = (x_0 + y_0)\,\mathrm{e}^{-\alpha(t)}.$$

It is observed that if $\alpha(t) \to \infty$, as $t \to \infty$, $\mathsf{E}[\xi_1(t) + \xi_2(t)] \to 0$, as $t \to \infty$. On the other hand, if $\alpha(t)$ tends to a finite limit, as $t \to \infty$, then, for any $\beta(t)$, the mean values of the variables $\xi_1(t)$, $\xi_2(t)$ as well as that of the sum $\xi_1(t) + \xi_2(t)$ will tend to a non-zero limit.

3.2.3.2. Colonization (or "invasion") represents an extreme case of migration. Considering the potential polymorphism of local populations, we see that when these populations are adapted to variable conditions of living on a small, ecologically dismembered areal, they form as soon as possible a sort of "compressed arch" and the species can quickly spread over large surfaces. [See DARWIN (1869, Ch. 11, p. 423; Ch. 12, p. 485): "Both in time and space, species and groups of species have their points of maximum development"].

A stochastic model for colonization was built by PYKE (1955, unpublished) and later presented in BHARUCHA-REID (1960). We shall recall here only the particular case of the Consael Type I Process when $y_0 = 0$, which is equivalent to a colonization process. In this situation we have instead of (3.12):

$$P(x, y; t) = \frac{x_0!}{x!\,y!\,(x_0 - x - y)!}\,(1 - \mathrm{e}^{-\alpha(t)})^{x_0} \times$$

$$\times \left[\frac{1 + \mathrm{e}^{-2\beta(t)}}{2(\mathrm{e}^{\alpha(t)} - 1)}\right]^x \left[\frac{1 - \mathrm{e}^{-2\beta(t)}}{2(\mathrm{e}^{\alpha(t)} - 1)}\right]^y,$$

and the mean values are

$$\mathsf{E}\,\xi_1(t) = \frac{x_0}{2}\,\mathrm{e}^{-\alpha(t)}\,(1 + \mathrm{e}^{-2\beta(t)}),$$

$$\mathsf{E}\,\xi_2(t) = \frac{x_0}{2}\,\mathrm{e}^{-\alpha(t)}\,(1 - \mathrm{e}^{-2\beta(t)}).$$

3.2.4. Taxis and kinesis as dispersion processes

3.2.4.1. The constitution of "coherent" (i.e., homogeneous) populations in a certain zone can be achieved under the condition that the B-objects considered should respond in a similar way to stimuli of physical, chemical or as yet unknown nature, existing in the respective zone. One may distinguish, according to the type of movement of iso-reactive B-objects, two efficient methods of aggregation (FRAENKEL and GUNN (1940)): (i) *taxis*, a directed movement, in which the B-object moves while orienting itself at an angle, although with errors, relative to the source or gradient of stimulation (i.e., a moth approaches a lamp along an approximately equiangular spiral), and (ii) *kinesis*, an undirected locomotor reaction, in which the speed of movement or turning frequency depends on the intensity of stimulation (see also WATSON (1965)).

The *taxis*-type movement could include, in our opinion, much more varied aggregation processes, such as, for example, the regrouping of different types of embryonic cells in coherent populations, after being dissociated and mixed (MOSCONA (1963), STEINBERG (1962)).

From the stochastic point of view, the model of aggregation through taxis is difficult to discuss. We recall here QUENOUILLE's (1947) research note on the problem of random flights. It is based on the expression derived by CHANDRASEKHAR (1943) for the probability density $p_n(\mathbf{x})$ of the position of an object, subjected to n displacements. When the displacements are random and of constant length l,

$$p_n(\mathbf{x}) = \frac{1}{2\pi^2 x} \int_0^\infty \sin xy \left(\frac{\sin ly}{ly}\right)^n y \, dy,$$

where $x = |\mathbf{x}|$.

3.2.4.2. A random model for *kinesis*-type movement of a random walk of the Brownian motion type was built by BROADBENT and KENDALL (1953). The prototype of the model was the behaviour of the larvae of the helminth *Trichostrongylus retortaeformis* which are hatched from eggs in the excretae of sheep or rabbits, and wander apparently at random until they climb and remain on blades of grass.

Let (x, y) be the coordinates of a larva, measured on a plane with origin at the point of release. Let us suppose that x and y are independent Gaussian variables with mean zero and variance σ^2. Their joint probability density is

$$\frac{1}{2\pi\sigma^2} \exp\left(-\frac{x^2 + y^2}{2\sigma^2}\right), \qquad -\infty < x, y < \infty.$$

Transforming to polar coordinates and integrating with respect to θ, one finds the marginal density at radius r to be

$$\frac{r}{\sigma^2}\exp\left(-\frac{r^2}{2\sigma^2}\right), \quad 0 \leqslant r < \infty$$

and thus we get the expected proportion of larvae contained in a circle of radius r (the radial cumulative distribution) as

$$g(r) = 1 - \exp(-r^2/2\sigma^2).$$

If we now assume that this distribution results from a random walk of the Brownian motion type, we have $\sigma^2 = \omega t$, where ω is the diffusion constant. Then, at time t the density at radius r will be

$$\frac{r}{\omega t}\exp\left(-\frac{r^2}{2\omega t}\right), \tag{3.13}$$

and the expected proportion in a circle of radius r will be $g(r; t) = 1 - \exp(-r^2/2\omega t)$.

For a complete description of the process we shall now introduce two assumptions:

1°. We assume that a larva, while performing the random walk may in any short time interval Δt, with probability $\alpha\Delta t$, come upon a blade of grass and climb it up. (Furthermore, having ascended the blade, it is unable to turn round and must stay there.)

2°. We assume that a larva is not necessarily trapped by the first blade of grass visited, but that it has probability p of being trapped when it visits a blade of grass, and that now $\beta\,\Delta\,t$ is the probability that a free larva will visit a blade of grass in a short interval Δt.

In case of assumption 1°, we write $P(t)$ for the probability that a larva is trapped on a blade of grass in the time interval $(0, t)$; if f is the probability density function of the time-to-trapping (the period of free motion), then

$$P(t) \equiv \int_0^t f(t)\,\mathrm{d}t, \quad 0 \leqslant t < \infty.$$

The probability that a larva is trapped in the interval $(0, t + \Delta t)$ is the sum of $P(t)$ and the joint probability that the larva is free at time t and trapped in the interval $(t, t + \Delta t)$, that is,

$$P(t + \Delta t) = P(t) + [1 - P(t)]\,\alpha\Delta t.$$

Hence,

$$f(t) = \alpha\left[1 - \int_0^t f(t)\,\mathrm{d}t\right],$$

with the solution

$$f(t) \equiv \alpha \mathrm{e}^{-\alpha t}, \quad 0 \leqslant t < \infty. \tag{3.14}$$

The probability density of the trapped larvae will be given by multiplying together the expressions (3.13) and (3.14), and integrating the result with regard to t. We obtain

$$I_0(\nu)\,\nu, \quad 0 \leqslant \nu < \infty \tag{3.15}$$

where $\nu \equiv r\sqrt{2\alpha/\omega}$ and $I_0(\nu)$ is a standard Bessel function. Using the differential equation

$$\frac{\mathrm{d}I_1(\nu)}{\mathrm{d}t}\nu + I_1(\nu) + \nu I_0(\nu) = 0,$$

we can integrate the ν-distribution to get the radial cumulative distribution

$$G(\nu) \equiv \mathrm{P}(r \geqslant \nu\sqrt{\omega/2\alpha}) = \nu I_1(\nu). \tag{3.16}$$

In case of assumption 2°, if we define $P_1(t)$ and $f_1(t)$ as in the first case, the probability that a larva is trapped in the interval $(0, t + \Delta t)$ is now the sum of $P_1(t)$ and the joint probability that the larva is free at time t, visits a blade of grass in the interval $(t, t + \Delta t)$ and is trapped there, i.e.,

$$P_1(t + \Delta t) = P_1(t) + [1 - P_1(t)]\,\beta\Delta t p.$$

The solution is now

$$f_1(t) \equiv p\beta \mathrm{e}^{-p\beta t}, \quad 0 \leqslant t < \infty.$$

The formulae (3.15) and (3.16) apply as before, with $\alpha \equiv p\beta$.

3.2.4.3. We shall complete our description of dispersion and aggregation by considering the behaviour of higher vertebrates which gather in flocks or in herds, migrate from one niche to another and separate at a given time in independent subgroups. This gregarious behaviour manifests intergroup (interindividual) relations. All forms of intraspecific relations, however well reglemented, are achieved through a balancing of actions, from cooperation and collaboration up to biologic competition and direct struggle. These phenomena are especially (or better) observed within small herds or flocks.

The model we shall discuss was set up by HOLGATE (1967 a) to describe the gregarious behaviour of elephants. Suppose that a given area contains n animals and that these animals are grouped into herds. The chance that two herds meet and join ("amalgamate") will increase with the number of herds, and it may, for instance, be taken to be proportional to the excess of the number over unity. Thus, the probability of an "amalgamation" in the interval $(t, t + \Delta t)$ given that there are m herds at time t is $d(m - 1)\,\Delta t + o(\Delta t)$.

The chance that a herd of size n_i, $1 \leqslant i \leqslant m$, separates into two will increase with n_i and may also be taken as being proportional to the excess of the size over unity. The probability that one of the herds in existence at time t will separate into two during the interval $(t, t + \Delta t)$, given that there are m herds at time t, is therefore

$$b \sum_{i=1}^{m} (n_i - 1)\,\Delta t + o(\Delta t) = b(n - m)\,\Delta t + o(\Delta t).$$

Finally, it will be assumed that the probability that a herd separates into more than two groups, or that more than two herds amalgamate during $(t, t + \Delta t)$ is $o(\Delta\, t)$.

It is easily observed that the fluctuations in the number of herds is a birth-and-death process. The stationary probability distribution $(p_i)_{1 \leqslant i \leqslant n}$ of our process will satisfy the equations (see I; 2.3.4.3.4):

$$b(n - m + 1)\, p_{m-1} - [d(m - 1) + b(n - m)]\, p_m + dmp_{m+1} = 0,$$

$$1 < m < n,$$

and,

$$- b(n - 1)\, p_1 + dp_2 = 0, \tag{3.17}$$

$$bp_{n-1} - d(n - 1)\, p_n = 0. \tag{3.18}$$

To solve the above equations, set $p_1 = c$. Then from (3.17)

$$p_2 = \frac{cb(n-1)}{d}.$$

It can be verified by induction that

$$p_i = c \binom{n-1}{i-1} \left(\frac{b}{d}\right)^{i-1}, \quad 1 \leqslant i \leqslant n.$$

The value of c can be deduced from the condition that $\sum_{i=1}^{n} p_i = 1$. Thus we finally have

$$p_i = \binom{n-1}{i-1} \left(\frac{b}{b+d}\right)^{i-1} \left(\frac{d}{b+d}\right)^{n-i}, \quad 1 \leqslant i \leqslant n, \qquad (3.19)$$

that is, in the equilibrium state the number of herds is distributed as "one plus a binomial variable". Notice that p_i can be interpreted as the proportion of time, during a long interval, that there are exactly $i\,(1 \leqslant i \leqslant n)$ herds. When there are i herds, the average herd size is n/i and hence the long term average herd size is

$$\frac{(b+d)^n - d^n}{b(b+d)^{n-1}}.$$

This has the limits n and 1, as $b/d \to 0$ and ∞, respectively, as it should.

3.3. Competition processes

3.3.1. Some introductory remarks

3.3.1.1. In community ecology, according to LEVINS (1965), a major question relates to the conditions of coexistence of species. The rate of change of each species in each niche is given by an equation whose terms are the rate of increase in an open environment, the carrying capacity of the environment in a niche, the density effects (e.g., the interaction coefficients) of the same and the other species, and the value of net migration. The first question that can be put by the biologist is: what are the conditions for equilibrium? There are several possibilities. First, if the rate of increase in an open environment is greater than zero, so that the species can maintain itself in a niche without immigration, and if all the coefficients of interaction are less than unity, so that the inhibiting effect of another species is less than the effect of individuals of the same species, a stable equilibrium is possible. Second, even if those conditions are not met, equilibrium is possible provided there is sufficient immigration.

The formula established by R. LEVINS is a good illustration of the problems we have tried to present in Section 3 and of their interdependence. The terms equilibrium and stability are often used in this section, a fact carrying implications for the structure of the random processes considered. On this occasion, we shall first recall FELLER's (1968, p. 295) remark that the expression *equilibrium distribution* "distracts attention from the important circumstance that it refers to a so-called *macroscopic equilibrium*, that is, an equilibrium maintained by a large number of transitions in opposite directions". Also, commenting their competition model, LESLIE and GOWER (1958) conclude that "it appears then, that, in a system of two competing species with a stable stationary state, the number of individuals over a relatively long period of time settles down to a type of distribution which is approximately normal in form, but with a degree of variation which may be greater than that expected for small deviations about the stable state. This greater degree of variation about the equilibrium level can only lead to an increased chance of random extinction of one or the other of the two species".

Let us examine the biological interpretation of this discussion. Treating the struggle for the habitat niche or the trophic niche, DARWIN (1869, p. 80) affirmed that "each species, even where it most abounds, is constantly suffering enormous destruction at some period of its life, from enemies or from competitors for the same place and food". The most frequent effect of the struggle for life was the expulsion of a species from an ecological niche and its occupation by another species. For example, the disappearance of carnivorous Marsupials from South America is due to the penetration on this continent, during the Myocene, following the rising of the Panama isthmus, of carnivorous mammals of the *Felidae* family. At present, there exist in South America only certain primitive types of Marsupials, the majority of this clas being represented by animals living in the Australian area. But even here, two different species, *Thylacinus cynocephalus* and *Sarcophilus satanicus*, have been eliminated because of the competition with the dingo *(Canis dingo)*, a half-wild dog brought in by man. We recall here the result of the introduction of the mongoose in Jamaica to destroy the rats damaging sugar-cane crops. The competition spread to other animals as well: the mongoose destroyed serpents, frogs, lizards, crabs, young domestic animals, etc. on the island. Therefore, a discussion of stability becomes particularly important, especially at present, as man continues to alter the biotype as well as the biocenosis.

3.3.1.2. Mathematical models of the competition process have been set up by A. LOTKA and V. VOLTERRA (see KOSTITZIN (1935)). The basic idea is to regard the population size $n(t)$ as a continuous differentiable function of time and express dn/dt as some function of n and of t. The approach to the problem also corresponds to Darwin's idea that for every species there are "many different checks, acting at different periods of life, and during different seasons or years ... some one check or some few being generally the most potent, but all concur in determining the average number or even the existence of the species" (1869, p. 86). When two or more interacting populations are involved, the differential equations will be functions of the several populations. Thus, even the deterministic models can rapidly become too difficult to solve in closed form (CHAPMAN's (1967) comment). Among the models set up in this field, competition models for food, prey-predator interaction and host-parasite interaction are well known. The three "laws" deduced by V. VOLTERRA and U. D'ANCONA for prey-predator interaction express, in fact, a stationary state (with both species persisting) which is approached by a series of damped oscillations. Recall the example of the periodic oscillations of rodents and carnivorous animals in northern regions: lemmings massively appear every four years and, at the same time, the populations of their predators (the polar fox, the polar owl, etc.) reach maximal sizes. See the book by C. S. ELTON, *Voles, Mice and Lemmings* (1942).

The first major attempt to provide stochastic versions of these deterministic models was made by FELLER (1939). But it was immediately obvious, CHAPMAN says, that the stochastic models are far more intractable than the deterministic models. The evident analytic approach was to replace the differential growth equation by transition probabilities which may depend on population size and time (see Kendall's model in the next subparagraph). However, as BAILEY (1964 a, p. 186) has observed, the transition probabilities are typically nonlinear functions of the population sizes, and all the difficulties which we will encounter in epidemic theory (in fact, a theory resulting from host-parasite interaction!), are present with additional complications. Such difficulties lead, among others, to a critical examination of both types of models, deterministic as well as stochastic (see WEISS (1963)) as in epidemic theory and, further, to Monte Carlo studies (BARNETT (1962), BARTLETT (1957 a, 1961), LESLIE (1958, 1962), LESLIE and GOWER (1958), etc.). See also the interesting paper of PARKER (1968) who deals with the simulation of the situation at the Kostenay Lake ecosystem, as well as the paper of GARFINKEL, MACARTHUR and SACK (1964).

However, the most important source of difficulties is the realism of the model. As stressed by KERNER (1962), who dealt with the statistical mechanics of interacting biological species, we are faced in ecology with a stunning, even paralyzing, array of variables and factual data. There is a series of questions of the following type: Where is account taken of the complex of breeding habits and peculiarities of the life cycles of the different species? Where is any description given of populations in space as well as time, of immigration currents and herding instincts, or of season, climate, geography, and topography? And so on. This is why the experimental models set up beginning in 1948 by T. PARK and his students on *Tribolium confusum* Duval and *Tribolium castaneum* Herbst (flour beetles) represent a new and realistic impulse to the creation of new competition models (see PARK (1954)). When these two species are brought together in a uniform environment (sifted wheat flour), one of the two species will always displace the other. At high temperatures and humidities, *T. castaneum* will win out; at low temperatures and humidities *T. confusum* will be the victor. Such a "competitive exclusion" was also noted by C. F. GAUSE, in 1934, on cultures of *Paramoecium caudatum* and *Paramoecium aurelia* (or *Didi-*

nium nasutum). Since the *Tribolium* has four stages in its life history, two active and two passive, and since the active stages are cannibalistic, it is seen that the model is of a prey-predator system but also at the same time a model of intraspecific competition. See in this sense ANDERSEN's (1960) paper. The model built by NEYMAN, PARK and SCOTT (1956) as well as that built by LESLIE (1962) represent two modalities of approaching the problem of the competition for *Tribolium* (see also CHAPMAN's comments).

3.3.1.3. In his 1952 paper D. G. KENDALL stressed the importance and difficulties of the problem of the existence of two biologically related species such that the growth law of the number of B-objects in the first species ("red") depends stochastically on the number of B-objects in the second species ("blue") and conversely. We shall not comment in detail this model, as it represents a special case of the competition processes set up by G.E.H. REUTER, a decade later.

3.3.2. Stochastic competition processes

3.3.2.1. We shall devote this paragraph to the theory of stochastic competition processes, as set up by REUTER (1961) and IGLEHART (1964 a, b)[4]. In the two-dimensional case we shall mean by a competition process a homogeneous Markov jump process $(\xi(t))_{t\geqslant 0}$ with state space $X = N^2 = \{(x, y) : x, y \in N\}$ and an intensity transition matrix $\mathbf{Q} = (q_{\mathbf{ij}})_{\mathbf{i},\mathbf{j}\in X}$ as follows

$q_{\mathbf{ij}}$	$\mathbf{j}$	
$r(x, y)$	$(x+1, y)$	
$u(x, y)$	$(x, y+1)$	
$l(x, y)$	$(x-1, y)$	or $\mathbf{i} = (x, y)$
$d(x, y)$	$(x, y-1)$	
$n(x, y)$	$(x-1, y+1)$	
$s(x, y)$	$(x+1, y-1)$	
0	other $\mathbf{j} \neq \mathbf{i}$	

where $r(x, y), \ldots, s(x, y) \geqslant 0$ and

$$q_{\mathbf{ii}} = -(r(x, y) + u(x, y) + \ldots + s(x, y)).$$

---

[4]) Taking Section 3 into account, it is possible to imagine setting up a general theory of dynamics processes as Markov jump processes with special state spaces and transition intensity matrices.

Here, the letters d, u, l, r were chosen (at IGLEHART's suggestion (1964 a)) to indicate motion down, up, left and right, respectively. The elements $n(x, y)$ and $s(x, y)$ allow motion on the northwest-southeast diagonal through the point $\mathbf{i} = (x, y)$. Also, because there are no states with $x < 0$ or $y < 0$, we must have

$$\begin{aligned} d(x, 0) &= s(x, 0) = 0, \qquad \text{for } x \in N, \\ l(0, y) &= n(0, y) = 0, \qquad \text{for } y \in N. \end{aligned} \tag{3.20}$$

More briefly, jumps from (x, y) always lead to one of the adjacent states $(x \pm 1, y)$, $(x, y \pm 1)$, $(x - 1, y + 1)$, $(x + 1, y - 1)$ but the boundaries $x = 0$ and $y = 0$ of the positive (x, y)-quadrant cannot be crossed.

A competition process is the natural analogue, in two dimensions, of the familiar one-dimensional birth-and-death process. Let us now examine the properties of two types of competition processes:

— *Type I*, where $r(x, y) = \ldots = s(x, y) = 0$ when $x = 0$ or $y = 0$, so that all states $(x, 0)$ and $(0, y)$ are absorbing.

Further

$$\begin{aligned} r(x, y) + u(x, y) &> 0, \\ l(x, y) + d(x, y) &> 0, \qquad x, y \in N^*, \end{aligned} \tag{3.21}$$

that is for each nonabsorbing state $\mathbf{i}$, some absorbing state $\mathbf{j}$ and every nonabsorbing state $\mathbf{j}$ is accessible from $\mathbf{i}$ [5].

— *Type II*, with irreducible transition intensity matrix $\mathbf{Q}$ (so that there are no absorbing states) and the relations (3.21) hold for all $(x, y) \neq (0,0)$. Note that necessarily $l(0,0) = d(0,0) = 0$, because of (3.20), but that $r(0,0) + u(0,0) > 0$ because $(0,0)$ is not absorbing.

A first example of Type II competition processes is precisely Kendall's two species process for which

$r(x, y)$	$u(x, y)$	$l(x, y)$	$d(x, y)$	$n(x, y)$	$s(x, y)$
αx	βy	γxy	δxy	0	0

and α, β, γ, and δ positive. This, as it stands, does not belong to Type I, but it will be noted that once one of the components becomes zero it remains zero; then one population becomes extinct and the other grows according to a linear pure birth process.

[5] State $\mathbf{j}$ is said to be accessible from $\mathbf{i}$ if there are states $\mathbf{i}_1 = \mathbf{i}, \mathbf{i}_2, \ldots, \mathbf{i}_{n-1}, \mathbf{i}_n = \mathbf{j}$, such that $q_{\mathbf{i}_k \mathbf{i}_{k+1}} > 0$, $1 \leqslant k \leqslant n - 1$.

The second example is BARTLETT's (1956) simple epidemic process for which

$r(x, y)$	$u(x, y)$	$l(x, y)$	$d(x, y)$	$n(x, y)$	$s(x, y)$
α	β	γx	δy	νxy	0

where $\alpha, \beta \geqslant 0$, $\gamma, \delta, \nu > 0$. (Compare with Subparagraph 5.3.1.5.).

Some results concerning competition processes are available. We shall give them without proof; for proofs the reader is referred to REUTER (1961). Put

$$M_k = \max\,(r(x, y) + u(x, y)),$$

$$m_k = \min\,(l(x, y) + d(x, y)),$$

the max and min being taken over (x, y) with $x + y = k$ and $x > 0$, $y > 0$ (Type I), $x \geqslant 0$, $y \geqslant 0$ (Type II). Thus M_k, m_k are defined and positive when $k \geqslant 2$ for Type I, defined when $k \geqslant 0$ and positive except for m_0 for Type II.

Then a sufficient condition for the uniqueness[6] of the competition process is

$$\sum_{k=2}^{\infty}\left(\frac{1}{M_k} + \frac{m_k}{M_k M_{k-1}} + \cdots + \frac{m_k \ldots m_3}{M_k \ldots M_2}\right) = \infty.$$

For a uniquely determined competition process of Type I, let A denote the set of absorbing states, and T the remaining set of states (x, y), with $x > 0$ and $y > 0$. Let $\pi_{\mathbf{ij}}$, $\mathbf{i}, \mathbf{j} \in X$, be the final probabilities (see I; Theorem 2.3.36). Then $\pi_{\mathbf{ij}} = 1$, if $\mathbf{i} = \mathbf{j} \in A$, while $\pi_{\mathbf{ij}} = 0$ if $\mathbf{j} \in T$; clearly, $\pi_{\mathbf{ij}}$ for $\mathbf{i} \in T$ and $\mathbf{j} \in A$ is the probability that the process, starting in $\mathbf{i}$, will ultimately enter (and then remain in) state $\mathbf{j}$.

Hence $\alpha_{\mathbf{i}} = \sum_{\mathbf{j} \in A} \pi_{\mathbf{ij}}$ with $\mathbf{i} \in T$ is the probability that some absorbing state is reached from $\mathbf{i}$. Either $\alpha_{\mathbf{i}} < 1$ for all $\mathbf{i} \in T$ or $\alpha_{\mathbf{i}} = 1$ for all $\mathbf{i} \in T$; the latter occurs if

$$\sum_{k=2}^{\infty} \frac{m_2 \ldots m_k}{M_2 \ldots M_k} = \infty.$$

If $\alpha_{\mathbf{i}} = 1$ for all $\mathbf{i} \in T$, let $\tau_{\mathbf{i}}$ be the mean time to reach A, starting in $\mathbf{i} \in T$. Then $\tau_{\mathbf{i}} < \infty$ for all $\mathbf{i} \in T$ if

$$\sum_{k=2}^{\infty} \frac{M_2 \ldots M_{k-1}}{m_2 \ldots m_k} < \infty.$$

[6] The method of proof is that used in I; 2.3.4.7.4.

Further, for a uniquely determined competition process of Type II either $\pi_{\mathbf{ij}} = 0$ for all $\mathbf{i}, \mathbf{j}$, or $\pi_{\mathbf{ij}} = \pi_{\mathbf{j}} > 0$ is independent of $\mathbf{i}$ and $\sum_{\mathbf{j} \in X} \pi_{\mathbf{j}} = 1$; the latter occurs if

$$\sum_{k=1}^{\infty} \frac{M_1 \ldots M_{k-1}}{m_1 \ldots m_k} < \infty.$$

The above results show that Bartlett's epidemic process is uniquely determined and positive recurrent (when $\alpha, \beta > 0$) while for Kendall's competition process, uniqueness holds and absorption is certain with finite mean absorption times.

IGLEHART (1964 a) gave the integral representation for a class of competition processes, similar in nature to those obtained by KARLIN and MCGREGOR (1957).

3.3.2.2. IGLEHART (1964 b) has studied a class of n-dimensional ($n > 2$) competition processes whose transition intensity matrix is defined by

$$q_{\mathbf{ij}} = \begin{cases} \lambda_k(\mathbf{i}), & \text{for } \mathbf{j} = (i_1, \ldots, i_{k-1}, i_k + 1, i_{k+1}, \ldots, i_n), \\ \mu_k(\mathbf{i}), & \text{for } \mathbf{j} = (i_1 \ldots, i_{k-1}, i_k - 1, i_{k+1}, \ldots, i_n), \\ \gamma_{kl}(\mathbf{i}), & \text{for } \mathbf{j} = (i_1, \ldots, i_{k-1}, i_k - 1, i_{k+1}, \ldots, i_{l-1}, i_l + 1, i_{l+1}, \ldots, i_n) \\ 0, & \text{for other } \mathbf{j} \neq \mathbf{i} \end{cases}$$

for $\mathbf{i} = (i_1, \ldots, i_n)$, where $1 \leqslant k, l \leqslant n,\ k \neq l$, and

$$q_{\mathbf{ii}} = -\left(\sum_{k=1}^{n} (\lambda_k(\mathbf{i}) + \mu_k(\mathbf{i})) + \sum_{\substack{k,l=1 \\ k \neq l}}^{n} \gamma_{kl}(\mathbf{i}) \right).$$

He assumes that $\mu_k(\mathbf{i}) = \gamma_{kl}(\mathbf{i}) = 0$ when $i_k = 0$ and that $\lambda_k(\mathbf{i})$, $\mu_k(\mathbf{i})$ and $\gamma_{kl}(\mathbf{i})$ are nonnegative. One also assumes that $\sum_{j=1}^{n} \lambda_j(\mathbf{i})$ and $\sum_{j=1}^{n} \mu_j(\mathbf{i})$ are positive when the state $\mathbf{i}$ is nonabsorbing with the exception that $\sum_{j=1}^{n} \mu_j(\mathbf{i}) = 0$ when $\mathbf{i}$ is the origin. The elements $\lambda_k(\mathbf{i})$ allow one birth in the kth coordinate, $\mu_k(\mathbf{i})$ one death in the kth coordinate and $\gamma_{kl}(\mathbf{i})$ a mutation from the kth to the lth coordinate. Clearly, all the states are stable.

The paper quoted gives conditions of uniqueness, recurrence, certainty for absorption and for finite mean absorption time. In addition to these, D. L. IGLEHART presents the characteristics of nine multivariate competition processes (irreducible or with absorbing states), among which three are extensions of linear birth-and-death processes.

3.3.3. Quasi-competition processes

3.3.3.1. This type of processes was generated by an important genetical problem. LYON (1961) made the hypothesis that in each individual cell of the normal female one of the two X chromosomes is genetically inactivated. The inactivation occurs only in embryogenesis, and it is a matter of chance whether the maternal or paternal X is inactivated in a given cell. However, once inactivation has occurred in a specific cell, all progeny of that cell maintain the same inactive X. Thus, the normal female ought to be a mosaic as far as the X chromosome is concerned. In approximately half of her cells only the X chromosome derived from her father will be functionally active, while in the other cells only the X chromosome of her mother is active (LYON (1961), BEUTLER et al. (1962)).

One of the genes located on the X chromosome controls (in man, equine species, Drosophila) the synthesis of glucose-6-phosphate dehydrogenase (G6PD), so that this enzyme becomes a fixed and natural cell marker. LYON's hypothesis was confirmed by the research work of DAVIDSON, NITOWSKY and CHILDS (1963), who demonstrated that clones of skin cells of heterozygous Negro females for a G6DP variant carried either one form of the enzyme, but not both (see also DAVIDSON (1968)). The phenomenon was also observed on cell cultures obtained from a benign tumour, leiomyoma of the uterus, from females heterozygous for the variant of G6DP (LINDER and GARTLER (1967)). The finding that the leiomyomas were of a single phenotype (as contrasted to the normal muscular uterus cells) is consistent with the concept that these tumors arose from single cells. Assuming a patch of 10,000 cells with a regular (cubic) shape, D. LINDER and S. M. GARTLER estimated the chance that a tumour arises from a given number of adjacent cells of like phenotype. The probability that a single tumour arises from two like cells is equal to the ratio of like pairs of adjacent cells to the total number of pairs of adjacent cells. This turns out to be 0.96 and the corresponding probability of their 121 runs is less than 5 per cent. The authors conclude that if no selection is at play during the growth of leiomyomas, their data rule out the possibility that the tumours arise from two cells of like phenotype.

One may, however, consider the assumption that the homogeneity of tumour cells — with respect to the identity of an active gene — ought to result from some kind of a competition-like process. More precisely, if every tumour starts with two kinds of cells, A and B, some chance mechanism would imply that large tumours containing both types of cells would be very rare (BÜHLER (1967 a)). This random mechanism is called "quasi-competition", provided that the word competition is not to be understood in its usual sense (BÜHLER (1967 b)). We have introduced this model beside the competition processes not only because of the close link between them (the quasi-competition process being a very particular case of Reuter's general scheme) but also to point out the interest and importance of research on cell ecology.

3.3.3.2. Let us assume that two populations of B-objects develop independently, each one according to a linear birth-and-death process $(\xi_1(t))_{t\geqslant 0}$ and $(\xi_2(t))_{t\geqslant 0}$. We want to study the conditional probability

$$\mathsf{P}\,(\xi_1(t) > 0,\ \xi_2(t) > 0 \mid \xi_1(t) + \xi_2(t) \geqslant n), \quad n \in N,$$

the condition $\xi_1(t) + \xi_2(t) \geqslant n$ reflecting the fact that one considers only tumours large enough to be detected and included in the study. Since, apparently, either one of the two cell types can be detected only if it is present in at least a certain proportion, say α, we may also consider

$$\mathsf{P}\,(\xi_1^*(t) \geqslant \alpha,\quad \xi_2^*(t) \geqslant \alpha \mid \xi_1(t) + \xi_2(t) \geqslant n),$$

where

$$\xi_1^*(t) = \frac{\xi_1(t)}{\xi_1(t) + \xi_2(t)}, \qquad \xi_2^*(t) = \frac{\xi_2(t)}{\xi_1(t) + \xi_2(t)}.$$

We mention that according to LINDER and GARTLER (1967), $0.05 \leqslant \alpha \leqslant$ $\leqslant 0.15$.

Under these assumptions, the following results hold:

Theorem 3.1. *Let two independent subcritical linear birth-and-death processes $(\xi_i(t))_{t\geqslant 0}$, $i = 1, 2$. Whatever $n \geqslant 1$ and whatever the fixed or random initial population sizes $\xi_1(0)$ and $\xi_2(0)$, we have*

$$\lim_{t\to\infty} \mathsf{P}\,(\xi_1(t) > 0,\ \xi_2(t) > 0 \mid \xi_1(t) + \xi_2(t) \geqslant n) = 0.$$

Proof. See BÜHLER (1967 b). ◇

Theorem 3.2. *Let $(\xi_i(t))_{t\geqslant 0}$, $i = 1, 2$, be two independent supercritical linear birth-and-death processes with the same parameters b, d. If $\xi_1(0) =$*

$= \xi_2(0) = 1$, *then for any* α *with* $0 \leqslant \alpha \leqslant 1/2$ *and* $n \in N^*$, *we have*

$$\lim_{t\to\infty} \mathrm{P}\,(\xi_1^*(t) > \alpha,\ \xi_2^*(t) > \alpha \mid \xi_1(t) + \xi_2(t) \geqslant n) =$$

$$= (1 - 2\alpha)\,\frac{1 - d/b}{1 + d/b}\,.$$

Proof. See BÜHLER (1967 b). ◇

In the case of the growth of benign tumours, at least the early stages of their growth are supercritical ($b > d$), so that this version (Theorem 3.2) seems to be closer to reality. An estimate of b can be obtained from studies using tritiated thymidine ($b \sim 0.01 - 0.1$ per day) and $(b - d)$ can be estimated from the sizes and ages of the tumours under study. Combining these estimates one can find that d/b may be of the order of 0.95 together with $\alpha = 0.1$. This combination would make the limit in Theorem 3.2 of the order of two per-cent (see also numerical estimates in BÜHLER (1967 b)). W. BÜHLER concludes that although there is evidence supporting the conjecture that leiomyomas originate from one single cell, the possibility cannot be excluded that the observed genetic homogeneity of tumours be caused by a competition-like chance mechanism.

3.3.4. Population excess and cannibalism

3.3.4.1. KENDALL (1952) has shown that two particular cases of stochastic growth processes are two possible models for population excess and cannibalism. It is about the birth-and-death processes with birth-and-death rates

$$b_n = \alpha\,(n_2 - n), \quad d_n = \beta(n - n_1), \quad 0 \leqslant n_1 \leqslant n \leqslant n_2,$$

and 0 otherwise,

and

$$b_n = \alpha(n_2/n - 1), \quad d_n = \beta\,(1 - n_1/n), \quad 0 < n_1 \leqslant n \leqslant n_2,$$

and 0 otherwise,

where α and β are constants. The second case is known as the Prendiville process and has been investigated in Paragraph 2.5.2.

We are not aware of models describing in any special way the biologic process of population excess, a process which is created when the correlation between population density and food conditions entails on interactions between B-objects growing together and results in elimination of a certain number of the same B-objects from the reproduction process. Population excess represents a particular case of the dynamics of species population, a paradoxical consequence of its prosperity since population excess leads to more rigid specialization, to a decrease of fecundity and to a reduction of the evolution rhythm.

In a paper meant for the lay reader, MARKERT (1966) presents the example of a small population of deer placed on a small island (of about 150 acres) in Chesapeake Bay about 44 years ago. The deer were kept well supplied with food. It was found that the colony grew, more or less exponentially, until it reached a density of about 1 deer per acre. Then the animals began to die off and did so in spite of adequate food and care. When these dead animals were examined there was marked evidence that they suffered from a variety of endocrinological disorders which can be described as due to the adrenal stress syndrome. The adrenal glands of these deer were hypertrophied and this led to deleterious physiological consequences, including the failure to reproduce.

Cannibalism is another mode of controlling the dimension of a population. This intraspecific process is rather wide spread as is well known in the case of worms, molluscs, insects, arachnids, fish, reptiles, etc. In all the cases when it occurs, it appears to be useful to the species, although it is achieved by the loss of number of destroyed B-objects: eggs, larvae or even adults. The stochastic model of cannibalism that we wish to present (MERTZ and DAVIES (1968)) is based upon the observations and experiments made on the genetic strain cIVa of *Tribolium castaneum*. Cannibalism is here exerted by larvae and adults on eggs and pupae.

Since the biologic process is little known, some details are deemed necessary to understand the parameters of the adequate stochastic model. For a population founded with adult beetles, there is an initial increase in eggs which tapers off in about four days when the first hatching begins. The young larvae immediately begin to destroy eggs and larval voracity increases with age. By the time a few hundred growing larvae are in the vial, almost every egg which is laid is destined to be eaten. Only when the first cohort of larvae has reached pupation, some 20 to 30 days later, is there again a relatively safe period for eggs. Then egg hatching resumes and a second cycle of larval abundance ensues. These cycles persist throughout the population history with the later cycles having only a slightly smaller amplitude than the earlier ones. In this way the cannibalistic behaviour of the larvae imposes a permanent and violent predator-prey cycle on the immature component of the population. The permanence of the cycle is insured by the voracity of the larvae, which are several times more predacious on eggs than are adults.

As a result of the cycling immature population, the pupae tend to appear in the population in evenly spaced waves about 35 to 40 days apart. Typically a wave consists of 100 to 400 individuals and every one of these pupae will become an adult unless it succumbs to cannibalism. Since the usual adult population consists of 70 to 100 individuals and the normal mortality of adults is only 10 to 20 per 30-day census period, the number of pupae eaten must necessarily be immense. Often every pupa in a wave is destroyed. However, when a wave is unusually large, survival may be high owing to the adult predators' inability to eat such large numbers of prey. Thus, one must take the advent of "satiety" into account.

3.3.4.2. Let us denote by:

A, the number of adult predators, which are assumed to be equally predacious and to act independently of one another in their predatory behaviour;

V, the initial number of prey (victims),

parameters fixed for the population considered. The random variables will be

$\alpha(t)$, the number of attacks generated up to time t by the adult population,

$\xi(t)$, the number of prey surviving at time t, where t is the time elapsed since the beginning of the collective vulnerable period.

All pupal prey and prey corpses will be assumed to be equally and synchronously vulnerable. A prey which has been attacked once is regarded a corpse and the corpses are assumed to be capable of sustaining an unlimited number of attacks during the collective period of vulnerability. The following probability distributions and probability generating functions will be considered:

$$p_x(t) = \mathsf{P}(\xi(t) = x), 0 \leqslant x \leqslant V,$$

with

$$F(z, t) = \mathsf{E}\, z^{\xi(t)} = \sum_{x=0}^{V} p_x(t) z^x,$$

and

$$q_a(t) = \mathsf{P}(\alpha(t) = a), \quad a \in N,$$

with

$$G(w, t) = \mathsf{E}\, w^{\alpha(t)} = \sum_{a \in N} q_a(t) w^a.$$

Let $g(w, t)$ be the probability generating function for the number of attacks made up to time t on V prey by a single adult. Because on the independence assumption,

$$G(w, t) = [g(w, t)]^A. \tag{3.22}$$

In order to obtain $F(z, t)$ from $G(w, t)$, consider the probability distribution for the number of survivors in the case that exactly a attacks have been made on a population of V prey. The attacks are distributed independently and with equal probabilities over the V prey: this is a version of classical occupancy problem (see e.g., FELLER (1968, p. 101)). Thus the probability that x prey survive at time t in case exactly a attacks have been made on a population of V prey will equal

$$\sum_{m=x}^{V} (-1)^{m-x} \binom{m}{x} \binom{V}{m} \left(1 - \frac{m}{V}\right)^a, \quad x \leqslant m \leqslant V,$$

consequently,

$$p_x(t) = \sum_{m=x}^{V} \binom{m}{x} \binom{V}{m} (-1)^{m-x} G(1 - m/V; t),$$

and

$$F(z, t) = \sum_{m=0}^{V} \binom{V}{m} (z-1)^m G(1 - m/V; t). \tag{3.23}$$

The mth factorial moment of $\xi(t)$ is given by the product of $m!$ and the coefficient of $(z-1)^m$ in $F(z, t)$, i.e.,

$$\frac{V!}{(V-m)!} G(1 - m/V; t).$$

Hence,

$$\mathsf{E}\,\xi(t) = V G(1 - 1/V; t),$$

$$\mathsf{D}\,\xi(t) = V(V-1)\, G(1 - 2/V; t) + VG(1 - 1/V; t)[1 - VG(1 - 1/V; t)].$$

Let us now consider the model of cannibalism itself and suppose that adults of *Tribolium* may exist in two states: "hunting" and "satiated". A satiated adult may be thought of as an adult who is occupied with a prey—either in attacking, handling, digesting, or in some other time-consuming way. At the beginning of the experiment all adults are assumed to be hunting. The rate of satiation μ is defined by assuming that the conditional probability that a given adult may attack a given pupa or pupal corpse (and thus become satiated) during the time interval $(t, t + \Delta t)$, given that the adult is hunting at time t, equals $\mu \Delta t + o(\Delta t)$.

The second transition from satiated to hunting will occur at some random time τ, after the onset of satiation. Suppose that at the beginning of the vulnerable period of the pupae, all the adults are hunting. The probability that an adult will rejoin the hunting component of the population during the interval $(t, t+\Delta t)$, given that the adult is satiated at time t, is $\nu\,\Delta t + o(\Delta t)$. The satiation time τ has an exponential probability density function with parameter ν. Suppose that at the beginning of the vulnerable period of the pupae all the adults are hunting.

From these assumptions we shall deduce[7] $G(w, t)$. Then we can solve for the distribution of surviving prey by applying equation (3.23).

In order to find the marginal distribution for the number of prey surviving at time t, consider a population consisting of one adult ($A = 1$). Let us define the following probabilities:

1°. $q_a^0(t)$, the probability that $\alpha(t) = a$ and the adult is not satiated at time t;

2°. $q_a^1(t)$, the probability that $\alpha(t) = a$ and the adult is satiated at time t.

We shall then define correspondingly the two generating functions $g_0(w, t)$ and $g_1(w, t)$ and then

$$g_1(w, t) + g_0(w, t) = \sum_{a \in N} q_a(t)\, w^a,$$

which is by definition equal to $g(w, t)$.

The differential equations for the transition probabilities become

$$\frac{\mathrm{d}\, q_a^1(t)}{\mathrm{d}t} = \mu\, V q_{a-1}^0(t) - \nu q_a^1(t),$$

$$\frac{\mathrm{d}\, q_a^0(t)}{\mathrm{d}t} = \nu\, q_a^1(t) - \mu V q_a^0(t).$$

---

[7] A more direct approach would be to try to find the joint probability distribution

$$\mathrm{P}(\xi(t) = x, \quad \eta(t) = y),$$

where $\eta(t)$ is the number of satiated adults at time t. However, since the Kolmogorov equations for the distribution in question lead to a soluble but not particularly tractable second order partial differential equation for the corresponding probability generating function, this approach was dropped. Nevertheless, one can obtain in this way the (marginal) distribution of $\eta(t)$: it is binomial with parameters A and

$$\frac{\mu V}{\mu V + \nu}\,(1 - \mathrm{e}^{-(\mu V + \nu)t}).$$

Multiplying both of these equations by w^a and summing over $a \in N$ yields

$$\frac{\partial}{\partial t} g_1(w, t) = \mu V w g_0(w, t) - \nu g_1(w, t),$$

$$\frac{\partial}{\partial t} g_0(w, t) = \nu g_1(w, t) - \mu V g_0(w, t).$$

Since at the beginning of the experiment the adult is assumed to be hunting and to have made no attacks, we have the initial conditions $g_1(w, 0) = 0$ and $g_0(w, 0) = 1$. We deduce

$$g_1(w, t) = \frac{\mu V w}{\rho_1 - \rho_2} (e^{\rho_1 t} - e^{\rho_2 t}),$$

$$g_0(w, t) = \frac{1}{\rho_1 - \rho_2} [(\nu + \rho_1) e^{\rho_1 t} - (\nu + \rho_2) e^{\rho_2 t}],$$

where ρ_1 and ρ_2 are the roots of the equation

$$\rho^2 + (\mu V + \nu) \rho + \mu \nu V (1 - w) = 0.$$

The probability generating function $g(w, t)$ is found by summing up g_0 and g_1 and further, $G(w, t)$ is obtained according to (3.22),

$$G(w, t) = \left[\frac{1}{\rho_2 - \rho_1} (e^{\rho_1 t}\{\mu V(1 - w) + \rho_2\} - e^{\rho_2 t}\{\mu V(1-w) + \rho_1\}) \right]^A.$$

We can obtain $F(z, t)$ using (3.23) and then the expectation and variance for $\xi(t)$.

The paper of Mertz and Davies (1968) also contains a development of the model without predator satiation and models in which the initial number of pupal prey is also a random variable.

3.3.4.3. In studies of the limiting egg population one observes that the possible effects due to the presence of the tunnels in the flour cannot be neglected. Under the assumption that no tunnels collapse, it is clear that the eggs are found in the tunnels only and not in the virgin flour. Therefore, the number of eggs found and eaten by the beetle will depend in particular on the time spent in retunnelling, i.e., moving through the existing tunnels. AHMED (1963) has built a stochastic model to describe the tunnelling and retunnelling of *Tribolium*; it is a three-state Markov chain.

3.4. Poikilopoiesis models

3.4.1. A stochastic approach to embryogenesis

3.4.1.1. We shall denote by poikilopoiesis (*poikilos*, various; *poiein*, to form) those stochastic models which represent the formation and behaviour of populations with alterexter B-objects generated by a precursor population of alteridem B-objects. Generally, it involves a category of cellular differentiation processes, a category belonging to the group of cytomorphosis (*morphosis*, shaping) processes. An example in this sense is formed by the population of histioblasts (young tissue cells) from which are derived a fairly large number of types of cells of the reticulo-histiocytic (or endothelial) system. Also, in embryogenesis, the blastomere S_2 in the nematode *Parascaris equorum* is the precursor of the cell populations, forming the endoderm, the mesenchyme and the worm's pharinx. It should be observed that the precursor population must hold relative omnipotency. This fact is clearly demonstrated in embryogenesis where this harmonious totipotence represents the ability of an early blastomere (or blastomeres) to give birth to various types of characteristic cells and structures and, moreover, the power to develop a perfectly organized animal body. The experiment of H. DRIESCH (1892), for example, is well known: the separation of blastomeres at the two-cell stage of sea urchin embryos provoked each blastomere to develop into an entire larva. The regeneration of iris and lens from pigmented retina in newts is also an example in this sense (STONE (1955)). Obviously, there is a difference between the potentialities of a cell and its actualities. It seems that the cells behave as if different sets of active genes (determinants) were parcelled out to them during embryonic development; however, such an interpretation would be mistaken, as the behaviour of chromosomes and their constituent genes is dependent upon external stimuli (MARKERT (1964)). During repeated cell division or under extreme environmental provocation, the differentiated state of the chromosome may be changed so as to express new patterns of gene function.

3.4.1.2. We shall now present a simple form of poikilopoiesis, i.e., the formation of the inner layer *(endoderm)* from the cells of the outer layer *(ectoderm)* in some inferior metazoans: medusae, calcareous sponges, etc. The generation of the new layer occurs through the migration of precursor cells from the outer part of the embryo to the inner cavity *(blastocoel)* and their transformation. The process was studied in details on some medusae (*Aegynopsis*, *Geryonia*, etc.) by I. I. METCHNIKOFF (1886).

To understand the construction of the adequate stochastic model (FIRESCU and TAUTU (1965 b)) we shall briefly describe the immigration process:

(a) Immigration is prepared by the gradual separation of some blastomeres from their link with the ectoderm and their existence as independent B-objects.

(b) The first stage of immigration is the shifting of some blastomeres (2—4 cells) followed by the immigration of a considerable number of cells. The immigration can be marked or weakly marked even within the same genus. Thus, with the medusa *Aurelia flavidula*, the immigration is strongly marked with the forms provided by the medusa *Eastport* and weakly marked with the forms provided by the medusa *Annasquama*.

(c) In the course of immigration, the nuclei of immigrating blastomeres are in a rest phase, immigration not being simultaneous with cell mitosis. Even when the first aggregate was made up, the freshly immigrated cells are quite rare in the mitotic state.

(d) In their further evolution, the endodermic cells suffer either a partial or a total removal or a qualitative transformation that will lead to the formation of the primordii of the organs of endodermic origin.

The stochastic model set up on the basis of these data is a bi-dimensional Markov jump process $(\xi(t), \eta(t))_{t\geqslant 0}$, the first component corresponding to emigration from the ectoderm, and the second to the birth, death and transformation of endoderm elements. The state space is $X = \{(x, y): x \in N^*, 0 \leqslant y \leqslant m\}$.

The modifications taking place in the system in the time interval $(t, t + \Delta t)$ are expressed [8] by means of transition intensities $q_{\mathbf{ij}}$ defined as follows:

$q_{\mathbf{ij}}$	$\mathbf{j}$	
$\nu x(m - y)$	$(x - 1, y + 1)$	
$\nu b y$	$(x, y + 1)$	$\mathbf{i} = (x, y)$,
$\nu y(d + r)$	$(x, y - 1)$	$b + d + r = 1$
$-\nu[x(m - y) + y]$	(x, y)	
0	other $\mathbf{j} \neq \mathbf{i}$	

where ν, b, d and r are immigration, birth, death and, respectively, transformation (removal) rates.

---

[8] Some small changes have been made here with respect to the original model.

3.4.2. Hematopoiesis models

3.4.2.1. The formation of blood cells from the hematogeneous bone marrow represents a complete process of poikilopoiesis: the precursor population (forming the so-called stem cell compartment) generates several cell populations (called *series* or *lines*) well differentiated from the morphological and physiological point of view and having their own evolution, from the young up to the mature cell entering the blood stream. These cell lines are well known: *erythrocytes*, anucleated cells, are the most specialized form of cells ladden with oxygen carriers; *neutrophils* are the "shock troups" primarily concerned with carrying noxious material and in secreting enzymes to aid in killing and digesting infectious material; *eosinophils* are concerned primarily with the antigen-antibody reactions; *basophils*, rich in histamine and heparin-like substances, may be important for initiating the inflammatory reaction; *thrombocytes* play an important part in the first stage of the blood clotting process as well as in the protection of the capillary endothelium, releasing a vasoconstrictor agent, serotonine.

We may thus consider that the hematogeneous bone marrow is represented by n populations of alterexter B-objects, each population going during its evolution through a system of compartments connected in series. To give an example, we mention that to become a mature red blood cell, an element from the erythrocyte line must pass through the following six compartments: pronormoblast → basophilic normoblast →

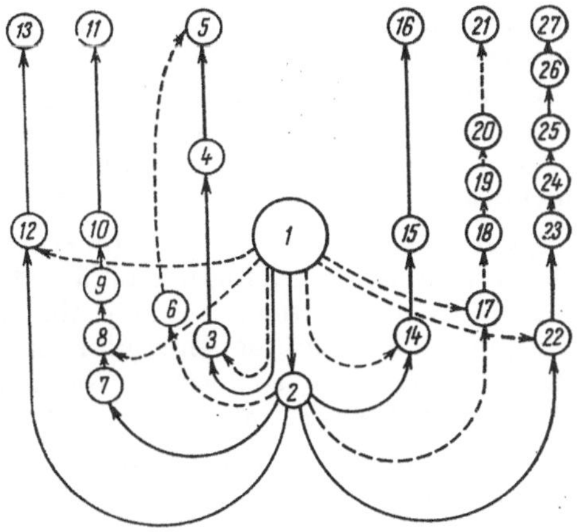

Fig. 4. — *1*: Undifferentiated reticulo-endothelial cell: *2*: Myeloblast; *3*: Reticulo-lymphocyte; *4*: Great lymphocyte; *5*: Lymphocyte; *6*: Immature lymphocyte (in leukemia); *7*: Leukoblast; *8*: Promyelocyte; *9*: Myelocyte (granulated); *10*: Metamyelocyte; *11*: Granulocyte; *12*: Monoblast; *13*: Monocyte; *14*: Megakaryocytoblast; *15*: Megakaryocyte; *16*: Thrombocyte; *17*: Promegaloblast; *18*: Basophile megaloblast; *19*: Polychromatic megaloblast; *20*: Orthochromatic megaloblast; *21*: Megalocyte; *22*: Pronormoblast: *23*; Basophile normoblast; *24*: Polychromatic normoblast; *25*: Orthochromatic normoblast; *26*: Reticulocyte; *27*: Erythrocyte.

polychrom normoblast → orthochrom normoblast → reticulocyte → erythrocyte. H. Downey's schema (1965; see Fuchs (1966)) illustrates the entire process of hematopoiesis (Figure 4).

If we therefore consider that each line S_i, $0 \leqslant i \leqslant n$, is formed by k compartments

$$S_i = \{S_{i0}, S_{i1}, \ldots, S_{ik}\}, \quad 0 \leqslant i \leqslant n; \qquad k \in N,$$

where S_{00} denotes the stem cell compartment, we may describe the evolution of a cell as a chain of transitions from the stem cell compartment through a series of multiplicative compartments, terminating in a non-mitotable (nondividing) compartment. Therefore, the adequate stochastic model of hematopoiesis is built by associating an n-dimensional birth-and-death process with an n-dimensional immigration-emigration process. The first will describe the multiplicative processes that take place within each line, while the second will describe the transitions from one compartment to the other, that is the cell differentiation process. We notice that the multiplicative compartments are characterized by receiving an inflow of cells from preceding compartments alongside intra-compartmental mitotic activity (see KILLMAN et al. (1963)).

Two simplifications are assumed in this model: (i) the number of compartments is considered as being the same for all lines; (ii) the transition from any compartment of line S_i to any compartment of line S_m, $1 \leqslant i, m \leqslant n$, is not taken into consideration (and this event does not seem truly essential).

FIRESCU and TAUTU (1965 a) have approached the process of hematopoiesis by building first of all a model in which the mitotable pool is considered as a whole and also the non-proliferating pool as a whole. The n cell lines thus have only two large compartments. An n-dimensional Kendall process is then associated to an n-dimensional emigration process. The role of immigration as a stabilizing factor is thus emphasized. Immigration influences not only the dimension of the population in the mitotable pool but also second or higher order central moments. However, the covariance matrix is independent of ν and its columns are identical.

The second model built by FIRESCU and TAUTU (1965 a) considers the entry in the non-mitotable pool and entry in the circulating blood of mature elements *(diabase)*. Thus they achieve an n-dimensional immigration — death and/or emigration process. Supposing that at the moment t there are x_i B-objects in the non-mitotable pool of line S_i, $1 \leqslant i \leqslant n$, in the time interval $(t, t + \Delta t)$ the following events may occur:

1°. Simultaneous entry into all n non-mitotable pools of one mature B-object with probability $\nu_{12\ldots n}(t)\, x_1 x_2 \ldots x_n \,\Delta t + o(\Delta t)$;

2°. Simultaneous disappearance, by death or emigration (diabase), from all n non-mitotable pools of a B-object, with probability $(d_{12\ldots n}(t) + r_{12\ldots n}(t))\, x_1 x_2 \ldots x_n \,\Delta t + o(\Delta t)$;

3°. The absence of any change with probability $1 - [x_1 x_2 \ldots x_n (\nu_{12\ldots n}(t) + d_{12\ldots n}(t) + r_{12\ldots n}(t))]\,\Delta t + o(\Delta t)$.

The Kolmogorov forward equations are easily written; by using classical methods one obtains equations of the moments for the cases $n = 1$, $n = 2$, $n = 3$ (see details in FIRESCU and TAUTU (1965 a).

3.4.2.2. FUCHS (1965, 1966) has presented a birth-and-death model for the evolution of a single cell line.

3.4.2.3. In a series of experiments with irradiated mice (see MCCULLOCH and TILL (1962)), it was demonstrated that the intravenous injection of bone marrow cells provokes the formation of macroscopic colonies in the spleen of the irradiated animals. The cells contributing to the formation of these colonies belong to the stem cell compartment and there is a linear relation between the number of injected cells and the number of formed colonies. The phenomenon was also observed with certain types of cancer cells, such as the lymphoma cells in the AKR mice (BUSH and BRUCE (1964)) and leukemia L1210 cells (WODINSKY et al. (1967)). TILL, MCCULLOCH and SIMINOVITCH (1964) studied the colonies-forming process using a birth-and-death process; here births mean the appearance of new colonies-forming cells and deaths mean the appearance of a differentiated cell that has lost the capacity to form colonies (in fact, a *heteromorphogeneous division*, i.e., a type of cell division where the change of characteristics coincides in time with mitosis).

3.4.2.4. We shall now present (TAUTU (1969)) another interpretation of the beginning of hematopoiesis, by studying the fluctuations in the stem cell compartment within the framework of dam processes (see GANI (1957), GANI and PYKE (1960), KENDALL (1957), MORAN (1956), etc.).

Let us consider the one-dimensional case, a situation in which the stem cell reservoir generates a single type of cell by heteromorphogeneous division. The general elements for such a model are the following:

1. The cell inflow. Let X_n denote the number of stem cells that, through homomorphogeneous division, enter the stem cell reservoir in the interval $(n, n + 1)$. By *homomorphogeneous division* we mean the type of cell division that results in the appearance of a daughter cell morphologically identical to the mother cell (see KILLMANN et al. (1963)). Let us suppose that $(X_n)_{n \in N}$ is a sequence of mutually independent and identically distributed random variables. From the data of TILL et al. (1964), it results that the number of macroscopic colonies formed in the spleen of irradiated mice can be approximated by a gamma distribution.

2. The dam content. The content fluctuations in the stem cell reservoir are given by the relation between the homomorphogeneous divisions (input) and the heteromorphogeneous divisions (output). Let W_n denote the number of cells found in the reservoir at time n,

exactly before influx X_n, with $W_0 = w_0 \geqslant 0$. We consider that the dam has an infinite capacity.

3. The cell outflow. The heteromorphogeneous division that gives birth to differentiated elements generates the exit of daughter cells from the stem cell reservoir. If we denote by M the outflow capacity of the reservoir, the number of cells set free just before time $n+1$ will be min $(M, W_n + X_n)$.

Under these assumptions we get

$$W_{n+1} = \max (W_n + X_n - M, 0), \quad n \in N.$$

The theory of such chains $(W_n)_{n \in N}$ has been extensively developed (see e.g. LLOYD (1963)).

3.4.2.5. The theory of reservoirs may also be applied to hematogeneous bone marrow as a whole. In this case, the cell outflow represents the number of cells passing into the blood flow. From the data presented by YOFFEY (1957), bone marrow is an extremely large reservoir for white cells series, forming about 100 times the number of granulocytes present in peripheral blood. Thus, the total number of blood granulocytes is 38×10^6, and the segmented neutrophils in the marrow $868 \times \times 10^6$, or a reserve in the marrow 23 times as large as that of the circulating blood. These are mature cells which can be discharged into the blood immediately, and may give rise to an "overflow" leucocytosis of 30,000 per mm^2. For the red cells series it is, however, possible that the bone marrow may contain a 24-hour supply of reticulocytes and probably slightly more.

Such a reservoir may be considered as having an infinite capacity, since the erythroid population can increase within the marrow (i) by displacing fat cells, (ii) by displacing other cells, (iii) by enlargement of bone cavity.

3.4.2.6. We shall close this paragraph by suggesting the study of the maturation process of cell lines using a multi-stage renewal process. The mechanism is identical to that of the promotion of hierarchical organization (SEAL (1945), VAJDA (1948)), called by BARTHOLOMEW (1963) "transfer of the oldest". See also POLLARD (1967) on hierarchical population models with Poisson recruitment.

4. Evolutionary processes

The stochastic models presented in this section are mostly generated by problems of statistical genetics; they will study the fluctuations of the genetic composition of populations, combining systematic mechanisms (mutation, migration and selection) with dispersion mechanisms (fluctuations of gene frequency by drift, variation of the intensity of systematic mechanisms). Since an enormous number of papers have dealt with these questions, we shall present here only stochastic models considered as being essential for understanding the phases of the evolutionary process.

In this sense we recall the considerations of WRIGHT (1964) in relation with the Darwinistic frame of the problem: "The idea that evolution comes about from the interaction of a stochastic and a directed process was the essence of Darwin's theory. The stochastic process that he invoked was the occurrence of small *random* variations which, he supposed, provide raw material for natural selection, a process *directed* by the requirements of the environment, and one that builds up, step by step, changes that would be inconceivably improbable at a single step". More generally, "the contribution of Darwin and Wallace was to perceive the relationship between intrapopulation variation and variation in time and space. Their theories amount, in short, to the realization that *intrapopulational variability is converted into spatial and temporal differentiation. The process of this conversion is the process of evolution*" (LEWONTIN (1963)).

In connection with our pointing out that the largest number of stochastic models of evolution are derived from problems of statistical genetics, we shall assume from the onset along with SEWALL WRIGHT, that "*change of gene frequency is treated as the elementary evolutionary process since it permits reduction to all factors to a common basis*".

4.1. Basic problems, models and methods

4.1.1. Paradigm for the stochastic evolutionary processes

4.1.1.1. As specified from the beginning, the fundamental quantity which is used in population genetics is the gene frequency. The adequacy of using gene frequencies to describe the genetic composition of a reproductive population stems from the fact that each gene is a self-reproducing entity and, therefore, its frequency changes almost continuously with time as long as the population is reasonably large (KIMURA (1970 b)). From our point of view, the population of B-objects considered will be here an aggregate of genotypes. Let us observe that FISHER (1953) held the opinion that one must consider a natural population "not so much as an aggregate of living individuals but as an aggregate of gene ratios. Such a change of viewpoint is similar to that familiar in the theory of gases, where the specification of the population of velocities is often more useful than that of a population of particles".

The arguments for such interpretation are, according to KIMURA (1970 b), the following: (i) each genotype does not necessarily reproduce its own kind under sexual reproduction. Therefore, each genotype possesses no continuity, as does each gene; (ii) the number of possible genotypes greatly exceeds the total number of individuals in a population. As a result, each individual is quite likely to represent a unique genotype in the entire history of the species (except for monozygotic twins). Thus, as an aggregate of genotypes, a population is an enormously complicated system. For example, with 100 segregating loci, each with a pair of alleles, the possible number of genotypes is about 5.2×10^{47}. Actually, for man, the number of possible genotypes is about $10^{4}-10^{9}$ rather than 10^{30}, because a haploid set of human chromosomes comprises 4×10^{3} nucleotide pairs, each able to exist in four different states, with respect to the purine-pyrimidine pairs.

Considering that the genotype represents the image of a certain genetic structure, we shall now speak about the g-pool (denoting the gene by g). The notion of "gene-pool" is equivalent, according to DOBZHANSKY (1967) with that of the "corporate genotype". By g-pool we understand the genes in the array of the gametes which give rise to the next generation; its composition can be described by means of the numbers of the frequencies of the genes and linked gene complexes. CROW (1968) considers that what is transmitted to the next generation is a sample of genes, not a genotype, since in a sexual population genotypes are formed and broken up by recombination every generation and a particular genotype is therefore evanescent.

However, J. F. CROW is right in stressing that despite the fact that the theory of evolution has been developed mainly in terms of gene frequency changes, the actual evolutionary observations have mainly concerned changes in phenotype. The increases in size of the horse and of its teeth are reasonably well documented over a period of tens of millions of years, but there is no reliable information on the number of gene substitutions that were required to effect this. Let us remark therefore that the g-pool or genotype aggregate is included within another population whose behaviour (respectively the mechanisms of transmitting a g-pool to the offspring) must be taken into account. Because of the obvious need to simplify, we shall consider a *Mendelian* population which is a reproduction community of sexual and either obligatorily or at least facultatively cross fertilizing organisms (DOBZHANSKY (1967)). This does not consider the asexual, exclusively self-fertilizing, parthenogenetic organisms, as well as some intermediate situations in which cross fertilization is rare.

What must be emphasized first of all is the kind of the mechanism of g-pool transmission, respectively if it is random or non-random. As one knows there are two mating systems in Mendelian populations, based either (i) on genotype similarity, or (ii) on phenotype similarity. Since we are here interested in the first system only, we shall distinguish two kinds of non-random matings. The first is called assortative mating, e.g., the mating between XX and XY genotypes in sexual reproduction. The second type of non-random mating is mating between closely related individuals, viz. various models of inbreeding (selfing, sib mating, etc.). Such models are clearly presented by KARLIN (1966, p. 382).

The random mating of I-objects (as we have denoted mature B-objects capable of reproduction in Subparagraph 1.2.1.5) is equivalent to random union of G-objects (denoting gametes). Thus, the frequency of a genotype, say a_1a_1, in the following generation is the frequency with which two G-objects drawn independently are both a_1. Because of these statements we are tempted to consider a population as an interbreeding group of I-objects, an inbreeding group thus being a line, a subpopulation. Within our models, the basic population being *panmictic*, where mating occurs at random, the probability of finding a partner of the opposite sex is the same for all I-objects, and fecundity is the same for all mates. The local populations we have dealt with in Subparagraph 3.2.3.1 are generally considered to be panmictic populations. Thus, panmixia is the first condition for an "ideal" population, in fact an abstract one. MALÉCOT (1948) has also considered isogamic populations (*isos*, equal; *gamos*, marriage).

Thus one must take into account the rules of generation succession. The next subparagraph will deal with this problem. The reader may find the necessary details in the first chapter of MORAN's (1962) book, where the structure of a breeding population is discussed.

4.1.1.2. The second condition for an ideal population is that the generations should be distinct and non-overlapping. Populations having such noninterbreeding generations are populations of I-objects which have a fixed breeding season and a lifetime which is not much longer, so that the parents never breed with their own offspring. Suppose that each generation consists of exactly x males and y females. We consider only a single autosomal locus, i.e., not sex-linked, at which there may be two alleles a_1 and a_2. The genotypes will be a_1a_1, a_1a_2, and a_2a_2. Suppose that i, $(x - i - j)$, and j are the numbers of male diploid individuals whose genotypes are a_1a_1, a_1a_2, a_2a_2, respectively. Similarly, we write k, $(y - k - l)$, l, for the numbers of corresponding females. When this generation dies, it is replaced by a new generation, with the same totals x and y, all the individuals being the offspring of matings between the parents. Ignoring selection and supposing that each individual is the result of a separate mating, the probability of a particular offspring being the offspring of a given male parent is x^{-1} and since the offspring are all formed independently, the number of offspring of a given male parent will have a binomial distribution with probability x^{-1} and index $x + y$. When x is not too small, this distribution is well approximated by a Poisson one with mean $(x+y)x^{-1}$. MALÉCOT (1955) calls this assumption "Poissonian fecundity" under conditions in which the extraction of useful gametes from the same generation is similar to the extraction from an urn with constant composition.

Let us now consider the case of overlapping generations. Suppose that there are x males and y females and the state of the system is defined as before by the integers (i, j, k, l). We now suppose that the individuals die at random one by one and are immediately replaced by a new object which is the offspring of a mating between l-objects of the opposite sex in the population immediately before death.

This model, as well as the one above, is a Markov chain since the probability of a transition from any state (i, j, k, l) to any other state depends only on (i, j, k, l). However, in case of overlapping generations, it is not possible to move to states for which the new i, j, k, and l differ from the old by more than one (whereas in the case of nonoverlapping generations it was possible, in general, to move to any other state (i, j, k, l)). Moreover, the time unit between transitions no longer corresponds to one generation.

Provided there is no selection, the probability of any given l-object dying at any event is $(x + y)^{-1} = d$. Thus the probability of any given l-object having a lifetime of n units is $d(1-d)^{n-1}$, $n \in N^*$. This is a geometric distribution with a modified first term. The expected lifetime is now $1/d = x + y$ and this is the number of time units in the chain which corresponds to one time interval in the previous model.

At each instant at which a death occurs, the probability of the l-object becoming a parent is x^{-1}, so that the probability generating function of the number of offspring of an l-individual is equal to

$$\sum_{n=1}^{\infty} d(1 - d)^{n-1} [(1 - x^{-1}) + zx^{-1}]^n =$$

$$= d(x - 1 + z) [dx + 1 - d - (1 - d)z]^{-1}.$$

This is a geometric distribution with a modified first term and may also be regarded as the convolution of a geometric distribution with a binomial with probability x^{-1} and index unity.

MORAN (1962, p. 14) notes that this is a distribution very different from the Poisson and there appears to be no particular evidence that it is a good fit to the distributions of natural populations. Thus, the above model sacrifices some empirical applicability in order to obtain a random process easier to deal with analytically. The above discussion is concerned with the distribution of the number of diploid offspring of a diploid individual. From it, we can easily obtain the distribution of the number of gametes which are found in the offspring and are descended from a specified factor in the diploid parent. This is obtained by replacing z in the above probability generating function by $1/2\ (1 + z)$.

This Markov chain can be embedded in a Markov process. One finally obtains the same modified geometric distribution as before.

4.1.1.3. The use of stochastic models in genetics has been justified from the beginning by the fact that the sampling of gametes introduces an element of chance into the change of gene frequencies. Of course, randomness exists even in the nature of the evolutionary process itself. Let us mention that this consideration of stochastic processes in population genetics leads us to problems of gene frequency distribution at equilibrium and the probability of gene fixation. These are of special importance, M. KIMURA says, for understanding the genetic structure of Mendelian populations. To emphasize random mechanisms in evolution, we now present the phases of the evolution process resulting from Wright's shifting balance theory.

(i) Random drift phase. In each deme, each set of gene frequencies drifts at random in a multidimensional probability distribution about the equilibrium point, characteristic of one of the multiple selective peaks. The set of equilibrium values for each gene frequency is the resultant of three sorts of "pressures": (a) those of recurrent mutation, (b) those of selection, and (c) those of recurrent immigration from other demes. The fluctuations, responsible for the random drift, may be due either to sampling accidents or to fluctuations in the values of the coefficients for the various pressures.

(ii) Phase of intrademic selection (mass selection). From time to time, the set of gene frequencies drifts across one of the many two-factor saddles in the probability distribution in a deme. There ensues a period of relatively rapid systematic change, dominated by selection among individuals (or families), until the set of gene frequencies approaches the equilibrium associated with the new selective peak, about which it now drifts at random, and thus returns to the first phase, but in general at a higher level.

(iii) Phase of interdemic selection. A deme that comes under control of a selective peak, superior to that controlling the neighbouring demes, produces a greater surplus population and by excess dispersion systematically shifts the positions of equilibrium in the neighbouring demes toward its own position, until the same saddle is crossed in them and they move autonomously to the same peak. This process spreads in concentric circles among the inferior demes. Two such circles spreading from different centers may overlap and give rise to a new center which combines the two different favorable interaction systems and becomes a still more active population source. The virtually infinite field of interaction systems may be explored in this way with only a small number of novel mutations as alleles which had been rare, largely displace the previously more abundant ones (WRIGHT (1964, 1970)).

The choice of Wright's theory is justified in our view by the mathematical results it generates. The reader can find extremely interesting comments about the biologic significance of Wright's and other theories in WADDINGTON (1969). The complex nature of the process was stressed by S. WRIGHT: the directed pressures of recurrent mutation, selection and immigration on the gene frequencies of demes in conjunction with random fluctuations due to accidents of sampling, and fluctuations in the coefficients describing the directed process, determine a stochastic distribution instead of a single equilibrium.

We must also add here the point of view of MAYR (1964), particularly attractive because of its clarity. He states that the opponents of the Darwinian theory have claimed that the conflict between the harmony of nature and the apparent haphazardness of evolutionary processes could *not* be resolved. In fact, this objection is valid only if evolution is a one-step process (see the same idea in WRIGHT (1964)). One must admit that every evolutionary change involves two steps. The first is the production of new genetic diversity through mutation, recombination, and related processes. On this level, randomness is indeed predominant. The second step, however, — selection of those individuals that are to make up the breeding population of the next generation, — is largely determined by genetically controlled adaptive properties. This is what natural selection means and, as is well known, this concept is the heart of the evolutionary theory — or, according to S. WRIGHT, an important cause of evolution. Thousands of experiments have proved that the probability that an individual should survive and reproduce is not a matter of accident, but a consequence of its genetic endowment. On the other hand, selective superiority gives only a statistical advantage. It increases the probability of survival and reproduction, other things being equal. Natural selection is measured in terms of the contribution a genotype makes to the genetic composition of the next generation.

We shall have to add to these theses also the importance of the ecologic factor (Wright's "ecologic opportunity"). Most species are restricted to progress, if any, along the line of increasing perfection in the niche that they occupy because of the occupation of all slightly different niches by other species. S. WRIGHT asserts that occasionally, however, an opportunity for evolution along one or more different lines may be presented (a) by arrival of a colony in a territory in which niches in which it can live are unoccupied, or (b) by some degree of accidental preadaptation to a drastically altered environment in which rival forms have become extinct, or (c) attainment in the course of specialization for one way of life, of adaptations that happen to open up an extensive new way of life.

4.1.1.4. Before presenting the fundamental assumptions of our models we shall consider, let us examine their general structure as presented in LEWONTIN (1963). The author starts from the idea that the relationship among the elements of an evolutionary model (i.e., entities, quantity variables, and rules) is a critical one. Thus, if the concept of Darwinian fitness refers to differential reproduction of diploid genotypes, the entities in the model must be genotypes and not genes, although it may be possible to derive expressions for changes in gene frequency from the genotypic model.

It is for this reason that models using populations or species as the basic physical entities are so difficult to construct, for there are no rigorous rules comparable to Mendelian laws and mating rules which apply to populations or species as wholes. However, one must not neglect the fact that genetic entities are, in turn, contained within a population or B-objects. An assertion of KIMURA (1964) corroborates our considerations in Subparagraph 4.1.1.1. A natural population which plays a significant role as *an evolutionary unit* should consist of a large number of individuals, and the gene frequencies for these behave practically as continuous variables. Also, any change of gene frequencies must in general be very slow measured by our ordinary time scale.The rate of change (evolution) in the size of the horse can be measured in terms of centidarwins – where a *darwin* is the unit of evolutionary rate at the phenotypic level, representing a change of one millionth per year, or a change by a factor e in a million years. This contrasts with the incredibly high rate of mutation: a mutation rate as low as 10^{-6} per gene yields three mutations per gamete per generation! Molecular considerations make it probable that under the relatively stable phenotype are a number of changes in individual genes (CROW (1968)). Following KIMURA (1969), the majority of *molecular* mutations due to base substitution must be neutral or almost neutral for natural selection. The remarkable constancy per year is most easily understoood by assuming that in various vertebrate lines the rate of production of neutral mutations per individual per year is constant. This inference is based on a simple principle that for neutral mutations the rate of gene substitution in a population is equal to the rate of production of new mutations per gamete, because for such a mutation the probability of gene fixation is equal to the initial frequency.

This is why the utilization of models considering an aggregate of genotypes largely attenuates all these difficulties in interpretation and confrontation, especially since the general laws of inheritance, of mutation and of natural selection are of such a form that they can be used in the construction of models. The rules that must be observed for the construction of an evolutionary model are, according to LEWONTIN (1963), as follows:

I. Relations between genotype and phenotype (dominance, pleiotropy, norm of reaction, especially of fitness, epistasis);

II. Environmental fluctuations;

III. Mating rules (panmixia, assortative mating, inbreeding, apomixis, etc.);

IV. Mechanisms of gene transmission (segregation ratio, number of alleles, number of loci, linkage, polyploidy, etc.[9]);

V. Mutation.

R. C. LEWONTIN thus presents the general diagram of a model for change in the composition of a population due to known evolutionary and genetic phenomena:

Proportions of adult genotypes at the moment of mating
↓ (Rules I, II)
Proportions of adult phenotypes at the moment of mating
↓ (Rule III)
Proportions of various matings
↓ (Rules IV, V)
Proportions of offspring genotypes at birth
↓ (Rules I, II)
Proportions of offspring phenotypes at birth
(Rules I, II (Fitness)) — returns to the top

Each rule in a stochastic model gives rise to a set of probabilities that particular events occur.

[9] *Dominance*: the parental allele manifested in the F_1 heterozygote; *Pleiotropy* (*pleion*, more; *trope*, turn): multiple phenotypic effect of a single allele; *Fitness*: the net productivity, for individuals; generally speaking, the probability of survival and reproduction; *Epistasis* (*epi*, upon; *stasis*, standing): dominance of a gene over another non-allelomorphic gene; *Apomixis*: a reproductive anomaly in plants akin to parthenogenesis; *Linkage*: gene tendency to remain associated through several generations; *Polyploidy*: reduplication of the chromosome number.

The diagram emphasizes, however, the complexity of the model and the actual difficulties for its representation due, on the one hand, to mathematical difficulties and, on the other hand, to the inability of biological science to rigorously express a number of situations and rules. Matters become still more complicated if the entire genotype (see CROW (1968)) or complex systems (LEWONTIN (1967)) are considered. At least during the first stage, one builds—as will be seen in the next subparagraph – a simpler model, rather severely characterized by DOBZHANSKY (1967) as "thoroughly unrealistic biologically" but "tractable mathematically, and, therefore, the favourite with some mathematical geneticists and genetical mathematicians".

4.1.1.5. The classical mathematical model has, according to DOBZHANSKY (1969), the following assumptions:

1°. The population is Mendelian;

2°. The population size is infinite or large enough to be treated as such;

3°. The environment is constant in time and in space for all members of the population;

4°. The variable gene loci are each represented by two or more alleles, one allele being "normal" and beneficial in the environment, and all other alleles disadvantageous in homozygotes;

5°. Most of the gene loci are occupied by identical alleles in all individuals, the variable loci are a minority;

6°. The genes produce their effects each independently of one another; epistatic interactions, especially as concerns fitness, do not occur or are negligible; the fitness variations caused by the alleles at the unfixed loci are simply additive or multiplicative.

According to these deductions, genetic uniformity is clearly the ideal state under the above model. All members of the population will be genetically identical, and all optimally fit. The model outlined makes all genetic diversity an unwelcome departure from the ideal optimal uniformity. Yet a good part of this genetic diversity is neither a sad accident, nor a departure from the Platonic eidos called the "optimum genotype" but a means whereby the population adapts to its environment and is able to deal with new environments.

We shall consider below the following fundamental assumptions:

I. The population is Mendelian (also: it is mono or dioecious; no differences in sexes);

II. Generally, it forms a deme, that is, a local, random breeding population;

III. The population size is a large but a fixed constant;

IV. Generations do not overlap;

V. The chromosomes are autosomes; it is considered that there are two alleles at each locus;

VI. Systematic pressures are exerted on the frequency of genes in a deme.

Assumptions 3° and 6° are implicitly contained; assumption 5° has no permanent character. In fact, these fundamental assumptions may be changed. LI (1967) has graphically represented the six simplifying assumptions for the "simplest case". (Fig. 5).

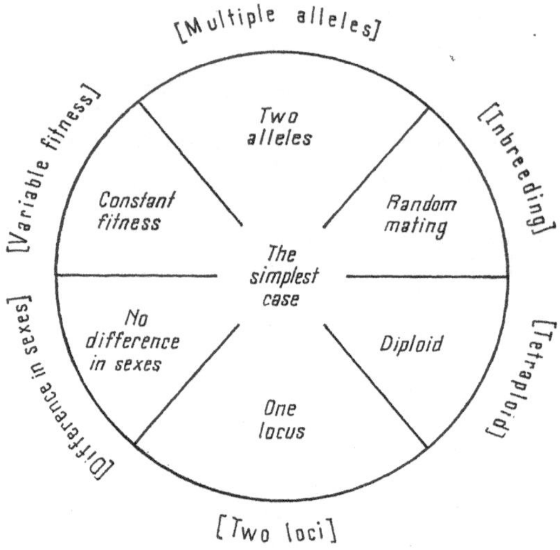

Figure 5

In Moran's simple model, assumption IV is discarded, while the multiallele Moran models adequately fulfill assumption V. Also, for the purpose of studying genetic drift one must discard assumption III. The simplest models ignore assumption VI which, in fact, was not shown in Figure 5.

4.1.2. Classical genetic stochastic models

4.1.2.1. We first discuss the stochastic models introduced implicitly by FISHER (1930) and explicitly by WRIGHT (1931) for investigating the effects of systematic pressures and random drift on the fluctuations of gene frequencies. In this field "the great trio Fisher, Haldane and Wright" (as they are called by KIMURA (1970)) has brilliantly dominated. The reader may find in KIMURA (1964, 1970) and EWENS (1969) the presentation of initial contributions and their meaning. According to FELLER (1968, p. 380), the formulation of Wright's model in terms of Markov chains is due to MALÉCOT (1944).

As the simplest possible case consider a haploid monoecious population having in each generation exactly n l-objects. The first five assumptions are admitted. According to (V) there will be two alleles,

a_1 and a_2, of a particular gene. This will therefore have $2n$ representatives in our population. We shall consider individuals having gene a_1 as I-objects of type a_1. The system is in state i ($0 \leqslant i \leqslant 2n$) if there are i objects of type a_1 and $n - i$ objects of type a_2. Under the conditions posed by our fundamental assumptions (random mating, no systematic pressures), the composition of the following generation is determined by $2n$ Bernoulli trials in which the I-objects will be of a_1-type with probability $i/2n$ and of type a_2 with probability $1 - (i/2n)$. Therefore, the transition probability from state i to state j will be

$$\binom{2n}{j}\left(\frac{i}{2n}\right)^{j}\left(1 - \frac{i}{2n}\right)^{2n-j}.$$

The fact that $2n$ is an even number has no mathematical significance and we shall write m for $2n$ for convenience.

By this procedure we generate a Markov chain $(\xi_k)_{k \in N}$ with state space $X = \{0, 1, \ldots, m\}$ and transition probabilities

$$p_{ij} = \binom{m}{j}\left(\frac{i}{m}\right)^{j}\left(\frac{m-i}{m}\right)^{m-j} = \binom{m}{j} p_i^j\, q_i^{m-j}, \quad 0 \leqslant i, j \leqslant m, \tag{4.1}$$

(with the convention $0^0 = 1$). We mention at the outset that Wright's model, characterized by (4.1), is the fundamental model in population genetics based on the scheme of Bernoulli trials, an ideal standard. In MORAN's (1958) model, assumption IV is abandoned and the assumption of overlapped generations is introduced. Thus, an additional element of randomness will appear due to the random choice of the I-object which is to die. The model is represented by a Markov chain embedded in a Markov process and is equivalent to the Ehrenfest urn model with absorbing states (KARLIN and MCGREGOR (1964)).

We remark that in both models we deal with I-objects, that is, with B-objects in the reproductive stage. Although for Wright's model this is realistic taking into account assumption IV — for Moran's model the treatment of each new B-object as an immediate potential parent is somewhat unrealistic. In this respect it is possible, as S. KARLIN suggests, to modify the model by associating with each new birth a probability f that the B-object will mature and be capable of producing offspring ($1 - f$ being the probability of no progeny). Analytically this has only the effect of adjusting the intensity of the rate of birth-death events.

4.1.2.2. Let us now introduce assumption VI, assuming that the systematic pressures are mutation, migration and selection.

(a) Mutation. Let us consider the forward mutation $a_1 \to a_2$ with probability α_1 and the reverse mutation $a_1 \leftarrow a_2$ with probability α_2. We get instead of the probabilities p_i and q_i from (4.1):

$$\begin{aligned} p_i &= \frac{i}{m}(1-\alpha_1) + \left(1 - \frac{i}{m}\right)\alpha_2, \\ q_i &= \frac{i}{m}\alpha_1 + \left(1 - \frac{i}{m}\right)(1-\alpha_2). \end{aligned} \tag{4.2}$$

(b) Migration (KARLIN and MCGREGOR (1965)). We take

$$\begin{aligned} p_i &= \frac{ai + \nu}{am + \nu + c}, \\ q_i &= \frac{a(m-i) + c}{am + \nu + c}, \end{aligned} \tag{4.3}$$

where $a, c, \nu > 0$. Here p_i represents the probability of reproducing an I-object of a_1-type which is proportional to the current frequency of a_1 genes plus a constant factor ν which represents an immigration rate. The similar probability for an I-object of a_2-type is q_i. The denominator is the normalizing constant so that $p_i + q_i = 1$.

(c) Selection. A selective advantage is assured by assuming that the relative number of offspring per object of I-objects of a_1-type and a_2-type have expectations proportional to $1 + \sigma$ and 1, respectively, where σ $(\sigma > 0)$ is the selective advantage which exists for allele a_1 over allele a_2. It is easy to understand that selection thus acts by excluding from reproduction some I-objects with a certain genotype, which has as result the modification of gene frequency. The mechanism of exclusion from reproduction is represented by the reduction of viability or of fertility (including the capacity of mating), both alternatives having the same result: the genotype against which selection acts will have a lesser contribution in gametes which form the zygotes of the future generation. We shall, therefore, consider a gametic and zygotic selection.

Let us dwell on the first type of selection. Suppose that I-objects of type a_1 and of type a_2 produce large numbers of offspring in the constant proportions c_1 and c_2, respectively, and that exactly m of these survive. Here, c_1 and c_2 have the following meaning: they are proportional to the relative survival values of gametes of type a_1 and a_2. Since only the ratio c_1/c_2 is relevant, we can write it as $1 + \sigma$. We have

$$p_i = \frac{(1+\sigma)i}{m + \sigma i}, \qquad q_i = \frac{m - i}{m + \sigma i}, \tag{4.4}$$

where the denominator is a normalizing constant.

4.1.2.3. Let us pass to the spectral analysis of the transition matrix **P** with entries given by (4.1). If the eigenvalues of **P** are λ_r, $0 \leqslant r \leqslant m$, we have $\lambda_0 = \lambda_1 = 1$ since there are two absorbing states, 0 and m. If the remaining eigenvalues are distinct we can write

$$\mathbf{P}^k = \mathbf{A} + \lambda_2^k \mathbf{P}_2 + \ldots + \lambda_m^k \mathbf{P}_m,$$

where **A** is a matrix having positive entries only in the first and last columns; these entries equal the absorption probabilities at 0 and m, for various values of the starting state. The matrices $\mathbf{P}_2, \ldots, \mathbf{P}_m$ are defined in terms of right and left eigenvectors (see I; 1.1.4.2.5). If

$$|\lambda_2| > |\lambda_3| > \ldots > |\lambda_m|,$$

then the rate at which $\mathbf{P}^k$ converges to the limiting matrix **A** is governed to a large extent by the value of λ_2. If λ_2 is very close to unity, this rate of approach will be slow and genetic variability will be lost slowly; if λ_2 is moderate or small, this rate will be quite rapid. The eigenvalues of the matrix **P** have been found by the method of generating functions (Feller (1951)) to be

$$\lambda_r = \binom{m}{r} r!\, m^{-r}, 0 \leqslant r \leqslant m.$$

The left eigenvectors corresponding to $\lambda_2, \ldots, \lambda_m$ are not known explicitly. The right eigenvectors are slightly more tractable and can be computed recursively, in terms of certain polynomial systems, although no analytic form is discernible. It seems likely that the eigenvectors cannot be expressed in terms of classical functions (Karlin and McGregor (1964)). We also note an interesting result due to Karlin and McGregor (1967 a) concerning the left eigenvector $(u_j(m))_{0 \leqslant j \leqslant m}$ of the matrix (4.1), corresponding to the eigenvalue $\lambda_2 = 1 - 1/m$. Let us impose the normalization $u_1(m) = 1$. Then, if $(u_j)_{j \in N^*}$ denotes the unique stationary measure normalized so that $u_1 = 1$ of the Poisson Galton-Watson chain with generating function $f(z) = e^{z-1}$ (see I; 1.2.2.4.2), we have

$$\lim_{m \to \infty} u_j(m) = u_j,\ j \in N^*.$$

In the mutation model the eigenvalues of **P** with entries given by (4.2) are

$$\lambda_r = (1 - \alpha_1 - \alpha_2)^r \binom{m}{r} r!\, m^{-r}, 0 \leqslant r \leqslant m.$$

Assuming $\alpha_1 > 0$, $\alpha_2 > 0$ and $\alpha_1 + \alpha_2 < 1$, the eigenvalues will be distinct. In this model there are no absorbing states but there is a limiting distribution

$$\lim_{k \to \infty} p_{ij}^{(k)} = \pi_j.$$

The difference between $p_{ij}^{(k)}$ and π_j decreases like $\lambda_1^k = (1 - \alpha_1 - \alpha_2)^k$, i.e., the rate of convergence does not depend on m.

For the selection model in which p_i are defined by (4.4) upper and lower bounds for the absorption probabilities at m were given by ARNOLD (1968) and MORAN (1960).

4.1.2.4. In his (1958) paper, MORAN modified Wright's model, by replacing assumption IV with the assumption of overlapping generations. Instead of assuming that all I-objects die at the same time, he assumed that at each instant at which the state of the model may change, one of the gametes, chosen at random, dies and is replaced by a new gamete which is a_1 or a_2 with probabilities p_i and q_i in (4.1), where i is the number of G-objects of type a_1 prior to the event. The two models, Wright's and Moran's, do not differ in the genetic nature of the I-objects considered, in the sense that the former would describe the modification of gene frequencies in populations with G-objects and the latter with Z-objects (which are diploid). The essential difference consists, as we have already specified, in the manner of succession of generations which introduces yet another random element. Since the large majority of diploid I-objects forms populations with overlapping generations, the Moran model is considered as a diploid model.

When the system is in state i and a birth-death event occurs, the subsequent state will be $i-1,\ i, i+1$, with probabilities

$$\frac{i}{m} q_i,\ \frac{i}{m} p_i + \left(1 - \frac{i}{m}\right) q_i,\ \left(1 - \frac{i}{m}\right) p_i, \tag{4.5}$$

respectively. The time intervals between successive events are identically distributed independent random variables with the same exponential density mde^{-mdt} where $1/d$ is the expected lifetime of an I-object. The process $(\xi(t) =$ the state at time $t)_{t\geqslant 0}$ is then a finite state Markov jump process. As in Wright's model, the homozygous states 0 and m act as absorbing barriers. The Moran model has several mathematical advantages and admits an extensive mathematical analysis.

In order to study this model, MORAN (1961) and KARLIN and MCGREGOR (1962) appealed to the standard method in the theory of birth-and-death processes. Since it appears as a special case of what we will discuss in Subparagraph 4.1.2.6, we shall not pursue their analysis here. In his initial paper, MORAN (1958) also presented this model as a Markov chain with the following transition probabilities

$$\begin{aligned} p_{i,\,i-1} &= \frac{i}{m} q_i = i\,(m-i)\,m^{-2}, \\ p_{i,\,i} &= \frac{i}{m} p_i + \left(1 - \frac{i}{m}\right) q_i = (i^2 + (m-i)^2)m^{-2}, \\ p_{i,\,i+1} &= \left(1 - \frac{i}{m}\right) p_i = (m-i)\,im^{-2}, \\ p_{i,\,j} &= 0 \text{ if } |\,i-j\,| > 1. \end{aligned} \tag{4.6}$$

Notice also that the unit time in Wright's model, which was equal to one generation there, must correspond to m units here. E. J HANNAN, in an appendix to MORAN's (1958) paper, deduces the eigenvalues of the transition matrix as

$$\lambda_r = 1 - r(r-1)\, m^{-2}, \quad 0 \leqslant r \leqslant m.$$

4.1.2.5. It is interesting to show how a bivariate model is set up whose components x_1 and x_2 represent the number of homozygotes a_1a_1 and a_2a_2 in a population with n diploid Z-objects (MORAN (1958)). Successive events will each consist of one of these Z-objects dying at random and being replaced by a new Z-object, whose precursors (two G-objects) are independently chosen to be of a_1-type or a_2-type with probabilities proportional to the numbers of G-objects of a_1 or a_2-type in the population before death. Once again, this model — which is clearly a Markov chain with $1/2\,(n+1)\,(n+2)$ states — could be regarded as embedded in a Markov process in which each Z-object has a negative exponential distribution of its lifetime. The states of the Markov chain are defined by the two variates x_1 and x_2, such that $0 \leqslant x_1, x_2 \leqslant n$, and $x_1 + x_2 \leqslant n$. The number of G-objects of type a_1 is $n + x_1 - x_2$ and thus we see that the transition probabilities $p_{\mathbf{ij}}$ are the following, if $\mathbf{i} = (x_1, x_2)$:

$\mathbf{j}$	$p_{\mathbf{ij}}$
$(x_1 - 1, x_2 - 1)$	$0,$
$(x_1 + 1, x_2 + 1)$	$0,$
$(x_1 + 1, x_2)$	$(n - x_1 - x_2)(n + x_1 - x_2)^2 (4n^3)^{-1},$
$(x_1, x_2 + 1)$	$(n - x_1 - x_2)(n - x_1 + x_2)^2 (4n^3)^{-1},$
$(x_1 - 1, x_2 + 1)$	$x_1 (n + x_2 - x_1)^2 (4n^3)^{-1},$
$(x_1 + 1, x_2 - 1)$	$x_2 (n + x_1 - x_2)^2 (4n^3)^{-1},$
$(x_1 - 1, x_2)$	$x_1 (n^2 - (x_1 - x_2)^2)(2n^3)^{-1},$
$(x_1, x_2 - 1)$	$x_2 (n^2 - (x_1 - x_2)^2)(2n^3)^{-1},$
(x_1, x_2)	$[x_1 (n + x_1 - x_2)^2 + 2(n - x_1 - x_2) \times (n^2 - (x_1 - x_2)^2) + x_2 (n + x_2 - x_1)^2](4n^3)^{-1}.$

There are two absorbing states $(n, 0)$ and $(0, n)$. In a similar way one can build Wright and Moran models by ignoring the supplementary assumption I, that is, for bisexual populations (see MORAN (1962, p. 144), EWENS (1969, p. 33)).

4.1.2.6. Let us now introduce assumption VI in the construction of the Moran model and suppose that mutation is the single systematic pressure. In this case, p_i and q_i are given by (4.2). Let us denote

$$b_i = d(m - i)\,[i(1 - \alpha_1) + (m - i)\,\alpha_2]\, m^{-1}, \tag{4.7}$$

$$d_i = di\,[i\,\alpha_1 + (m - i)\,(1 - \alpha_2)]\, m^{-1}.$$

The continuous time model can be recognized as a finite state birth-and-death process with birth and death rates b_i and d_i, $0 \leq i \leq m$. The transition matrix $\mathbf{P}(t) = (p_{ij}(t))$ satisfies the equation

$$\mathbf{P}'(t) = \mathbf{Q}\mathbf{P}(t),$$

where $\mathbf{Q}$ is the Jacobi matrix

$$\mathbf{Q} = \begin{pmatrix} -b_0 & b_0 & & & & \\ d_1 & -(d_1+b_1) & b_1 & & & \\ & d_2 & -(d_2+b_2) & b_2 & & \\ & & & \ddots & & \\ & & & d_{n-1} & -(d_{n-1}+b_{n-1}) & b_{n-1} \\ & & & & d_n & -d_n \end{pmatrix}.$$

KARLIN and MCGREGOR (1962, 1964) have found the eigenvectors and the eigenvalues of the matrix $\mathbf{Q}$. The eigenvectors can be expressed in terms of Hahn polynomials and the eigenvalues in case $\alpha_1, \alpha_2 > 0$, $\alpha_1 + \alpha_2 < 1$ are given by

$$\lambda_r = -\frac{d(1-\alpha_1-\alpha_2)\, r\,(r+y_1+y_2+1)}{m},$$

where

$$y_1 = \frac{n\alpha_2}{1-\alpha_1-\alpha_2} - 1, \quad y_2 = \frac{n\alpha_1}{1-\alpha_1-\alpha_2} - 1.$$

The rate of convergence to the limiting distribution is like $e^{-\lambda_1 t} = e^{-d(\alpha_1+\alpha_2)t}$. To compare this with the result for the Wright model it is natural to take $t = 1/d$ as comparable to one generation time and we have, as for the Wright model, the rate of convergence $e^{-(\alpha_1+\alpha_2)} = 1 - (\alpha_1+\alpha_2)$ approximately, per generation.

For the treatment of the considered model as a birth-and-death process with balanced non-linearities see BATHER (1963).

4.1.2.7. In case of the introduction of selection as systematic pressure, the Moran model causes the same difficulties as the Wright model.

4.1.2.8. Multi-allele Moran models with and without mutations have been studied with rare insight by KARLIN and MCGREGOR (1964). We draw the reader's attention to two interesting aspects: the consideration of the three-allele case with mutation (where $\alpha_1 + \alpha_2 + \alpha_3 < 1$) as a simplification to a three-urn Ehrenfest model and the consideration of cases in which the deduced limit process may be a mixed birth-and-death and diffusion process.

4.1.3. Tendency to homozygosity

4.1.3.1. We shall study here the problem of reaching the homozygosity state with or without assumption II. The amount of genetic diversity within a single interbreeding population is regulated by a balance of mechanisms that favour inbreeding and such that favour outbreeding (MAYR (1964)). The extremes, in this respect, are much greater among plants and lower animals.

In fact, no natural population of higher animals will normally breed completely at random; this fact is most usual within animal breeding farms. We shall now study the tendency to homozygosity under conditions of nonrandom mating as well as under conditions of random mating. Selection pressure is admitted. We have also added the case of subdivided populations for the biologic conclusions that can be drawn. The effects of migration have been taken into account: the population is divided; however, demes are not completely isolated.

4.1.3.2. The stochastic model for sib mating is well known and has already been presented in Subparagraph 1.2.3.1, with the corresponding matrix of transition probabilities. We add that in this case assumption III is also not wholly respected: the population is not a large-sized one, rather it represents a line.

Let us therefore consider a set of families maintained through a mating of two full sibs, male and female: that is, with the maximum possible consanguinity. The limitation to two individuals per family keeps the total number of animals constant from one generation to another. The individuals are subject to breeding or artificial selection by elimination of those presenting undesirable characteristics, which could be present in both sexes, or just in one sex. According to assumption V, we shall consider the existence of two alleles a_1 and a_2, and the following selective values shall be given to the corresponding genotypes:

$$\left.\begin{array}{ll}\text{male:} & \sigma_1, 1 \text{ and } \eta_1 \\ \text{female:} & \sigma_2, 1 \text{ and } \eta_2\end{array}\right\} \begin{array}{l}\text{for genotypes } a_1a_1,\ a_1a_2 \text{ and} \\ a_2a_2, \text{ respectively.}\end{array}$$

This model is a finite absorbing Markov chain, the absorbing states being the homozygosity states of mates $a_1a_1 \cup a_1a_1$ and $a_2a_2 \cup a_2a_2$. We emphasize the fact that the problem of reaching the homozygosity state is a particularly important one in models of population genetics. The spectral analysis of the Markov chain corresponding to the sib mating model can be found in BAILEY (1964 a, p. 53).

By using I; Theorem 1.1.52 (for details see BOSSO, SORARRAIN and FAVRET (1969)) one finally deduces the results given in Table 3. In all cases the starting state is the heterozygous state $(a_1, a_1, a_2\, a_2)$.

Table 3

Assumptions about selection	σ_1	σ_2	η_1	η_2	Mean time to homo-zygosity	Variance of the time to homozygosity
No selection	1	1	1	1	6.67	22.67
Selection on both sexes	1/2 2	1/2 2	1/2 2	1/2 2	13.85 4.5	144.42 6.75
Selection on one sex	1/2 2	0 0	1/2 2	0 0	15.15 2.92	138.65 4.13
Selective advantage for male genotype $a_2 a_2$	1/2 2	1/2 2	1/2 2	0 0	19.77 5.23	355.04 23.22

Thus, it may be observed that if selection pressure acts more strongly in favour of homozygotes (i.e., for $\sigma_1 = \sigma_2 = \eta_1 = \eta_2 = 2$), the mean time up to homozygosity is small as was to be expected. Under these conditions, selection on one sex is more advantageous to breeders. If selection pressure favours heterozygotes (i.e., for $\sigma_1 = \sigma_2 = \eta_1 = \eta_2 = 1/2$), the absorbing state is reached in a longer time that in the absence of selection ($\sigma_1 = \sigma_2 = \eta_1 = \eta_2 = 1$). In this situation, selection on both sexes is to be preferred. Attention is drawn to the large values of variance of the time to homozygosity.

Kemeny and Snell (1960, p. 176) present a model of phenotype selection, the selection factor being expressed through the intensity with which "opposites attract one another" (heterogeneous mating). While the attraction of opposites increases, the probability of obtaining a pure line with dominant genotype a_1a_1 decreases. If allele a_2 is recessive, a pure recessive line will be obtained earlier, which in fact means the effacing of phenotypical dissimilarity, since the allele which does not manifest itself phenotypically is the favoured one.

4.1.3.3. We note that I; Theorem 1.1.52 can also be applied to discrete time Moran's model allowing selection when p_i and q_i are given by (4.4). Ewens (1963 c) computed the fundamental matrix of the corresponding Markov chain as

$$\mathbf{N} = \begin{pmatrix} n_{11} & n'_{12} & n'_{13} & \dots & n'_{1,m-1} \\ n_{21} & n_{22} & n'_{23} & \dots & n'_{2,m-1} \\ \cdot & \cdot & \cdot & \cdot & \cdot \\ n_{m-2,1} & n_{m-2,2} & n_{m-2,3} & \dots & n'_{m-2,m-1} \\ n_{m-1,1} & n_{m-1,2} & n_{m-1,3} & \dots & n_{m-1,m-1} \end{pmatrix},$$

where

$$n_{ij} = \frac{(1+\sigma)^{m-i}-1}{(1+\sigma)^m-1} \cdot \frac{(1+\sigma)^j-1}{\sigma p_{j,\,j-1}},$$

$$n'_{ij} = \frac{(1+\sigma)^m-(1+\sigma)^{m-i}}{(1+\sigma)^m-1} \cdot \frac{(1+\sigma)^m-(1+\sigma)^j}{[(1+\sigma)^m-(1+\sigma)^{m-1}]\,p_{j,\,j+1}}.$$

Consequently, the absorption probabilities at 0 and m (i.e., the probability that l-a_1-objects or l-a_2-objects are eventually eliminated), starting in state i are respectively given by

$$\frac{(1+\sigma)^{m-i}-1}{(1+\sigma)^m-1}, \text{ and } 1-\frac{(1+\sigma)^{m-i}-1}{(1+\sigma)^m-1};$$

the mean time to absorption, under the same assumption, equals

$$\frac{(1+\sigma)^{m-i}-1}{(1+\sigma)^m-1} \sum_{j=1}^{i} \frac{(1+\sigma)^j-1}{\sigma p_{j,\,j-1}} +$$

$$+\frac{(1+\sigma)^m-(1+\sigma)^{m-i}}{(1+\sigma)^m-1} \sum_{j=i+1}^{m-1} \frac{(1+\sigma)^m-(1+\sigma)^j}{[(1+\sigma)^m-(1+\sigma)^{m-1}]\,p_{j,\,j+1}},$$

and an expression for the variance of the time to absorption is also available.

To obtain the model without selection we let $\sigma \to 0$. This case was considered by WATTERSON (1961). The corresponding absorption probabilities, starting in state i, will be $(m-i)/m$ and i/m respectively, and the mean time to absorption will be given by

$$(m-i) \sum_{j=1}^{i} (1-jm^{-1})^{-1} + im \sum_{j=i+1}^{m-1} j^{-1}. \tag{4.8}$$

It is interesting to note that for Wright's model without selection (4.1), the absorption probabilities at 0 and m, starting in state i, also equal $(m-i)\,/\,m$ and i/m. See FELLER (1968, p. 399) and also KHAZANIE and MCKEAN (1966 a).

4.1.3.4. Let us now consider the effect of the geographic distribution of an l-object population. We shall take into account the modifications of assumptions II and VI: many populations are spread out over a plane and mating is more likely to occur between l-objects whose parents were close together; under these conditions the effect of systematic pressures may be different. In fact, it has been shown (MORAN (1962, p. 175)) that as long as selection operates in opposite directions in two subpopulations, a stable polymorphism is possible and this is very similar to the situation in a single population in which the heterozygote is more favoured than either of the homozygotes. The effect of intermigration has also been studied, and MORAN (1959, 1962) considered a Wright model with subdivided populations in which migrants into any subpopulation come equally from all the other subpopulations. The conclusion is that under the conditions imposed on the model, the effect of subdivision is small and becomes negligible when the number of migrants in each generation and in each subpopulation is larger than one or two. The conclusion reached by EWENS (1969, p. 38) is that, provided that no selective differences exist between the subpopulations, the collection of subpopulations can normally be regarded for practical purposes as a simple large random-mating population.

MARUYAMA (1970) has built a more realistic model. He considers a population consisting for r demes of finite size, say with n diploid l-objects each, located linearly. Assume that migration between adjacent demes occurs at rate ν per generation, that is, the i-th deme receives migration from the $(i-1)$-th and the $(i+1)$-th demes at rate $\nu/2$ if $i \neq 1$ or $i \neq r$. Each terminal deme receives migration from only the corresponding subterminal deme at rate ν which is twice the rate at which it would be receiving if it were not at the border. Finally, one must suppose that no other kind of migration nor mutation occurs in any colony. The most significant result from the biological point of view is that the genetic variability decreases asymptotically at rate $\nu\pi^2/(2r^2)$, provided that the number of demes is large. One immediately observes that the rate is proportional to the reciprocal of the square of r and is independent of the number of l-objects in a single deme. If we have the inequality $(\nu\pi^2)/(2r^2) > 1/2\,rn$, the asymptotic rate of decline of heterozygosity is approximately equal to $1/2\,rn$, being thus equal to the rate in a panmictic population of rn diploid l-objects.

4.1.4. Diffusion approximations

4.1.4.1. Let us consider Wright's model (4.1). Putting $\eta_n = \xi_n m^{-1}$, where n is the time measured in units of generations, it is easily seen that

$$\mathsf{E}\,[\eta_{n+1} - \eta_n] = 0,$$

$$\mathsf{E}\,[(\eta_{n+1} - \eta_n)^2 \mid \eta_n] = m^{-1}\eta_n(1 - \eta_n).$$

This suggests (see I; 2.3.4.3.2) the possibility of approximating η_n, when m is large, by a diffusion process with drift 0 and diffusion coefficient proportional to $x(1 - x)$. Similar considerations can also be made for (discrete time) Moran's model. Before describing a few rigorous results concerning diffusion approximations, we shall quote FELLER's (1966, p. 325) opinion, according to which the genetical implications of the diffusion process are somewhat dubious because of the assumption of the constant population size (assumption III), the effect of which is not generally appreciated. For a description depending on an equation in two space variables see FELLER (1951).

The use of diffusion approximations requires the assumption of an infinite population: although the state space is discrete, it is treated as though it were continuous. The tacit assumption in such an approach is that for large population sizes, the approximation of the discrete time, discrete space process by the continuous diffusion process does not result in serious error. A rigorous justification of the diffusion approach was given by WATTERSON (1961, 1962, 1964) with respect to the models considered by S. WRIGHT and P. A. P. MORAN.

Diffusion approximations are used when exact expressions of various quantities of interest are not available. To consider an example, for Wright's selection model (4.4) the absorption probabilities at 0 and m are not known. Passing to the diffusion approximation, one finds (see e.g. EWENS (1963 a)) that the absorption probability at m starting in state i could be approximated by

$$\frac{1 - e^{-2\sigma i}}{1 - e^{-2\sigma m}}$$

(in case $\sigma = 0$, this value is exact). Clearly, for small values of σ, the diffusion approximation is very close to the true value. The exact values were obtained simply by taking powers of a transition matrix and by a matrix inversion. It will be noted that in all cases the diffusion approximation always exceeds the true absorption probability; that this will always be so for positive σ is shown by MORAN (1962). The problem of justifying diffusion approximations and estimating total expected discrepancies has been studied by EWENS (1963 a, d; 1965).

4.1.4.2. The interpretation given by M. KIMURA starts directly from the supposition that two alleles a_1 and a_2 have frequencies x and $1 - x$ respectively in a large population of I-objects. Denote by $p(t; i, x)$ the conditional probability density that the gene frequency is x at time t, given that the initial frequency, at $t = 0$, is i. Various models are introduced by writing down a Fokker-Planck equation for $p(t; i, x)$.

The Fokker-Planck equations which appear in population genetics usually have singularities at the boundaries and a deep mathematical investigation of these was carried out by FELLER (1951) who clarified the nature of the boundaries by using the semi-group theory. We also mention, in the same direction, KOLMOGOROV's (1959) paper on the necessity for "fusing" the branching process results with the diffusion results in the neighbourhood of the boundaries. MILLER (1962) has developed a method for estimating eigenvalues of the Fokker-Planck equation in genetics. For details in a series of genetic situations, as well as random fluctuation of selection intensities, probability of gene fixation and average number of generations until fixation, stationary gene frequency distribution or number of heterozygous nucleotide sites, see CROW and KIMURA (1955), KIMURA (1955—1970), and KIMURA and OHTA (1969—1971).

4.1.4.3. We shall close this paragraph with a brief discussion following KARLIN and MCGREGOR (1964) on limiting diffusion processe as $m \to \infty$ for (continuous parameter) Moran models. It may be observed, for example, that in the Moran model with mutation there is a slight fluctuation of the population size of I-a_1-objects about the mean equilibrium level, so that the states near $i = m\,\alpha_1/(\alpha_1+\alpha_2)$ seem to be favoured. Indeed, from (4.7) we see that $p_i > q_i$ or $p_i < q_i$ according as $i > m\,\alpha_1/(\alpha_1+\alpha_2)$ or $i < m\alpha_1/(\alpha_1+\alpha_2)$, so there is attraction toward the special value. When m is large, the system tends to spend long periods of time in the neighbourhood of the pseudo-equilibrium state $i = m\,\alpha_1/(\alpha_1+\alpha_2)$. The study of fluctuations about the pseudo-equilibrium state leads us to consider a limit procedure. With α_1, α_2 fixed, one assumes the initial population to be

$$i = \left[\frac{m\alpha_1}{\alpha_1+\alpha_2} + x\sqrt{mc}\right] = s(x),$$

where $c = 2\,\alpha_1\,\alpha_2/(\alpha_1+\alpha_2)^3$, x is a fixed real number and $[u]$ denotes the greatest integer $\leqslant u$.

One can prove that for any fixed $y \in R$,

$$\lim_{m\to\infty} \sum_{j\leqslant s(y)} p_{s(x),\,j}(t) = P(t;\, x,\, (-\infty, y)),$$

where P is the transition function of the Ornstein-Uhlenbeck process (see I; 2.3.5.5.3) governed by the backward equation

$$\frac{\partial P}{\partial t} = d\left[\frac{1}{2}\,\frac{\partial^2 P}{\partial x^2} - x\,\frac{\partial P}{\partial x}\right].$$

The existence of the limit is interpreted by S. KARLIN and J. MCGREGOR as follows: When the population is large, the gene frequencies are given appropriately by a deterministic model, since the fluctuations in state i (the number of I-a_1-objects) are only of order $\sqrt{m}$ and hence do not affect the gene frequencies. In the study of the fluctuations in state i, the above diffusion process ought to give a "reasonable" approximation.

For another example of limit procedures, the reader is referred to the remarkable paper by S. KARLIN and J. McGREGOR already quoted here.

4.1.5. Non-Mendelian situations

4.1.5.1. The genetic mechanisms studied until now have been based upon three laws stated by GREGOR MENDEL: (I) the law of the uniformity of the first generation, (II) the law of gene segregation in the second generation, (III) the law of gene free combination. Mendel's fundamental structural assumption is that each gene is found on each single chromosome; organisms of the same species will therefore present a number of character differences equal to the pairs of homologous chromosomes.

Later research in experimental genetics and cytogenetics has shown that, besides other types of segregation, there is also a mechanism of transmission of genetic structures which does not confirm Mendel's second law: the linked transmission of genes from the same chromosome. This phenomenon was called "linkage" by THOMAS H. MORGAN (Nobel prize 1933), and is equivalent to that of "coupling" mentioned as early as 1906 by W. BATESON and R. C. PUNNETT. However, linkage can only be admitted if Mendel's structural fundamental assumption is replaced by the double assumption that there are several genes on one chromosome and that they form a line.

Linkage may be complete or uncomplete. When it is only partial, it may be possible that during meiosis (or even mitosis) there is an exchange of linkage groups between homologous chromosomes, i.e., an exchange of genes located at large distances from one another. This is considered as a recombination process. The mixture and reassortment of the genetic material given by parental organisms is carried out by this process. It is quite unnecessary to stress the importance of this process for the evolution of populations (see BODMER and PARSONS (1962)). Mathematical problems of gene linkage have been studied by BAILEY (1961, 1964 a, p. 224; 1967 b, Ch. 10).

4.1.5.2. We shall present, following KARLIN, McGREGOR and BODMER (1967), a Markov model for the production of recombinant G-objects in a finite (captive) population.

Consider two linked loci in a population of diploid I-objects with alleles a_1, a_2, and b_1, b_2, at the first and second locus, respectively. The normal G-object is of a_1b_1 type but types a_1b_2 and a_2b_1 also exist.

The gametic output of a double heterozygote $a_2b_1 \cup a_1b_2$ contains the four types a_1b_2, a_2b_1, a_1b_1, a_2b_2 in the proportions

$$\frac{1-r}{2}, \quad \frac{1-r}{2}, \quad \frac{r}{2}, \quad \frac{r}{2}, \qquad 0 \leqslant r \leqslant 1/2,$$

where r is the recombination fraction. However, let us consider that in a population of n diploid I-objects, we have only genotypes $a_1b_2 \,||\, a_1b_2$, $a_1b_2 \,||\, a_2b_1$, $a_2b_1 \,||\, a_2b_1$ (here, $||$ indicates the linkage group on the two homologous chromosomes; one can easily observe that this situation is a replica of the classical situation of three genotypes for a single gene). In the population considered, there are thus two types of G-objects, type a_1b_2 with frequency i, and type a_2b_1 with frequency $2n - i$. We determine the next generation by random sampling from the gametic output above. If n is not too small, the genotype frequencies of the next generation can be obtained with sufficient accuracy from the gene frequencies of the present generation by assuming a Hardy-Weinberg distribution: the genotypes $a_1b_2 \,||\, a_1b_2$, $a_2b_1 \,||\, a_1b_2$, $a_2b_1 \,||\, a_2b_1$ occur in the present generation with the frequencies

$$\left(\frac{i}{2n}\right)^2, \quad 2\frac{i}{2n}\left(1-\frac{i}{2n}\right), \quad \left(1-\frac{i}{2n}\right)^2.$$

A G-object selected at random from the output of the present generation is therefore of type a_1b_2 with probability

$$p_i = \left[2n\left(\frac{i}{2n}\right)^2 + 2n\,\frac{i}{2n}\left(1-\frac{i}{2n}\right)(1-r)\right] \Big/ 2n =$$

$$= \frac{i}{2n} - r\,\frac{i}{2n}\left(1-\frac{i}{2n}\right),$$

and of type a_2b_1 with probability

$$q_i = 1 - \frac{i}{2n} - r\,\frac{i}{2n}\left(1-\frac{i}{2n}\right),$$

and is of recombinant type (a_2b_2 or a_1b_1) with probability $1 - p_i - q_i$. When $2n$ G-objects are chosen by repeated sampling forming the n individuals of the next generation, the probability that j G-objects of type a_1b_2 and $2n - j$ G-objects of type a_2b_1 are obtained is given by

$$p_{ij} = \binom{2n}{j} p_i^j\, q_i^{2n-j}, \qquad 0 \leqslant j \leqslant 2n. \tag{4.9}$$

The probability that no recombinant G-objects are obtained is therefore

$$\sum_{j=0}^{2n} p_{ij} = (p_i + q_i)^{2n} = \left[1 - 2r\frac{i}{2n}\left(1 - \frac{i}{2n}\right)\right]^{2n},$$

and the probability that one or more recombinants appear is $1 - \sum_{j=0}^{2n} p_{ij}$.

The Markov chain with transition probabilities (4.9) ends either with the appearance of a recombinant type or with fixation. A main objective is to determine the probability of recombination *before fixation* as a function of the initial gametic frequencies, the recombination fraction r, the population size n, selective values and the mating system. This objective, however, can be only reached by calling on with an approximating diffusion process. It was shown by S. KARLIN, J. MCGREGOR and W. BODMER that the probability of a recombinant appearing prior to fixation given the initial frequency of the G-object of type a_1b_2 is $x = i/2n$, is approximately

$$1 - \frac{\sinh 4xn\, r^{1/2} + \sinh 4\,(1-x)\, n\, r^{1/2}}{\sinh 4n\, r^{1/2}}.$$

We see immediately that when $r \gg 1/n^2$, then the probability of recombination before fixation is approximately 1. On the other hand, if rn^2 is small, then this probability is approximately zero and fixation almost certainly occurs before recombination.

If we are interested in the creation of the a_1b_1 recombinant G-object, while G-objects of type a_2b_2 are discarded when formed, then the corresponding probability will approximately be

$$1 - \frac{\sinh x\, n\,(8r)^{1/2} + \sinh (1-x)n\,(8r)^{1/2}}{\sinh n\,(8r)^{1/2}}.$$

The preceding model can also be analyzed taking into account selection effects or random mating with and without selection.

4.1.6. Direct product Galton-Watson chains

4.1.6.1. Genetic models based of gene frequency may also be studied by means of Galton-Watson chains. For this purpose, it is first of all necessary to consider instead of a single population of alterexter l-objects, each independently multiplying according to Galton-Watson chains, $(\xi_n)_{n\in N}$ and $(\tilde{\xi}_n)_{n\in N}$.

If I-a_1-objects and I-a_2-objects reproduce according to offspring distributions with probability generating functions $f(z)$ and $g(z)$ respectively, then the joint probability generating function for the generation after starting with i I-a_1-objects and $m - i$ I-a_2-objects is

$$f^i(w) \cdot g^{m-i}(z).$$

This explains the term direct product branching process given by KARLIN and McGREGOR (1964) to this type of two-dimensional chain.

We add the assumption that each generation has the same number of offspring. Let m be this number. We can write

$$p_{ij} = \mathsf{P}(\xi_1 = j, \tilde{\xi}_1 = m - j \mid \xi_0 = i, \tilde{\xi}_0 = m - i, \xi_1 + \tilde{\xi}_1 = m) =$$

$$= \frac{\mathsf{P}(\xi_1 = j, \tilde{\xi}_1 = m - j \mid \xi_0 = i, \tilde{\xi}_0 = m - i)}{\mathsf{P}(\xi_1 + \tilde{\xi}_1 = m \mid \xi_0 = i, \tilde{\xi}_0 = m - i)} = \tag{4.10}$$

$$= \frac{\text{coefficient of } w^j z^{m-j} \text{ in } f^i(w) g^{m-i}(z)}{\text{coefficient of } z^m \text{ in } f^i(z) g^{m-i}(z)}.$$

The equality of the denominators rests on the fact that $f^i(z)\, g^{m-i}(z)$ is the probability generating function for the total number of progeny under the assumption that I-a_1-objects and I-a_2-objects are indistinguishable from one another. We can regard (4.10) as the transition probabilities of a finite Markov chain with state space $X = \{0, 1, \ldots, m\}$.

Let us now assume that $f(w)$ and $g(z)$ are given by

$$f(w) = e^{\lambda(w-1)}, \; g(z) = e^{\mu(z-1)}, \; \lambda > 0, \; \mu > 0.$$

Then, (4.10) becomes

$$p_{ij} = \binom{m}{j} \left(\frac{i\lambda}{i\lambda + j\mu}\right)^j \left(\frac{j\mu}{i\lambda + j\mu}\right)^{m-j}, \quad 0 \leqslant i, j \leqslant m. \tag{4.11}$$

If we put $\lambda = \mu$ in (4.11) we shall again find (4.1), i.e., the transition probabilities for Wright's model.

4.1.6.2. To build a mutation model, one first assumes that mutation occurs first, followed by growth ("branching multiplication") or vice versa. These lead to different Markov chains. One retains the assumption of mutation $a_1 \to a_2$ with probability α_1 $(0 \leqslant \alpha_1 \leqslant 1)$ and of mutation $a_1 \leftarrow a_2$ with probability α_2 $(0 \leqslant \alpha_2 \leqslant 1)$.

It is easier to describe the mechanism of the mutation model directly in terms of the corresponding probability generating functions (KARLIN (1966, p. 394), KARLIN and McGREGOR (1965)).

(a) Mutation follows reproduction. Let $f(w)$ and $g(z)$ represent the probability generating function of the numbers of progeny due to one parent of type a_1 and one of type a_2, respectively. Suppose that following reproduction, each offspring of an I-a_1-object (or I-a_2-object) can produce further objects of both kinds with generating functions $A(w, z)$ and $B(w, z)$. In this way one can describe a reproduction process in two stages: The first stage corresponds to the usual multiplication process (with alteridem generations), the second one to a process of growth and mutation: the coefficient a_{kl} in $A(w, z)$ is the probability that in the second stage of reproduction, an offspring of type a_1 will produce an additional of k objects of type a_1 and l objects of type a_2. The final probability generating function culminating both states of reproduction depicting the offspring population stemming from one I-a_1-object is $f[A(w, z)]$, and that due to one I-a_2-object is $g[B(w, z)]$.

To obtain the mutation process we must take

$$A(w, z) = (1 - \alpha_1) w + \alpha_1 z,$$

$$B(w, z) = \alpha_2 w + (1 - \alpha_2) z.$$

Therefore, the probability generating function $f(w, z)$ of the offspring population resulting from reproduction of one I-a_1-object effected afterwards by mutation pressures will be

$$f(w, z) = f[(1 - \alpha_1) w + \alpha_1 z].$$

Similarly, we see that the progeny population due to one I-a_2-object taking account of mutation pressures is

$$g(w, z) = g[\alpha_2 w + (1 - \alpha_2) z].$$

(b) Mutation precedes reproduction. In this case,

$$f(w, z) = (1 - \alpha_1) f(w) + \alpha_1 g(z),$$

$$g(w, z) = (1 - \alpha_2) g(z) + \alpha_2 f(w).$$

We can devise a more general construction. Let $f(w, z)$ be the probability generating function of the offspring population. Similarly, we assume that an I-a_1-object may produce objects of both types and let $g(w, z)$ denote the probability generating function of his offspring population.

Let $h(w, z)$ represent the probability generating function of the number of I-a_1-objects and I-a_2-objects immigrating into the system during each period.

Let $(\xi_k, \tilde{\xi}_k)_{k\in N}$ denote the resulting two-dimensional chain. The probability generating function of $(\xi_k, \tilde{\xi}_k)$ under the initial condition $\xi_0 = i$, $\tilde{\xi}_0 = j$ is

$$[f(w, z)]^i\ [g(w, z)]^j\ h(w, z).$$

The transition probabilities for the induced Markov chain obtained under the condition that the population size is fixed can be calculated as above. We get

$$p_{ij} = \mathsf{P}(\xi_1 = j, \tilde{\xi}_1 = m - j \mid \xi_0 = i, \tilde{\xi}_0 = m - i;\ \xi_1 + \tilde{\xi}_1 = m) =$$

$$= \frac{\text{coefficient of } w^j z^{m-j} \text{ in } f^i(w, z)\, g^{m-i}(w, z)\, h(w, z)}{\text{coefficient of } z^m \text{ in } f^i(z, z)\, g^{m-i}(z, z)\, h(z, z)}. \quad (4.12)$$

For the special choice

$$f(w, z) = \mathrm{e}^{-\lambda} \exp(\lambda[(1 - \alpha_1)\, w + \alpha_1 z]),$$

$$g(w, z) = \mathrm{e}^{-\mu} \exp(\mu[\alpha_2 w + (1 - \alpha_2) z]),$$

$$h(w, z) = \exp(\gamma_1 (w - 1) + \gamma_2 (z - 1)),$$

the transition probabilities of the induced Markov chain reduce to

$$p_{ij} = \binom{m}{j} \frac{[(1-\alpha_1)\, i\lambda + \mu(m-i)\,\alpha_2 + \gamma_1]^j\, [\alpha_1 i\lambda + \mu(m-i)(1-\alpha_2) + \gamma_2]^{m-j}}{[i\lambda + \mu\,(m - i) + \gamma_1 + \gamma_2]^m}, \quad (4.13)$$

which is the Wright model with selection, mutation, and migration.

For the mathematical treatment of induced Markov chains including eigenvalues and eigenvectors we refer the reader to KARLIN (1966, Ch. 13), KARLIN and MCGREGOR (1965).

4.1.6.3. The above theory can be extended to multi-allele models by making use of multitype Galton-Watson chains. The system will be in state $\mathbf{i} = (i_1, \ldots, i_s)$ if there are i_k B-objects of type a_k, $1 \leqslant k \leqslant s$. In the sequel we shall no longer denote the alleles of gene a by $a_1, \ldots, a_k$, but we shall simply say, for example, B-objects of type k.

The state space $X = N^s$ will consist of all s-tuples of nonnegative integers $\mathbf{i} = (i_1, \ldots, i_s)$ which, according to assumption III, obey the constraint $\sum_{k=1}^{s} i_k = m$. We shall also keep our assumption II and IV.

Obviously, assumption V is modified in the sense that at one locus there are s alleles. Suppose each B-object of type k in one generation yields progeny of all types with generating function $f^k(z_1, \ldots, z_s)$, $1 \leqslant k \leqslant s$. These B-objects are assumed to act independently.

Let $\boldsymbol{\xi}_n = (\xi_n^1, \ldots, \xi_n^s)$ denote the associated chain, that is, ξ_n^k represents the number of B-objects of type k in the n-th generation. Let us now introduce assumption VI on the existence of a systematic pressure, immigration. The probability generating function of the progeny in one generation given i_k objects of type k, $1 \leqslant k \leqslant s$, is

$$[f^1(z_1, \ldots, z_s)]^{i_1} \ldots [f^s(z_1, \ldots, z_s)]^{i_s} h(z_1, \ldots, z_s), \tag{4.14}$$

where h represents the probability generating function of the number of B-objects of various types immigrating into the system.

As in the case of the model with two alleles, we can write

$$p_{\mathbf{ij}} = \mathrm{P}\left(\boldsymbol{\xi}_{n+1} = \mathbf{j} \mid \boldsymbol{\xi}_n = \mathbf{i}, \sum_{k=1}^{s} \xi_{n+1}^k = m\right) =$$

$$= \frac{\text{coefficient of } z_1^{j_1} z_2^{j_2} \ldots z_s^{j_s} \text{ in } h(\mathbf{z}) \prod_{k=1}^{s} [f^k(\mathbf{z})]^{i_k}}{\text{coefficient of } w^m \text{ in } h(\mathbf{w}) \prod_{k=1}^{s} [f^k(\mathbf{w})]^{i_k}}, \tag{4.15}$$

where $\mathbf{z} = (z_1, \ldots, z_s)$ and $\mathbf{w} = (w, \ldots, w)$.

The multi-allele version of Wright model allowing mutation, migration and selection is a special case for which, following KARLIN and MCGREGOR (1965),

$$f^l(\mathbf{z}) = \exp\left(\sigma_l\left(\sum_{k=1}^{s} \alpha_{lk} z_k - 1\right)\right), \; 1 \leqslant l \leqslant s,$$

$$h(\mathbf{z}) = \exp\left(\sum_{k=1}^{s} \nu_k (z_k - 1)\right),$$

$$\alpha_{lk} \geqslant 0, \; \sum_{k=1}^{s} \alpha_{lk} = 1, \; \sigma_l > 0, \; \nu_k > 0. \tag{4.16}$$

The parameters can be interpreted as follows: α_{lk} represents the probability that a B-object of type l will mutate after birth into a B-object of type k; σ_l measures the relative fertility (as selection coefficient) of a B-object of type l and ν_k is the average rate at which B-objects of type k are immigrating into the population. The probabilities are

$$p_{ij} = \frac{m!}{j_1! \dots j_s!} \cdot \frac{\prod_{k=1}^{s} \left[\sum_{l=1}^{s} i_l \alpha_{lk} \sigma_k + \nu_k \right]^{j_k}}{\left[\sum_{k,l} i_l \alpha_{lk} \sigma_k + \sum_k \nu_k \right]^m}. \tag{4.17}$$

Conversely, one can show that if the matrix $(\alpha_{lk})_{1 \leqslant l,\, k \leqslant s}$ is irreducible and has rank at least 2, and (4.15) is identical to (4.17), then the generating functions f^k, $1 \leqslant k \leqslant s$, and h have the Poisson form (4.16). This implies that the Wright multi-allele model (4.17) is restricted in that the fertility distribution per individual is necessarily Poisson.

KARLIN and MCGREGOR (1965) also investigate the limiting processes for $m \to \infty$ associated with these multitype Galton-Watson chains. The multi-allele version of Moran model is studied, using other methods, by KARLIN and MCGREGOR (1964).

4.1.6.4. Other uses of multitype Galton-Watson chains in genetics have been suggested by EWENS (1968) and POLLARD (1968 a).

MODE (1966 a) considered a variation of multitype Galton-Watson chains in order to study the following situation: Let us suppose a population having genotype A_1 or genotype A_2 (for example, pathogenic organisms in an avirulent or a virulent form). Let us further suppose that only a forward mutation $A_1 \to A_2$ is possible. In addition, suppose that the numbers of offspring contributed to the population in the nth generation by a B-object of genotype A_i, $i = 1, 2$, are independent Poisson random variables. Let us also consider that there are no immigrants (e.g. no outside source of infection). The following questions are answered: If one starts with a population containing i_1 B-objects of genotype A_1, then what is the probability (a) that the genotype A_2 has not appeared by generation n but the genotype A_1 is still present, (b) genotype A_1 becomes extinct in generation n but genotype A_2 has not yet appeared and (c) a B-object of genotype A_2 appears for the first time in generation n and genotype A_1 is still present.

4.1.6.5. We have shown in Subparagraph 4.1.1.4 that genetic models of population structure must take the diversity of environments into account (rule II). In this case it is not only a matter of meteorological variations but, particularly, of modifications appearing in the ecosystem to which any population may belong, that is in the habitat and in biocenosis. One can take the assumption of uniform environmental conditions as valid when the habitat is a specially equipped laboratory or the blood stream of warmblooded vertebrates, in cases of parasitism. Even in such situations a modification of the ecosystem is possible such as, for example, the passing of parasites into different parts of the host's body or the infecting of another host. We further mention here the problem of population growth in a randomly varying environment such that the finite rate of increase per generation is a random variable (LEVINS (1969), LEWONTIN and COHEN (1969)).

It is possible to introduce the idea of a diversity of environments into our models if we impose two new assumptions, i.e.: (i) the modifications of the environment can be reflected in the modification of the size of generations, (ii) the probability generating functions for two sequential generations is a compounding generating function and can be used to study each generation as a sequence of separate stages, each stage studied with a separate probability generating function.

Suppose that there are r possible environments. According to the first assumption mentioned above, each environment has its own offspring distribution for which the probability generating function is $f_i(z)$, $1 \leqslant i \leqslant r$. The ith environment occurs each generation with probability γ_i. Starting with one individual in generation zero, the probability generating function for the number of I-objects in the kth generation satisfies

$$\varphi_k(z) = \sum_{i=1}^{r} \gamma_i \varphi_{k-1}[f_i(z)].$$

The compouding of the $f_i(z)$ into φ represents, as usual, the passage of a generation. WILKINSON (1967) dealt with such random fluctuations in the offspring distribution which he identifies as the result of stochastic environments. We will present some aspects of his thesis using interpretations given by SCHAFFER (1970).

The expected number of offspring from one I-object in one generation is

$$m = \sum_{i=1}^{r} \gamma_i f_i'(1) = \sum_{i=1}^{r} \gamma_i m_i,$$

which is the arithmetic mean of the m_i, the means for each environment. It can be shown by mathematical induction that the expected number of I-objects in the kth generation is $\varphi_k'(1) = m^k$. However, this result is exactly the same as that expected from constant environments. In the determination of the ultimate survival probability of an I-object in a varying environment, the value of m does not play exactly the same role as with a constant environment.

On the basis of known considerations in the theory of Galton-Watson chains, one then introduces a new quantity which governs the ultimate survival probability

$$\overline{m} = \prod_{i=1}^{r} m_i^{\gamma_i},$$

that is, the weighted geometric mean of the distribution. When $\overline{m} \leqslant 1$, then the ultimate survival probability is zero, and also $m \leqslant 1$. The dependence on the geometric mean illustrates the multiplicative relationship of the means of the individual distributions when they are compounded. In order that the mean of compounded distributions be greater than unity, their geometric mean (the kth root of the product of k of them) must be greater than unity. For the random selection of a sequence of different possible environments, each of them should occur in the compounding with their relative frequencies γ_i, and so they are weighted in $\overline{m}$ according to these frequencies. From this relationship between the geometric and arithmetic means it can be seen why the ultimate survival probability is lowered if the offspring distribution is a random variable in successive generations —which can be interpreted as the result of stochastic environments.

WILKINSON (1967) also discusses the case in which the environments are not chosen entirely at random for each generation, but instead form a Markov chain, with each environment having (possibly) different probabilities of being succeeded by each of the others. It was thus demonstrated that the ultimate survival probability is zero whenever $\overline{m}$ of the environmental means is <1, regardless of the initial environment. In this case, m is computed with the weights, the γ_i, being the probabilities of each environment in the steady state distribution for the Markov chain. While this result is similar to that of randomly chosen environments, it is still an unanswered question whether $\overline{m} > 1$ is sufficient to give an ultimate survival probability greater than zero.

4.1.6.6. Galton-Watson chains have also been used for representing some non-Mendelian situations. Thus, KOJIMA and SCHAFFER (1964) studied the chain for two linked mutant genes and characterized each generation by three stages, which requires a compounding of three probability generating functions. The survival of the descendents with two linked mutant genes can thus be studied as a Galton-Watson chain with probability generating function

$$\varphi(z) = F[R(W(z))],$$

where $F(z)$ is the probability generating function of the number of offspring, $R(z)$ is the probability generating function of the distribution of an offspring with a mutant chromosome; $R(z)$ generates the probabilities of the outcomes which depend on segregation and recombination, and $W(z)$ is the probability generating function of the distribution of an offspring carrying the mutant chromosome surviving till sexual maturity.

A problem similar to that considered in Subparagraph 4.1.6.2 has been posed by KOJIMA and SCHAFFER (1967) and studied using multi-type Galton-Watson chains.

4.1.6.7. Another problem of interest is that of inversion. It has been studied by OHTA (1966), OHTA and KOJIMA (1968). The origin of each inversion chromosome is probably unique and nonrecurrent, because a given inversion first appears as a result of chromosome breakages, and the likelihood of repeating the same set of breakages and reunion in a short interval of time is extremely small. On the other hand, from structural considerations, one must take into account the fact that the investigation of the survival of an inversion is the study of the survival of gene blocks completely linked together. The situation is analogous to that of the survival of a mutation where its selective advantage is changing over time according to a certain function. Because of the changing fitness values of inversions over time, the process of compounding the probability generating functions must be heterogeneous with respect to the fitness parameter. The probability generating function for the number of inversion heterozygote progeny per one inversion heterozygote parent is expressed as exp $(m(z-1))$, where m is the average number of inversion heterozygote progeny per one inversion heterozygote parent, and the number of progeny follows a Poisson distribution.

4.2. Random drift and systematic evolutionary processes

4.2.1. Random drift

4.2.1.1. The problem of genetic drift cannot be left out of any book dealing with the application of stochastic processes in biology, because this phenomenon represents one of the random mechanisms of evolution. WRIGHT (1964) has outlined the idea that "random drift is a significant evolutionary process only when there is superficial control by selection". The rearrangements of genes in chromosomes, the mutual effect of some genes, the association of genes in polygenic systems, are nothing but gene interactions that have a selective value and reduce the accidental fluctuations of gene frequency. Genetic drift essentially represents the rapid change of gene frequency in small populations — that is by removing assumption III specified in Subparagraph 4.1.1.5. A biochemical example is the variant form of serum albumin found in as many as 25% of members of a group of North American Indians known as the Naskapi and in several closely related tribes. This variant does not appear to occur elsewhere in the world. Another example is the very high incidence of a "silent" serum cholinesterase allele in a population of Eskimos in Alaska (HARRIS (1970 b, p. 240)).

One way of approaching the matter is by the ascertainment that the chance of meeting any type of gametes is proportional to the population size. If, for example, each pair of chromosomes of *Zea mais*

($2n = 20$) contained a single pair of alleles, the number of combinations that would be achieved in the second generation, according to the Mendelian laws would be around 4^{10}. The number of male G-objects being very great, it is possible that all the calculated combinations could be achieved even by a small number of plants. However, the number of female G-objects is very small for each plant so that in fact one achieves only a small part of the possible combinations. This exhibits the mechanism of accidental losses of genes in small populations. Owing to the same mechanism, one may explain why two genetic characteristics with the same adaptive value are kept in a species: skin with dark spots or with vertebral line for brown frogs, both characters having the same protective function against enemies (see STUGREN (1966)). According to WRIGHT (1964), relatively pure random drift is most conspicuous in laboratory experiments in inbreeding. Random drift implies a cumulative process (WRIGHT (1970)).

FISHER (1922) considered the random sampling of G-objects as the factor causing random fluctuation in gene frequency and assuming no selection, he investigated its effect on the decrease of variability in a species. He called this the "Hagedoorn effect", the term "drift" (= infinitesimal velocity) being imposed by WRIGHT (1929). It was also S. WRIGHT who showed that random drift is most likely to be conspicuous in nature in the case of heterotic characters but the resulting variation in gene frequencies obviously has little evolutionary significance, unless the locus is associated with others in an interaction system. There is also little evolutionary significance in the case of degenerative changes that may occur in very small completely isolated natural populations.

The mathematical treatment of the problem leads to the results already stated in Paragraph 4.1.4. In case of two alleles and in the absence of any systematic pressure, it is demonstrated that as time goes on the gene frequency tends to deviate from the initial frequency and the probability that both alleles coexist within a population gradually decreases due to chance fixation of one of the alleles. Eventually, this probability decreases at the rate of $1/m$ in each generation. In the mathematical formulation for Wright's intrademic phases, random drift is defined as the portion of the multidimensional probability distribution of the set of gene frequencies in the neighborhood of the currently controlling selective peak.

Linkage disequilibrium due to random genetic drift when two loci are considered simultaneously has also been studied (HILL and ROBERTSON (1968), OHTA and KIMURA (1969 a, b)). No precise solution was obtained for the general case of an arbitrary number of alleles, only the expression of asymptotic behaviour (KIMURA (1955)).

4.2.1.2. KIMURA (1964) also considers the random drift in the wide sense, introducing the assumption of linear pressure of mutation and migration. By studying the drift in a subdivided population with a random number of migrants, POLLAK (1966) finds that if the mean number of migrants per generation from one subpopulation to another is at least as large as 1, the population behaves almost as if it were not subdivided (see e.g. Subparagraph 4.1.4.3).

4.2.1.3. Modern research in biochemistry has emphasized the existence of isozymes and lead to detailed studies of genetics and the appearance of the notion of "phenotypic drift" (also called inter-allelic interaction). The segregation of phenotypes that is observed in *Tetrahymena pyriformis* heterozygotes does not result because of genetic assortment, but occurs as a result of factors which affect gene expression in the macronuclear subunit (ALLEN (1968)).

4.2.2. Selection

4.2.2.1. Experimental research, especially in chemotherapy and immunology, as well as some ecologic investigations, have given rise to new interpretations of the word selection—a word which even today has no exact formulation. The fundamental theorem of natural selection given by R. A. FISHER ("the rate of increase in [mean] fitness of any organism at any time is equal to its [additive] genetic variance in fitness at that time"), discovered again by MORAN (1967), also presents difficulties of interpretation. It seems that the latest discussions have reached the conclusion that the Darwinian principle of natural selection is the link which permits us to go from the microlevel of genes and chromosomes to the macroproperties of multi-species systems (LEVINS (1965)). Yet, if *intra*populational selection has been better studied, the mathematics of the *inter*-population selection has been largely ignored (LEVENE (1967)).

Selection presents itself as a system of contradictory forces. Thus it is known there is a selection pressure towards the development of more complex ecosystems: when there is only a small number of interacting species, there is a tendency for violent fluctuations in numbers, which may lead to the extinction of some species and the complete collapse of the whole system (see WADDINGTON (1969)). The result of the interaction between genetic drift and selection is decided by the population size, the succession rate of generations, the ecologic niche, etc. In each population, in each new biologic situation, there exist a wealth of relations. Selection also interacts with evolutionary material and the diversity of genotypes is also influencing the values of recombination probabilities which exist in a population (SHIKATA (1967)). We must understand that when speaking about selection, we must deal with certain structures and not with the material of evolution. According to S. WRIGHT there are five types of structures: (i) the single gene, (ii) gene system within a genotype; (iii) genotype systems within populations with a single selective peak (uniform and unitary adaptation), (iv) genotype system within the population, with several selective peak (adaptation by genetic polymorphism), (v) deme (isolated Mendelian population).

4.2.2.2. MORAN (1967) has given a formal description of the genetic composition of G-objects, considering selection dependent on several loci. Unfortunately, he considered only populations large enough for stochastic variations to be ignored.

We mention recent interest in allowing for random selective advantages of a gene (see e.g. JENSEN and POLLAK (1969)) as well as the treatment of the problem of natural selection using multitype age-dependent branching processes (MODE (1968)) when the relative selective advantage is understood as being dependent on the distribution of lifetime of an I-object reproducing asexually.

4.2.3. Mutation

4.2.3.1. We have studied the process of mutation within the growth processes of population with alterexter B-objects as well as in presenting Wright and Moran models. This paragraph deals with another aspect of the mutational process which precises its exact significance as systematic pressure. We shall use for this purpose the comments made by KARLIN and MCGREGOR (1967 b) concerning Fisher's theory of balance of mutation and genetic drift. The biologic importance of the discussion will be evident.

Radiogenetics research carried out in the last decades is well on the way to alter the biologic conception of the role of mutations in the evolutionary process. As some biologists are fond of saying, a new species does not appear through mutation but through the selection of mutations. It is not the variability as a whole which is important in the analysis of evolution but the configuration of variability. FISHER (1930a) was interested in the problem of determining the rate of mutation necessary in order to maintain a sufficient degree of heterozygosity which thus contributes toward the genetic variance. The increase of the number of heterozygous genotypes stresses genetic polymorphism as a frequent phenomenon in natural populations. However, the origin of polymorphism not only resides in the mutational process but also in the appearance of adaptive phenotypes with temporary value. (Research work carried out by N. P. DUBININ and J. CLAUSEN is fundamental to biologists in this field.)

Consider a fixed number l, $l \in N^*$, of loci at each of which there is initially one mutant and $2n - 1$ nonmutant genes. R. A. FISHER implicitly assumes that the fluctuations of these specific mutant populations, relative to a given locus, follow the laws of a Markov chain with transition probabilities of type (4.1). Each locus will eventually become homozygous (i.e., lost or fixed) due to random elimination if no further mutation occurs.

Let l be chosen so that on the average, one locus becomes homozygous per generation. To compensate for this, a new locus is introduced at each generation having one mutant and $2n - 1$ nonmutants. The problem is to evaluate l and to determine the asymptotic mean number l_i^* of loci having i mutants, $1 \leqslant i \leqslant 2n - 1$.

KARLIN and MCGREGOR (1967 b) consider that it is possible to alter the model as follows. Instead of feeding in one new mutant in each generation, one starts each new locus with a random number of mutants with possible values $1, \ldots, 2n - 1$ with probabilities $u_1, \ldots, u_{2n-1}$, where $(u_i)_{1 \leqslant i \leqslant 2n-1}$ is the conditional limiting distribution of the Markov chain, given fixation has not occurred.

[It is not difficult to see that the vector $(u_i)_{1 \leqslant i \leqslant 2n-1}$ is the left eigenvector associated with the eigenvalue $\lambda_2 = 1 - \dfrac{1}{2n}$ of the matrix defined in (4.1) with $m = 2n$, normalized so that $\sum_{i=1}^{2n-1} u_i = 1$].

In our induced Markov chain, approach to absorption occurs at rate $1/2\,n$, that is, in each generation $1/2\,n$ of the existing populations on the average become fixed. Choosing $l = 2n$, then since each new locus is started with a size following the distribution law $(u_i)_{1 \leqslant i \leqslant 2n-1}$ the expected number of loci becoming fixed in each generation is one.

4.2.3.2. Let us now examine an essentially equivalent formulation. We shall start from the same assumption that initially there is one mutant at a single locus and, consecutively, $2n - 1$ nonmutants. After a variable number of generations, fixation of either mutant or nonmutant occurs. Let the mean number of generations for this to happen to be k and let g_i be the mean number of generations for which the number of mutants is i. By definition, $k = \sum_{i=1}^{2n-1} g_i$ and it is not difficult to see by a standard ergodic argument that $k = l$, $g_i = l_i^*$, as well.

These considerations must be linked to a model of mutation described in Subparagraph 2.3.5.7. This model also contains KIMURA and CROW's (1964) model on the number of alleles that can be maintained in a finite population. The computation of the asymptotic expected number of mutant lines with k representatives starting initially with a given number of lines is of course the same as evaluating the quantity l_k^*. R. A. FISHER succeeds in estimating l_k^* by passing to a related Galton-Watson chain. He determines l_k^* approximately as a/k (a is a suitable constant). This agrees with the result of Theorem 2.1.

Thus it may be seen that the problem of ascertaining the number of loci with k heterozygotes out of a population of n individuals is mathematically equivalent to the problem of determining the number of different alleles maintained by a balance of mutation and random elimination as set forth in this last model (see also Subparagraph 2.3.5.9.).

4.3. Problems of molecular genetics

4.3.1. Growing point of donor DNA attachment model

4.3.1.1. We now pass from stochastic models of "organismic" populations to stochastic models of molecular populations. Owing to the outstanding discoveries of modern molecular genetics we know the hierarchical organizing of the genetic material and its physical and chemical structure but we do not yet know the genetic laws governing each structural level. Obviously, events are not reducible from one level to the other. As ZUCKERKANDL and PAULING (1962) have shown, many phenotypic differences may be the result of changes in the pattern of timing and rate of activity of structural genes rather than of changes in functional properties of the polypeptides as a result of changes in amino acid sequence.

Probabilists who wish to deal with stochastic modelling in molecular genetics face great difficulties springing from three primary sources. First of all, there is the complexity of the processes involved. As LEWONTIN (1970) has stressed, our present picture is of nucleotide bases, three of which make a codon; of codons, 200 or so of which make a gene; of genes, 2000 or so of which make a chromosome; and of chromosomes, a dozen or so of which make a genome. And of all these units of hierarchy, it is only the chromosomes that obey Mendel's law of independent assortment, and only the nucleotide base is indivisible. R. LEWONTIN shows that population geneticists are often accused of having failed to incorporate the findings of modern molecular genetics in their models, but the situation is far worse than that; they have not even incorporated the findings of T. H. MORGAN. (One must interpretate Paragraph 4.1.5 in the same sense.)

Finally, there are the difficulties of mathematics itself. Stochastic models of the synthesis of biologic polymers, such as proteins and nucleic acids, raise numerous problems of mathematical technique. We refer to the application of GOOD's (1955) results on multitype Galton-Watson chains to the statistics of polymer distributions (GORDON (1962)) but particularly to the model of protein formation (THIEBAUX (1967)).

4.3.1.2. We shall now present the stochastic treatment of a sequence from the transformation process. As is known, genetic variability may also be obtained by means of mechanisms other than those known from population genetics. One of these is *transfer*, which can be represented by four possibilities: transformation, conjugation, transduction and sexduction. Transformation is achieved by the penetration of a DNA fragment from the donor into the genome of the recipient; the exogenous DNA fragment may replace, by a recombination process, a homologous nucleotidic sequence in the genome of the recipient cell[10].

BODMER (1967) considers that the whole process of transformation can be formally divided into at least five stages, not all of which are necessarily independent: (1) the initial attachment of donor DNA molecules to the recipient competent cell; (2) the entry of the DNA into the receptive cell; (3) synapsis between the donor DNA and the recipient chromosome; (4) recombination (or integration); (5) expression of the transformed state. The transformant is a recombinant (see Paragraph 4.1.5) into which the functional units of DNA are linearly integrated in the chromosome of the transformant, owing to some recipients. These have a certain degree of specificity, in the sense that DNA belonging to a remote species is irreversibly fixed, but is not integrated into the chromosome, does not form transformants and is not phenotypically expressed.

We must also consider the fact that not all cells are able to irreversibly fix DNA and enter the transformation process. In a cell population, 1—2 per cent may appear as transformants. The ability of a cell to be transformed is called *competence*; it is related to a certain phase of the life cycle. A recent example was given by GREEN (1970).

However, one cannot as yet build a stochastic model sufficiently detailed to include all the phases as mentioned above. Following BODMER (1967), we shall only present the growing point attachment model. As is presently known, the donor DNA is integrated at a stationary growing point of DNA synthesis. The growing point is probably associated with the cell wall membrane in such a way as to make it readily accessible to exogeneous DNA. Therefore one can consider the fate of a given attachment site under the following assumptions:

1°. Donor molecules arrive at the site according to a Poisson process with parameter λ;

2°. There is only one attachment site (or at most perhaps two or three);

3°. The proportion of acceptable donor molecules is c, (because only those DNA molecules containing regions homologous to the portion of the recipient chromosome surrounding the growing point associated with any given attachment site can be integrated at that site);

4°. Acceptable molecules fix the sites, preventing any further attachments;

---

[10] This phenomenon was pointed out in 1926 on streptococci by I. CANTACUZINO, on pneumococci by F. GRIFFITH in 1927, and proved in 1944 by O. T. AVERY, C.M. MACLEOD and M. MCCARTY.

5°. Unacceptable molecules stay at the site for a random length of time τ with probability density function g, and while at the site they prevent any further attachments.

The probability density function f of a site becoming fixed will then satisfy the integral equation

$$f(t) = c\lambda e^{-\lambda t} + \int_0^t \int_0^{t-\theta} (1-c)\,\lambda e^{-\lambda\theta}\, g(\tau) f(t-\theta-\tau)\, d\tau\, d\theta. \quad (4.18)$$

If we take Laplace transforms then the above integral equation becomes

$$\mathscr{f}(s) = \frac{c\lambda}{s+\lambda} + \frac{(1-c)\,\lambda}{s+\lambda}\, \mathscr{g}(s)\mathscr{f}(s), \quad (4.19)$$

where $\mathscr{f}(s)$ and $\mathscr{g}(s)$ are the Laplace transforms of f and g, respectively. Solving (4.15) we have

$$\mathscr{f}(s) = \frac{c\lambda}{(s+\lambda) - (1-c)\,\lambda \mathscr{g}(s)}.$$

If we assume that g is an exponential density function with parameter μ, then $\mathscr{g}(s) = \mu / (s+\mu)$, and

$$\mathscr{f}(s) = \frac{c\lambda(\rho_1+\mu)}{(\rho_1-\rho_2)(s-\rho_1)} + \frac{c\lambda(\rho_2+\mu)}{(\rho_2-\rho_1)(s-\rho_2)},$$

givin g, on inversion,

$$f(t) = \frac{c\lambda}{\rho_1-\rho_2}\left[(\rho_1+\mu)\, e^{\rho_1 t} - (\rho_2+\mu)\, e^{\rho_2 t}\right],$$

where ρ_1 and ρ_2 are the roots of the quadratic equation

$$s^2 + s(\lambda+\mu) + c\lambda\mu = 0.$$

The probability of a site not being fixed by time t is therefore

$$F(t) = \int_t^\infty f(\tau)\, d\tau = \frac{c\lambda}{\rho_2-\rho_1}\left[\left(1+\frac{\mu}{\rho_1}\right) e^{\rho_1 t} - \left(1+\frac{\mu}{\rho_2}\right) e^{\rho_2 t}\right].$$

Since ρ_1 and ρ_2 are both negative, as $t \to \infty$, $F(t) \to 0$, leaving no unfixed cells. When t is small:

$$F(t) = \frac{c\lambda}{\rho_2 - \rho_1}\left[\frac{\rho_1 + \mu}{\rho_1}(1 + \rho_1 t) - \frac{\rho_2 + \mu}{\rho_2}(1 + \rho_2 t)\right] + O(t^2),$$

so that the proportion of transformed sites is

$$1 - F(t) = c\lambda\, t + O(t^2),$$

since $\rho_1\rho_2 = c\lambda\mu$. The existence of two attachment sites and the phenomenon of competence are also considered in BODMER's (1967)paper.

4.3.2. An example of a random system with complete connections

4.3.2.1. One of the first examples of a chain with complete connections given by ONICESCU and MIHOC (1937) was the spreading of tuberculosis cases in several generations as related to the heredity of individuals. The probability of an individual's catching tuberculosis depends not only on whether one or the other of his parents was ill with it, but also on the probability that his parents had caught tuberculosis.

IOSIFESCU and TAUTU (1968) gave an example of a random system with complete connections in molecular genetics. This is the modification of the structure of polypeptide chains during biochemical evolution, because, as observed by EDSALL (1964), modern proteins chemistry is closely connected with evolutionary biology. The evolutionary process can be envisaged as having involved a successive series of gene duplications followed in each instance by the divergent evolution of the products formed as a result of point mutations causing different amino acid substitutions (HARRIS (1970 b, p. 77)). The best known examples are the hemoglobin polypeptide chains — denoted α, β, γ and δ. If we consider structurally different hemoglobin chains in man, we may divide them in two groups: (i) chains differing from each other by more than one amino acid substitution; these have all been found to be controlled by distinct loci and are present in all normal human beings; (ii) chains differing by only one amino acid substitution — here the gene that controls such a chain has been found to be an allele of the gene in control of the nearby identical chain; this allele occurs in small proportions of the population and controls "abnormal" hemoglobins.

KIMURA (1969 a) put forward the hypothesis that at least in the hemoglobin α and β chains, amino acid substitutions and the underlying nucleotide substitutions have proceeded at a constant rate and in a fortuitous manner throughout the diverse lines of vertebrate evolution during the past 500 million years. He proposed a new measure, the *pauling*, a unit of evolutionary rate at the molecular level, defined as the rate of substitution of 10^{-9} per amino acid site per year. Accordingly, the hemoglobin rate is near one pauling, while the rates for seven proteins range from 33 centipauling (insulins) to 4.3 paulings (fibropeptide A). In the same sense, differences between horse and human globins suggest the idea that different sequencing could be the result of nine stable mutations that may have taken place during the course of 150 million years of evolution that separate horse and man from a common ancestor (ZUCKERKANDL and PAULING (1962)).

It is also significant that the α- and β-chain genes, among the most different within the species, have been shown to be on separate chromosomes, or at least not to be closely linked, while the β- and δ-chain genes, which resemble each other most, appear to be linked. The finding that the α-locus is not closely linked to the β- or δ-loci presumably implies that the original gene duplication was associated with separation of the loci or that a translocation of a chromosomal segment causing the separation of these loci occurred at some subsequent stage in their evolutionary history (HARRIS (1970 b, p. 78)).

These genetic considerations lead naturally to the construction of an adaptive (learning) model, by making use of a homogeneous random system with complete connections (see I; 1.3.4.2.4). We shall denote by W the set of those linkage groups which contain genes controlling the synthesis of a polypeptide chain (or of other structural units, as in the case of the synthesis of isozymes). The elements w of W will be the possible states of this genetic system, e.g. duplication, mutations, crossing over, position effect, etc. The responses of this system are expressed through the modifications of the structural unity the synthesis of which it controls. We shall denote by A the set of all possible sequences of amino acids in a polypeptide chain (or in another structural unit that is considered).

We must also suppose that the external situation to which the genetic system is submitted is formed by environmental modifications. Each exposure can alter the response tendencies of the genetic system. We shall denote by M the set of all possible environmental changes which certainly include cytoplasm modifications. The sets W, A and M are finite. Let us assume we are also given:

(1) for every $w \in W$, a probability distribution $p(w; \cdot)$ on A;

(2) a probability distribution $q(*)$ on M, which is independent of $w \in W$;

(3) a mapping u of $W \times X$ into W, where $X = A \times M$.

Define also a transition function from W to X by

$$P(w; \cdot\,; *) = p(w; \cdot)\, q(*).$$

The proposed model finds application in processes of cell differentiation as well as in the process of the formation of isozymes (see HARRIS (1970 b, Ch. 2)). For instance, during the first stage of embryonic development, before the differentiation of erythrocytes, none of the genes for hemoglobin synthesis appear to be active, though they are surely present in the cells of the embryo (MARKERT (1964)). Then, as erythrocytes begin to develop, the α and γ genes are activated and fetal hemoglobin is synthetized. Later, the γ gene is suppressed and the β gene activated, although for a time both function together, producing mixtures of adult and fetal hemoglobin. Thus, paraphrasing C. L. MARKERT, cell differentiation becomes an expression of an antecedent pattern of the states of a certain genetic system. The most fundamental expression of cell differentiation is the appearance of a new specific protein, which must be regarded as a product involving some particular relations between genes and environment (also cytoplasm). During development, genes and environment (also cytoplasm) interact to produce new patterns of active genes, which lead to the synthesis of new varieties of proteins.

5. Models in physiology and pathology

In this section we shall restrict ourselves to the presentation of several stochastic models grouped according to extremely general criteria. Space considerations do not allow more than a bird's-eye-view of most of them here, although they will certainly find their way into another work. In what follows we shall emphasize the modelling of the clinical process, the control of some biologic processes and the essential aspects of the mathematical theory of epidemics.

5.1. Stochastic models in physiology

5.1.1. Models of the appearance and the transmission of the neural flux

5.1.1.1. A basic characteristic of B-objects is their ability to respond to stimuli emanating from their environment. This property, in its elementary form, is called *excitability*; the neuron is a highly specialized cell in this respect. The change provoked by an adequate stimulus is not only limited to the application site of the stimulus but also propagates along the nerve fiber. This propagated modification is called an *impulse*, and its electric manifestation is called the *action potential*. E. DU BOIS-RAYMOND states as early as 1848 in his book on animal electricity, that by establishing the fact that the nerve impulse is an electric phenomenon, the secular dream of physicists and physiologists had been realized.

Using FRANCK's (1956) considerations in our introduction to the study of models for biological excitation processes, we can state that the production of the excited state is subject to the *all-or-none principle* (E. D. ADRIAN in 1914), i.e., a stimulus acting on the cell will bring about either the maximum excitation, or no excitation at all, depending on whether or not it exceeds a critical threshold. Accordingly, the stimulus causes only the release of the response and does not affect its intensity. This behaviour represents, according to BRAZIER (1967), "a beautiful digital signal"; it is due to a specific instability of the cell membrane with respect to certain adequate stimuli.

The best known mathematical model for the conduction of the nerve impulse belongs to A. L. HODGKIN and A. F. HUXLEY (in 1952, 1958): it describes the movement of ions along the axon membrane when impulses pass along the fiber in the form of electrical action potentials. The mathematical treatment is based on the probabilities that these ions are concentrated at the relevant place on the membrane at the critical time. Essentially they consist of a partial differential equation describing the current distribution along the nerve fiber and some additional differential equations describing the current-voltage time characteristics of the nerve membrane (see e.g. BRAZIER's (1967) comments).

These changes of the membrane of excitable B-objects (and particularly on the neuron membrane) obviously have a series of consequences. We shall recall the most important ones (following FRANCK (1967)):

I. The threshold properties of the excitation release (= validity of the all-or-none principle);

II. The spontaneous decrement-free propagation of the locally-released excitation state;

III. The appearance of refractory behaviour (reduction of excitability as a result of the previous excitation);

IV. The accomodation phenomena, that is, the reduction of excitability as a result of a previous sub-threshold stimulus;

V. Rhythmical behaviour, i.e., the ability to respond under certain conditions to a steady stimulus with rhythmical recurring excitation.

Excitation of an excitable B-object is thus due to a sudden change in the fiber membrane by which electrical energy is released in an explosive manner. The transition from the unexcited (or *resting*) to the excited (or *active*) state occurs when the potential difference which normally exists across the fiber membrane is rapidly reduced to a critical level (the threshold). At this point, the potential changes, becomes self-reinforcing and undergoes a rapid automatic cycle, known as the "spike" or action potential. Thus, during activity, the membrane potential reverses sign, and its total oscillation is several times greater than the threshold change initially required for excitation (CASTILLO and KATZ (1956)).

5.1.1.2. CAIANIELLO (1961) has built a mathematical model of the neuron by considering a linear threshold element with n input lines ($n \in N^*$) and (constant) coupling coefficients $a_k (1 \leqslant k \leqslant n)$. The observation times are considered to be a discrete sequence of equidistant intervals τ, 2τ, ..., where τ denotes the neuron's delay. At any time, each of these input lines can be in one of two possible states, 0 and 1, corresponding to the physiological situations of activity and inactivity respectively. The inputs are connected to an adder element by means of the coupling coefficients which transform the state η_k of the kth input line into $a_k \eta_k$ so that for $a_k > 0$ there is excitation, while for a_k negative there is inhibition.

At the output of the adder, we shall consider the appearance of net excitation. It is expressed by the sum

$$\xi_n = \sum_{k=1}^{n} a_k \eta_k .$$

A binary decision element follows the adder. Denoting the constant threshold of the neuron by θ, the neuron fires iff $\xi_n > \theta$. Firing means that the output line becomes active after a neuron's delay interval τ.

In order to give a probabilistic description of the neuron, DE LUCA and RICCIARDI (1968) assumed the input variables η_k are independent with $\mathsf{P}(\eta_k = 1) = p_k$, $\mathsf{P}(\eta_k = 0) = q_k$, $p_k + q_k = 1$, $1 \leqslant k \leqslant n$. They first analyze the conditions under which use can be made of the central limit theorem for calculating the probability density function of the net excitation arriving at the decision element of the neuron. A sufficient condition for the central limit theorem to apply is the divergence of the series $\sum_{k \in N^*} p_k q_k$. If this condition holds and n is large, then the firing probability approximately equals $1 - \Phi(u)$, where

$$u = \frac{\theta - \mathsf{E}\xi_n}{\sqrt{\mathsf{D}\xi_n}},$$

and $\Phi(\cdot)$ is the distribution function of the standard normal distribution.

The probability for the neuron to fire k times in the time interval $m\tau$ is given by

$$\binom{m}{k} [\Phi(u)]^{m-k} [1 - \Phi(u)]^k .$$

5.1.1.3. DE LUCA and RICCIARDI (1968) also deal with the problem of determining the probability for the neuron to exhibit a preassigned "histogram of intervals" between successive spikes in the interval $m\tau$. They studied a model which takes into account the existence of a refractory period effective after each firing and a finite summation time of the signals coming along the input lines. These assumptions suggest the use of electronic counter models, which represent well-known examples of recurrent phenomena (see e.g. COX and MILLER (1965, p. 341), FELLER (1968, p. 306; 1966, p. 189), PRABHU (1965 a, p. 191), PYKE (1958), TAKÁCS (1956), etc.). We also mention LESLIE's (1967) paper on recurrent composite events. This category of events is the discrete analogue of a model he proposed for the firing of cone cells in the retina of the eye for which a Poisson input of photons results in a succession of output signals along the optic nerve. See also RUNNENBURG (1969) and TEN HOOPEN (1966).

5.1.1.4. It results from the above that the realization of a sequence of nerve impulses can be studied within the theory of streams of events (point processes). Cox and LEWIS (1966) have analysed the data obtained by FATT and KATZ (1952) on time intervals between approximately 800 impulses along an axon. The logarithmic survival function of these intervals is quasi-linear, so that intervals have a marginal law close to an exponential. The series of events may thus be considered as the superposition of a number of sequences of almost regular impulses. In the case of a central neuron we clearly are concerned with a superposition of point processes (see Cox and MILLER (1965, p. 363)). By a judicious interpretation of electrophysiological phenomena it is possible also to consider the electrocardiogram as a superposition of two (or three) point processes.

In their previous papers, Cox and SMITH (1953, 1954) considered in detail the distribution of the interval between successive events in the pooled process. In their 1953 paper these authors considered sources at each of which events occur at strictly regular intervals with mutually irrational periods. In the 1954 paper, the intervals between successive events were assumed to be independent random variables, all with the same distribution so that each source constitutes a renewal process of a familiar type (see FELLER (1941)). Expressions were derived for the output interval distribution and for the variance-time function, defined as the variance of the number of events occurring in time t, considered as a function of time t. This is a measure for distinguishing the pooled output of a sequence of events from a random series.

Further investigations have been made by TEN HOOPEN and REUVER (1965, 1966). See also DIETZ (1968), POGGIO and VIERNSTEIN (1964), and SRINIVASAN and RAJAMANNAR (1970).

5.1.1.5. Among other classes of stochastic models used in the study of nerve activity we mention a random walk model (GERSTEIN and MANDELBROT (1964)), diffusion models (GLUSS (1967), JOHANNESMA (1968)) and random telegraph signal model (FIRESCU and TAUTU (1969)). Some electrophysiologists have shown that in cells in the striate cortex of some animals there is a strong tendency for discharges to occur in repetitives bursts, or clusters, and that the length and frequency of appearance of these bursts are functions of the nature and intensity of the stimulus and also of the wakeful state of the animal. It seems, therefore, that a study of possible mechanisms for grouped firing may be useful in analysing stimulus-response characteristics, at least in cortical units. Models for the clustered firing were considered by SMITH and SMITH (1965) and THOMAS (1966).

For a comprehensive survey see MOORE et al. (1966), SEGUNDO et al. (1966), and STEIN (1967).

5.1.2. Chemical mediation processes

5.1.2.1. The overwhelming majority of connections that are known to occur between neurons or between neurons and other excitable B-objects called "effectors" (e.g. gland cells, muscle fibers, electroplaques) are formed by synaptic junctions. The term "synapse" was initially used by C. S. SHERRINGTON to designate a state of anatomical contiguity between nerve cells. The terminal fibers (presynaptic) of a neuron have at their extremities small formations called (because of their shape) synaptic buttons. These buttons are attached to the surface of the body or on the fibers (postsynaptic) of another neuron, but between the two structures there is a space of approximately 200 Å (synaptic lacuna). The synaptic buttons contain a series of vesicles (having a size of 200—600 Å) which in their turn contain the chemical transmitter of the neural influx: acetylcholine (ACh). This is the "key substance" (WASER (1962)) responsible for these changes of membrane properties leading to depolarization. This neurotransmitter is formed in the mitochondria of the nerve axon and accumulates on carrier proteins located in small vesicles of the presynaptic axoplasm. The impulse travelling along the nerve fiber liberates ACh from its inactive ("bounded") form and puts it in direct contact with the receptors in the post-synaptic membrane, triggering the well-known events of depolarization. "Quanta" of 1000—10,000 ACh molecules are liberated from the vesicles. The neuro-transmitter is inactivated by an enzyme, acetylcholinesterase, while the synaptic membrane is repolarized.

It follows that to excite the synaptic membrane the transmitter must cross the synaptic space by diffusion. The synaptic membrane is probably restricted to sites of innervation since histochemical preparations of cholinoceptive synapses show that only the regions in the immediate vicinity of presynaptic terminals possess esterase activity.

5.1.2.2. In the absence of nerve propagated activity, quanta of ACh are spontaneously liberated at random intervals. This event is well observed where the two excitable B-objects in contiguity are a neuron and a muscle fiber. Action potentials thus developed are called miniature potentials, and these occur randomly in different fibers of the same muscle. The time intervals at which they appear follow the negative exponential distribution, and at least for moderate length of observation, the process is well approximated in all its characteristics by the Poisson process.

According to VERE-JONES (1966), the system must be understood as a "two stage" queueing process in which the first stage representing the passage of vesicles through the region of immediate access onto the neuron membrane has the logical structure of the queueing system $M/M/\infty$, and feeds into the second stage, the discharge of quanta of ACh through the neuron membrane itself, which has the stucture $M/G/\infty$. The result we are required to establish is that the output to the queueing system $M/G/\infty$ is also Poisson. This has been proved by MIRASOL (1963).

Here is the intuitive proof according to KENDALL (1964): If we sort the incoming ACh molecules according to the "service times" that they will subsequently require (i.e., for the forming of the ACh-receptor complex), we obtain a family of independent Poisson inputs. It is possible to consider the "server" as a protein molecule which acts in two ways: first as a receptor for ACh causing depolarization and ion flux through the membrane by structural deformation or change of electric charge, and secondly as an esterase splitting ACh. Because there is no limit on the number of "servers", these components of the input pass independently through the system, and for each component the service time is *fixed*, so that the output for that component is a shifted version of the input, and thus again Poisson. Finally we have to combine the component outputs, which are independent Poisson streams obtaining a gross output which is a Poisson stream, as required.

5.1.2.3. The model of the response to impulses applied at regular short intervals of time is, as we shall see, a positive-recurrent Markov chain (VERE-JONES (1966)). Let us consider the region of the nerve terminal at the instants immediately following an impulse, thus extracting a process in discrete time from the continuous process in which it is imbedded. The vesicles diffuse freely through the region of immediate access, and are unhindered by overcrowding effects.

Let n_j denote the number of vesicles in the region of immediate access at the time of the jth impulse, $j \in N$. Let u_j denote also the number released by impulse and i_j the number coming into the region between the jth and $(j+1)$st impulses. Then the progress of the system is governed by the basic equation

$$n_{j+1} = n_j + i_j - u_j = n_j + r_j, \tag{5.1}$$

where r_j is the number remaining after the jth impulse.

D. VERE-JONES considers that it is plausible to suppose, at least as a first approximation, that each of the vesicles in the region of immediate access has the same probability p of releasing its ACh at the time of the impulse, and that they act independently. In general, p will be much larger than the probability that in a period comparable to the duration of the impulse, a vesicle will give rise to a "miniature" potential.

For a given number $m \in N^*$ of vesicles in the region of immediate access, the distribution of the number of quanta emitted will then have a binomial distribution with parameters p and m. Even when the number m is itself a random variable, the binomial assumption enables a simple calculation to be made for the distribution of emitted quanta in terms of the distribution of m. Indeed, let us denote by $N_j(z)$ the probability generating function for n_j and, respectively, by $U_j(z)$ and $R_j(z)$ the probability generating functions of u_j and r_j. We have

$$U_j(z) = N_j(q + pz) \tag{5.2}$$

$$R_j(z) = N_j(p + qz), \tag{5.3}$$

where $q = 1 - p$. In particular,

$$\mathsf{E}u = p\mathsf{E}n, \qquad \mathsf{D}u = p^2\mathsf{D}n + pq\mathsf{E}n,$$

$$\mathsf{E}r = q\mathsf{E}n, \qquad \mathsf{D}r = q^2\mathsf{D}n + pq\mathsf{E}n,$$

$$\mathrm{Cov}\,[n,\ r] = q\mathsf{D}n,$$

$$\mathrm{Cov}\,[n,\ u] = p\mathsf{D}n,$$

where the index j has been omitted for brevity.

Let us now denote by $G(z)$ the probability generating function of the input i_j. We suppose that for each j, i_j has a fixed distribution independent of n_k and r_k, $k \leqslant j$. This assumes that the interior reserve is large, so as to be essentially unaffected by the drain due to excitation. For such an input, then, we shall have the recurrence relation

$$N_{j+1}(z) = G(z)\, N_j(p + qz), \tag{5.4}$$

according to the relations established before. Solving these equations recursively yields

$$N_j(z) = G(z)\, G[1 - q(1 - z)]\, G[1 - q^2(1 - z)] \ldots G[1 - q^{j-1}(1 - z)] \times$$
$$\times N_0[1 - q^j(1 - z)]. \tag{5.5}$$

As $j \to \infty$, the probability generating function $N_j(z)$ approaches the infinite product

$$\prod_{j=0}^{\infty} G[1 - q^j(1 - z)],$$

which is readily shown to be convergent and a solution of the equation

$$\varphi(z) = G(z)\, \varphi(p + qz),$$

if i_j has a finite mean. The limit (representing the generating function of the stationary distribution) is then independent of the initial distribution of n_0 and thus the model constitutes a positive-recurrent Markov chain.

From relations (5.4) or (5.5) one easily obtains

$$\mathsf{E}n_j = \frac{\mathsf{E}i_0}{p} + q^j \left(\mathsf{E}n_0 - \frac{\mathsf{E}i_0}{p}\right).$$

Then from (5.2) the expected values of the outputs will equal

$$p\,\mathsf{E}\,n_j = \mathsf{E}i_0 + q^j(p\mathsf{E}n_0 - \mathsf{E}i_0).$$

Thus, the output means tend geometrically to their equilibrium value $\mathsf{E}i_0$, the common ratio of the geometric series being equal to q. The calculation of variances leads to similar conclusions. The variance of n_j also converges geometrically to its equilibrium value, which equals $(\mathsf{D}i_0 + q\,\mathsf{E}i_0)\,/\,(p + pq)$. The variance of the output converges geometrically to its equilibrium value which equals $(p\mathsf{D}i_0 + 2q\mathsf{E}i_0)/(1 + q)$.

If we suppose that the input has a Poisson distribution, the equilibrium distribution of n_j is also Poisson with parameter $\mathsf{E}\,i_0 p^{-1}$. Further, substituting from (5.2), we see that the equilibrium distribution of the output is Poisson with the same parameter as the input. For studies of Poisson distributions in the case of nerve impulses see Lewis (1965) and McCulloch (1948).

5.1.2.4. D. Vere-Jones also considers the case when the synaptic vesicles are not freely diffusing throughout the outer region (the region of immediate access to the terminal membrane) and only a finite number of sites, n, is available. Equation (5.1) does not change its form if we make the obvious changes in interpretation. The distribution of i_j will depend on the number of free sites available, and hence on r_j and n_j. The simplest way of taking this dependence into account is to suppose that each free site has fixed probability ω of being occupied in the interval between two impulses, where this probability is independent of what occurs at the other sites, and of the past history of the process.

With this assumption, r_j and u_j are obtained from n_j by binomial sampling with parameters q, p respectively, and i_j is obtained from $n - r_j$ by binomial sampling with parameter ω.

Considerations of the same type as in Subparagraph 5.1.2.3 lead to the conclusion that the limiting distribution of n_j shall be, in this situation, binomial with parameters $\omega^* = \omega/\,(1 - q\,[1 - \omega])$ and n, and the limiting distribution of the output is binomial with parameters $p\omega^*$ and n.

5.1.2.5. Bartoszyński et al. (1962) have built a stochastic model of acetylcholinesterase transportation from the place of formation (microsomes) to the terminal fiber. The associated process is a multidimensional birth-and-death process.

5.1.2.6. We shall now consider another example of chemical mediation of the nerve impulse, i.e., the appearance of the visual stimulus. The nervous structures of the retina are represented by three cell layers followed by two synaptic layers. The first layer of the retina consists of pigmented cells which make and store photochemical substances; the next layer consists of two kinds of neurons called, according to their shape, rods and cones. Research work carried out by G. WALD has clarified the question of the photochemical basis of sight, respectively the mechanism by which a quantum of light is transformed into a visual response. Rods contain a red pigment, rhodopsin, which bleaches under light, into retinene and opsin (a protein). These two substances, when placed in the dark, will regenerate into rhodopsin.

WALD (1954) suggested that the rod is composed of a number of independent compartments, each containing a number of rhodopsin molecules. This compartment model corresponds, mathematically, to that of electronic counters. Since the rod is composed of many such compartments, each acting independently, this is equivalent to saying that the rod is a group of independent particular counters. The mathematical treatment of the process has been given by BENTLEY(1963a, b).

The counter model and Bentley's model can be compared by noting that time between counts is equivalent to time between visual responses being sent to the brain. The probability that the process carries on at $T + t$ given no quanta arriving after time T is just the probability that a visual response is recorded at $T + t$ if an adaptation light is used on a subject till time T and then the subject is placed in the dark for a period of length t. The reciprocal of this probability is an indication of visual threshold. Arrival rate is based upon factors such as intensity of light and number of compartments within the rod.

5.1.3. Some problems of stochastic networks

5.1.3.1. This paragraph will deal with some problems of general interest deriving from the study of the stochastic models of neural networks. This is natural as a neural network is a particular form of cell assemblies, with the following properties (see CANE (1967)):

I. At a given time, a cell can be active (firing) or inactive;

II. Cells are connected by paths; a pair of cells A, B may be unconnected, or connected by any number of paths; these paths are directional;

III. If there is a path $A \to B$, the firing of A at some time may contribute to the firing of B at a later time; A may be described as an input cell for B;

IV. A cell fires either as the result of an external stimulus or because of the firing of its input cells.

These four properties must be completed by a series of other details. Thus, the "inactive" state admitted by property I must be dichotomized in "used inactive" when the cell is in a refractory period and "sensitive inactive" when the cell can be reactivated. For property II, the number of paths, or a probability distribution for the number, must be given. The existence of a path $A \to B$ (property III) supposes that it is crossed in a certain time (synaptic delay), if this time is h, then it is usually assumed that the firing of A at time t can only affect the state of B at time $t + h$. (In discrete time models h is taken as the unit of time and the refractory period is taken either as less than h or equal to some multiple of it.) For property IV we need to know the minimum number θ of input cells which must fire together if a cell is to become active: θ ($\geqslant 1$) is the threshold of the receiving cell and may also be described by a probability distribution.

5.1.3.2. In the paper mentioned, VIOLET CANE discusses the two stochastic models in which neurons are randomly connected: SHIMBEL and RAPOPORT (1948) and ASHBY, VON FOERSTER and WALKER (1962). In the first case, the authors use a discrete time model without inhibitory effects and there is a given distribution of thresholds. In the second model the threshold is fixed, as is the number of inputs to each cell. Results are worked out in terms of the proportion of cells firing at a given time.

If we include in the argument the fact that the number of cells is n, $n \geqslant 2$, and fixed (which can mean either that we consider the whole net at successive times or successive pieces of the net, each of the same size), then if j cells fire at time t,

$$\mathrm{P}\,(\text{cell in set next available fires}) = \varphi_j = \begin{cases} \displaystyle\sum_{i=\theta}^{n} \binom{n}{i} \left(\frac{j}{n}\right)^i \left(1 - \frac{j}{n}\right)^{n-i}, & j \geqslant \theta, \\ 0, & j < \theta, \end{cases}$$

where θ is the fixed threshold number. Clearly, $\varphi_n = 1$.

The process is then a Markov chain with transition probabilities

$$p_{jk} = \binom{n}{k} \varphi_j^k (1 - \varphi_j)^{n-k}, \quad 0 \leqslant j, k \leqslant n, \tag{5.6}$$

with the convention $0^0 = 1$. The reader may easily discover by himself that if $n = \theta = 1$, the model is the same as Wright's model (4.1). For this case (and clearly also for the case of other values of n and θ), states 0 and n are absorbing. ASHBY et al. (1962) conclude that, whatever the input, the "brain" would ultimately be in a state of coma or epilepsy, and regard this as a paradox.

If one considers that following an active state there is a used state, only state 0 will be an absorbing one (see BEURLE (1956)). Let a, u, s, denote the number of active, used and sensitive neurons. The Markov process corresponding to the situation in which there are three types of neurons and transitions: active → used, used → sensitive, and sensitive → active, are easily built. The equations of moments, but especially their corresponding deterministic analogue, could represent the course of an endemic infection in which a was the proportion infected, s the proportion of susceptibles, $1 - a - s$ the proportion recovered and temporarily immune. We should still expect the infection to die out ultimately but can nevertheless investigate whether a temporary state of stable equilibrium is possible.

5.1.3.3. These considerations (CANE (1967)), point out the internal link between the models of neural networks and those known in genetics and epidemiology. The correspondence with genetic models goes even further; the case where the number of input paths is greater and threshold higher would correspond in genetics to multiple parentage with a dominance rule for the case in which offspring have had several parents (multiple parentage) and eventually a dominant allele. This case has, perhaps naturally (as V. CANE says), not been considered by geneticists. Spontaneous firing or spasmodic failure of a cell would correspond to the two sorts of mutations possible.

The model of BEURLE (1956) or that of GRIFFITH (1963) where connections are nonrandom resemble models for the spread of infection when the effect of geographical location is taken into account. In particular, Beurle's equations leading to plane waves are essentially the equation for a linear epidemic.

Finally, as regards absorbing state, it appears in general that, in a randomly connected network with only excitatory inputs, the effect of an external stimulus will ultimately die out if there is a long recovery time for cells, and will either die out or flare up if there is a recovery time less than the synaptic delay. If inhibitory inputs are included in the model, it would seem plausible that this would simply increase the probability that a process would die out.

5.1.3.4. The general problem which also includes the study of neural nets is that of populations of "synaptic" B-objects, that is of B-objects on which one observes a certain form of junctional transmission of a particular signal. This explains V. CANE's interest in the behaviour of coral colonies and clutches of eggs. If a single polyp is electrically stimulated it retracts along with the others in its neighbourhood (see HORRIDGE (1956)); eggs of game birds incubated in a clutch tend to hatch all at the same time; even eggs added to the clutch 24 hours late will hatch at about the same time as the rest of the clutch. The two main questions that must be answered by studying this type of populations are: (i) the modality of the junctional transmission, (ii) the behaviour of these populations as a whole. The only attempts to deal with the second question are concerned with learning. The model for a "brain", proposed by CANE (1967), is based on some of the facts known about vision.

5.1.4. A model of muscle contraction

5.1.4.1. A striated muscle fibril as seen under the electron microscope is formed of two sets of filaments, one being thicker (double) than the other. The muscle striations (or "bands"), arise from the characteristic arrangement of the thick and thin filaments. The dark *A*

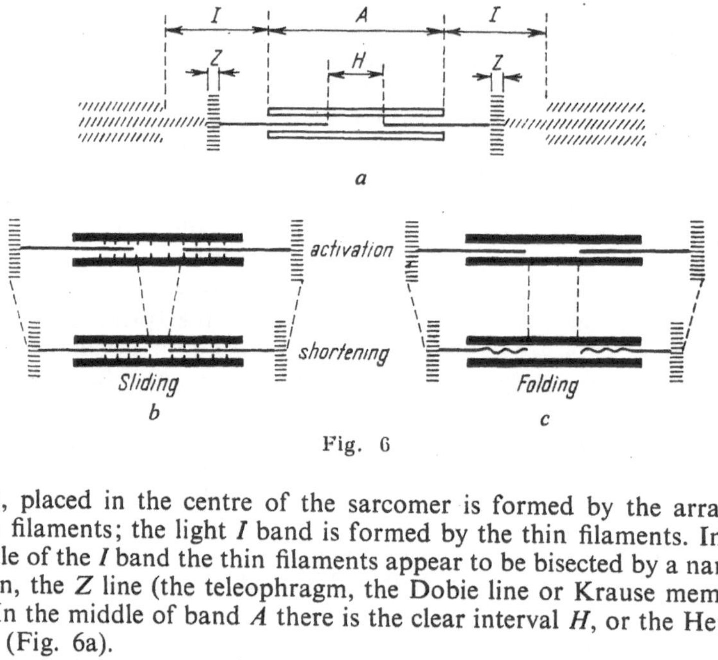

Fig. 6

band, placed in the centre of the sarcomer is formed by the array of thick filaments; the light *I* band is formed by the thin filaments. In the middle of the *I* band the thin filaments appear to be bisected by a narrow region, the *Z* line (the teleophragm, the Dobie line or Krause membrane). In the middle of band *A* there is the clear interval *H*, or the Hensen zone (Fig. 6a).

At normal muscle lengths, the thin filaments extend into the *A* band, interdigitating there with the thick filaments. It was shown in 1954 by A. F. HUXLEY and R. NIEDERGERKE, that when living muscle shortens, the decrease in length takes place almost entirely in the light *I* bands, while the width of the dark *A* bands remains constant. Two mechanisms of contraction were suggested: the filaments could *slide* along each other (Fig. 6b) or, after anchoring its ends, the thin filament could shorten by *folding* (Fig. 6c). Details on the two physiological models may be found, for instance, in PODOLSKY (1961).

One knows today that the thick filaments are formed from a fibrillar protein, myosin (consisting of two molecules of "light" meromyosin and one molecule of "heavy" meromyosin) while the thin filaments largely consist of another protein, actin. Each thick filament is surrounded by six thin filaments, each thin filament having three thick filaments around it. One may thus imagine that the reactive sites of the two proteins are linearly placed along the two filaments. The reversible complex actomyosin ⇌ myosin + actin represents the contractile protein in the muscle tissue; actomyosin (as well as myosin in fact) has an enzymatic function, being capable of splitting the terminal phosphate group in the molecule of adenosinetriphosphoric acid (ATP). Thus, it appears that the actomyosin → ATP reaction is the fundamental event of muscle contraction.

According to PODOLSKY (1961), the sites D, which are distributed along the myosin (thick) filaments are active centers, with enzymatic activity, and the complementary sites K which are distributed along the actin (thin) filaments can be identified with nucleotide binding (Fig. 7). One assumes that interaction may take place when these sites pass each other. If each interaction is associated with the using up of one substrate molecule M (that is, ATP or creatine phosphate), the interaction rate can be equated to the rate of chemical reaction driving the contractile mechanism.

For the purpose of building a stochastic model on the basis of these data from physiology and biochemistry (GREENHOUSE (1961)) we must make the following assertions:

(a) The myosin filament may be interpreted as a *fixed* line;

(b) Along this line there are special (enzymatic) points equally spaced, denoted as *positions*;

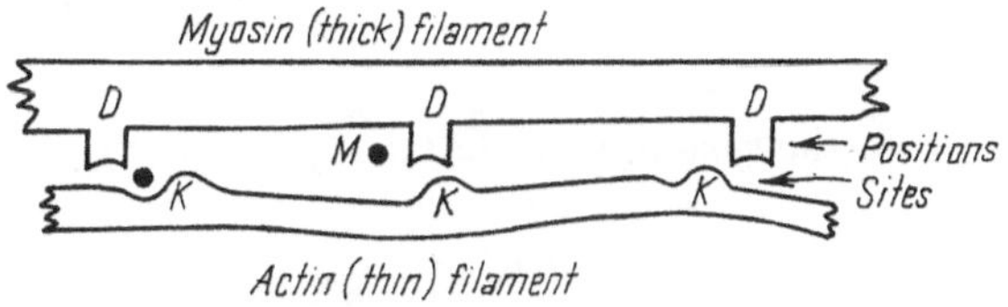

Fig. 7

(c) The actin filament may be interpreted as a *moving* line; it moves in a linear direction, parallel to the fixed line;

(d) On the moving line there exist equally spaced points, denoted as *sites*;

(e) A site may be in either of two states: *vacant* or *filled*;

(f) The changing of the state of a site can only take place following interactions with the positions;

(g) Consequently: a vacant site can become filled at certain (enzyme) positions, denoted as *load* positions; a filled site can become vacant at *release* positions;

(h) The movement starts from an arbitrary fixed starting position and, further, the release and load positions alternate so that the odd numbered positions are release positions, and the even numbered ones are load positions.

One may easily see that to build our model, some simplifications must be introduced. We shall consider as "event", the occurrence or nonoccurrence of an interaction between sites and positions and we shall make the following probabilistic assumptions:

1°. An event between a site and a position is independent of any previous events involving that site and position;

2°. Only one event can occur between a specific site and a specific position;

3°. Taking into account assertion (c), the moving line moves with uniform velocity past the fixed line;

4°. A filled site becomes vacant at a release position with a constant probability α which is the same for all release positions; similarly, a vacant site becomes filled at a load position with a constant probability β which is the same for all load positions.

According to these assumptions, the model of muscle contraction is a nonhomogeneous Markov chain with two states: 0 (vacant site) and 1 (filled site). The probabilities of passing from one state to the other depend on the characteristics of the positions: release (odd numbered, $n = 2r + 1$) and load (even numbered, $n = 2r$, $r \in N$), as it results from assertions (g) and (h). Namely, we have the following transition matrices:

$$^{2r}\mathbf{P} = \mathbf{A} = \begin{pmatrix} 1 & 0 \\ \alpha & 1-\alpha \end{pmatrix}, \quad ^{2r+1}\mathbf{P} = \mathbf{B} = \begin{pmatrix} 1-\beta & \beta \\ 0 & 1 \end{pmatrix}, \quad r \in N.$$

Let

$$p_j^{(n)} = \mathrm{P}\,(\text{the system is in state } j \text{ at position } n),\ j = 0,1,$$

$$\mathbf{p}^{(n)} = \begin{pmatrix} p_0^{(n)} \\ p_1^{(n)} \end{pmatrix}, \quad n \in N.$$

Then,

$$\mathbf{p}^{(n)} = \mathbf{B}'\,\mathbf{A}'\,\mathbf{p}^{(n-2)} = (\mathbf{B}'\,\mathbf{A}')^{\frac{n}{2}}\,\mathbf{p}^{(0)}, \quad n = 2,\ 4\ldots \tag{5.7}$$

$$\mathbf{p}^{(n)} = \mathbf{A}'\,\mathbf{B}'\,\mathbf{p}^{(n-2)} = (\mathbf{A}'\,\mathbf{B}')^{\frac{n-1}{2}}\,\mathbf{p}^{(1)}, n = 3,\ 5\ldots \tag{5.8}$$

In (5.7) $\mathbf{p}^{(0)}$ has different values according to the initial condition of the process : if the site starts filled, $p_0^{(0)} = 0$ and $p_1^{(0)} = 1$; if it starts vacant, $p_0^{(0)} = 1$ and $p_1^{(0)} = 0$. Similarly, for (5.8), the initial conditions will determine the structure of $\mathbf{p}^{(1)}$: if the site starts filled, $p_0^{(1)} = \alpha$ and $p_1^{(1)} = 1 - \alpha$; if the site starts vacant, $p_0^{(1)} = 1$ and $p_1^{(1)} = 0$.

From (5.7) we deduce

$$p_1^{(n)} = \beta \frac{1 - [(1-\alpha)(1-\beta)]^{\frac{n}{2}}}{1-(1-\alpha)(1-\beta)}, \quad n = 0,2,4, \dots$$

if the site starts in state 0, and

$$p_1^{(n)} = [(1-\alpha)\,(1-\beta)]^{\frac{n}{2}} + \beta \frac{1 - [(1-\alpha)(1-\beta)]^{\frac{n}{2}}}{1-(1-\alpha)(1-\beta)}, \quad n = 0, 2, 4,$$

if the site starts in state 1.

Similarly, one may deduce from (5.8)

$$p_1^{(n)} = \beta(1-\alpha) \frac{1 - [(1-\alpha)(1-\beta)]^{\frac{n-1}{2}}}{1-(1-\alpha)(1-\beta)}, \quad n = 1, 3, 5, \dots$$

if the site starts in state 0, and

$$p_1^{(n)} = (1-\alpha)\,[(1-\alpha)\,(1-\beta)]^{\frac{n-1}{2}} + \beta\,(1-\alpha) \frac{1 - [(1-\alpha)(1-\beta)]^{\frac{n-1}{2}}}{1-(1-\alpha)(1-\beta)}, \quad n = 1,3,5,\dots$$

if the site starts in state 0, and

$$p_1^{(n)} = (1-\alpha)[(1-\alpha)(1-\beta)]^{\frac{n-1}{2}} + \beta(1-\alpha) \frac{1 - [(1-\alpha)(1-\beta)]^{\frac{n-1}{2}}}{1-(1-\alpha)(1-\beta)}, \quad n = 1,3,5,\dots$$

if the site starts in state 1.

5.1.4.2. S. W. GREENHOUSE has also considered the case in which each site can be in any one of $m + 1$ states. When a site in state i, $0 \leqslant i \leqslant m$, on the moving line encounters a release (odd) position on the fixed line segment, any number $i - j$, $0 \leqslant j \leqslant i$, of interactions can occur resulting in the site changing from state i to state j. Similarly, when a site in state i encounters a load (even) position on the fixed line segment, any number $j - i$, $i \leqslant j \leqslant m$, of interactions can occur resulting in the site changing from state i to state j. It is assumed that the number of interactions occurring at the juxtaposition of a site and a position has a binomial distribution with parameter α at a release position and parameter β at a load position. Thus we shall write for the transition probabilities at an odd position

$$ {}^{2r}p_{ij} = \begin{cases} \dbinom{i}{i-j} \alpha^{i-j}(1-\alpha)^j, & j \leqslant i,\ 0 \leqslant i \leqslant m, 0 \leqslant j \leqslant i, \\ & r \in N, \\ 0, & j > i, \end{cases}$$

and for the transition probabilities at an even position

$$
{}^{2r+1}p_{ij} = \begin{cases} \binom{m-i}{j-i} \beta^{j-i} (1-\beta)^{m-j}, & j \geqslant i, 0 \leqslant i \leqslant m, i \leqslant j \leqslant m, \\ & r \in N, \\ 0, & j < i. \end{cases}
$$

For details, see GREENHOUSE (1961).

5.1.4.3. We shall conclude by mentioning that assumptions 3° and 4° stated at the beginning of this paragraph deviate from the real situation in the muscular contraction problem in that, during shortening, a site on the thin filament moves with a varying velocity. The probability of an interaction between a site and a position will clearly be greatly affected by the speed with which a site moves past a position: the slower the speed, the greater the probability of an interaction. One may suppose that an adequate model of this situation would be a two-state semi-Markov process (see as a suggestive example, COX and MILLER (1965, p. 360)).

5.1.5. Renewal processes in pharmacology

5.1.5.1. The data from neurophysiology presented in Subparagraph 5.1.2.1 will greatly facilitate the presentation of the problem of the action of drugs and pharmacoreceptors. Indeed, the fundamental idea of J. N. LANGLEY concerning the existence of a specialized "receptive substance", an idea presented as long ago as 1905, is based on the remarkable sensitivity of synaptic transmission to pharmacological agents. Langley's thesis was essentially (following SCHILD's (1962) comments) that adrenaline cannot act on nerve endings because it still produces effects after complete muscle denervation and it cannot act directly on the contractile element since some smooth muscles respond to adrenaline strongly, some weakly and some not at all. Therefore, it must be acting on a special receptive substance. Other facts confirm this thesis. After denervation, acetylcholine receptors appear all over the muscle-fiber surface, increasing their total number some 1000-fold. On the other hand, the concentration of cholinesterase at the end-plate does not increase, but on the contrary decreases.

The most widely accepted approach to drug action is that invoking receptors, more precisely "pharmacoreceptors", that is, receptors connected with a pharmacological response. Naturally, considering the above, more detailed models of pharmacoreceptors will feature the receptors of excitable B-objects. Synapses may be further classified pharmacologically according to the nature of the *pharmacon* (= drug) to which they respond. The synaptic pharmacons may be classified as "activators" which excite the synaptic membrane, mimicking the action of neural stimuli, or as "inactivators" which hinder or prevent excitation of the membrane by the neural stimuli or by activator drugs. The different action components of electrically excitable membrane may be affected differentially by pharmacological agents, also membranes of the same electrogenic type in different parts of the same cell may have different pharmacological properties (GRUNDFEST (1961)).

5.1.5.2. The model of pharmacologic response we shall now present is part of the renewal processes related to the counter models. Following TAKÁCS (1956), a counter can be considered as a device for transforming the sequence of arrival instants (of some physical particles) into a sequence of recording instants. The transformation obviously depends upon the type of counter considered and the nature of the sequence of pulse length. Suppose that an impulse arrives at time $t = 0$ and that the counter registers this impulse. This registration causes a dead time, and so the impulses arriving during this dead time will not be registered by the counter. In the simpler case, the impulses which arrive during a dead time do not cause any dead time, so that each dead time is caused only by a registered impulse. This is a Type I counter. In a Type II counter, each arriving impulse causes a dead time, so that arrivals during a dead time prolong that dead time (PRABHU (1965 a, p. 175)). According to PYKE's (1958) terminology, a Type I counter is one in which dead time is produced only after an event has been registered, while a Type II counter is one in which dead time is produced after each event (an emitted impulse) has occurred.

Let $(X_i)_{i\in N^*}$, $(Y_j)_{j\in N}$ be two independent sequences of independent random variables such that X_i and Y_j have distribution functions F and G, respectively. Set $X_0 = 0$ and $S_k = \sum_{i=0}^{k} X_i$ for $k \in N$. Assume throughout this discussion that $F(0) = G(0-) = 0$, F is a non-lattice distribution and all distributions are right continuous. Define for a Type I counter

$$n_0 = 0,$$
$$n_j = \min_k \{S_k > Y_{j-1} + S_{n_{j-1}}\}, \quad j \in N, \tag{5.9}$$

and for a Type II counter

$$n_0 = 0,$$
$$n_j = \min_k \{k > n_{j-1}, S_k > S_i + Y_i, n_{j-1} \leqslant i \leqslant k-1\}. \tag{5.10}$$

Clearly, $(n_j - n_{j-1})_{j \in N^*}$ is a sequence of identically and independently distributed random variables. Define

$$Z_j = S_{n_j} - S_{n_{j-1}}, \quad j \in N^*,$$

and denote by H the common distribution function of the Z_j.

To understand the connection between the above notations and the counter problem, let X_i be the time between the ith and $(i + 1)$ st impulses, and let Y_j represent the dead time caused respectively by the registration of an event at time S_{n_j} in the Type I model and the occurrence of an event at time S_j in the Type II model (time being measured from the registration of some event). The random variables Z_j denote the time between successive counts or registrations. Finally, define the lengths of time the counter is free during successive registrations, $W_j = Z_j - Y_{j-1}, j \in N^*$, and denote by K their common distribution function.

5.1.5.3. Let us now interpret the above within the framework of the pharmacological process considered. Clearly, what has been called an impulse represents for us the contact between a pharmacon molecule and its cell receptor. Drugs act on pharmacoreceptors only at the moment of impact. Also, Y_j as dead time represents for us the time of forming the pharmacon-receptor complex; the appearance of this complex leads to the liberation of one quantum of excitation (e.g., a pharmacological response). If the drug remains on the receptor site, no further excitation occurs. A stimulant drug is now one which dissociates rapidly from the receptors, a strong antagonist is one which dissociates very slowly but nevertheless gives high receptor occupancies (PATON (1962)). The sequence $(Z_j)_{j \in N^*}$ shall be considered in this situation as the process of pharmacological response.

FIRESCU, SAVU and TAUTU (1970) started from the assumption that the pharmacoreceptor may be identified with a Type II counter, with some special characteristics. Thus, it was considered that if F is an exponential distribution function, one can assume that G is a hyperexponential distribution function. By means of this assumption one tried to point out the variable duration of the pharmacon-receptor complex in relation to the number and/or the kind of drug molecules. It was also supposed that the pharmacoreceptor also has a regeneration period (technically called *excess dead time*). Considering these precautions, it is difficult to criticize the fact that the Type I counter has not been chosen as a pharmacoreceptor model in which the counter and relations between the entry and the registering of impulses would be similar to those of dose and effect in pharmacology.

The main result of the paper mentioned was the derivation of expressions for the Laplace-Stieltjes transforms of H and K in terms of the parameters of F and G.

5.1.6. Multicompartment systems

5.1.6.1. The problem of compartments used in a rough sense has been discussed in Subparagraph 1.2.5.6. Examples were also given in Subparagraphs 3.2.2.5 and 3.4.2.1.

We shall now approach this problem from a more restricted point of view, the general kinetic model being reduced to the study of the distribution of a radioactive substance in a closed system of compartments under certain conditions. Thus, according to the definition given by SCHOENFELD (1963), compartmental systems are mathematical models used to describe measurements made with radioactive isotopes of physiological substances. The general model assumed consists of a finite number of compartments with turnover rates proportional to the amounts of material in the compartments, and with the behaviour of the isotope reflecting that of the unlabelled substance. Thus one finds that the distribution of the isotope is by itself a model of the distribution of the corresponding physiological substance (see e. g. BERMAN (1963) on the formulation and testing of models). We shall consider, therefore, the four assumptions presented by CORNFIELD, STEINFELD and GREENHOUSE (1960):

I. The fraction of substance in compartment i going to compartment j in unit time is the same for labelled and unlabelled substances. (This is the fundamental assumption of isotope work and is sufficient to determine the equations describing the movement of labelled substances.)

II. The fractional amount of unlabelled substance going from compartment i to compartment j in unit time is independent of time.

III. The volume of the vascular compartment is a constant independent of time.

IV. With respect to unlabelled substances, the body is in steady state and is producing a time-independent amount of such substance in each compartment.

In general, the structure of the multi-compartmented system is fixed. In his stochastic theory of configurations, KOŽEŠNÍK (1965) implies that one may take into account all possible connections between compartments. In a detailed paper, BERNARD, SHENTON and UPPULURI (1967) have studied a compartmental system in which the connections between the compartments are random variables.

5.1.6.2. The model built by WIGGINS (1960) represents one of the models starting from radioisotope experiments: radioactive particles migrate from one compartment into the other and the problem is to calculate the number of these particles in compartment i after a long administration. The model is a migration model "with chronic feeding". The concrete situation is the incorporation of some highly noxious radioisotopes such as, for example, strontium 90. The contamination of the environment with Sr^{90} is the factor of continuous feeding of the biological system.

Suppose that a biological system consists of r compartments, $r \in N^*$, and let a single particle be introduced into compartment 1 at time $t = 0$. We assume that the behaviour of the particle within the system is random and can be described in terms of a vector-valued Markov process $\boldsymbol{\eta}(t) = (\eta_1(t), \ldots, \eta_r(t))_{t \geqslant 0}$. Here $\eta_i(t)$, $1 \leqslant i \leqslant r$, equals 1 or 0, according as the particle is or is not in compartment i at time t. If $\eta_i(t) = 1$, then necessarily all other $\eta_j(t)$, $j \neq i$, have the value 0. We wish first to determine the probabilities

$$p_i(t) = \mathsf{P}\,(\eta_i(t) = 1 \mid \eta_1(0) = 1), \quad 1 \leqslant i \leqslant r,$$

which denote the transition probability of the system from compartment 1 to compartment i during time t.

Let λ_{ij}, $1 \leqslant i, j \leqslant r$, $i \neq j$, be a system of nonnegative constants and assume that in the time interval $(t, t + \Delta t)$ the probability of transition from compartment i to compartment j is given by

$$\mathsf{P}\,(\eta_j\,(t + \Delta t) = 1 \,|\, \eta_i\,(t) = 1) = \lambda_{ij}\,\Delta t + o\,(\Delta t).$$

Thus we obtain the system of differential equations

$$p'\,(t) = \sum_{i=1}^{j-1} \lambda_{ij}\, p_i\,(t) - \lambda_j . p_j\,(t) + \sum_{i=j+1}^{r} \lambda_{ij}\, p_i\,(t), \; 1 \leqslant i, j \leqslant r, \qquad (5.11)$$

where

$$\lambda_j . = \sum_{\substack{k=1 \\ k \neq j}}^{r} \lambda_{jk} .$$

We assume the system to be closed (that is without excretion compartment), and hence we must have

$$\sum_{i=1}^{r} p_i\,(t) = 1. \qquad (5.12)$$

it is well known that the solutions of the system (5.11) are expressed in terms of linear combinations of exponential functions. However, in most cases of interest one is less interested in obtaining explicit solutions for the probabilities $p_j\,(t)$ than in obtaining information in case a large number of particles are introduced in the system.

Let $f(t)$ be a monotone nondecreasing piecewise continuous function over the nonnegative time axis. The function f will have the interpretation of a feeding function which measures the total quantity of particles introduced into the system up to time t. The probability generating function of $(\eta_i(t_i))_{1 \leqslant i \leqslant r}$ is

$$G_1(z_1, \ldots, z_r; t) = \sum_{i=1}^{r-1} p_i(t) z_i + \left[1 - \sum_{i=1}^{r-1} p_i(t)\right] z_r.$$

Suppose now that not one but n particles are introduced into compartment 1 at $t = 0$, and these particles behave independently within the system. Then the corresponding probability generating function will be

$$G_n(z_1, \ldots, z_r; t) = \left[\sum_{i=1}^{r-1} p_i(t) z_i + \left(1 - \sum_{i=1}^{r-1} p_i(t)\right) z_r\right]^n.$$

Let the time interval $[0, t)$ now be divided into m subintervals each of length $\Delta t = tm^{-1}$ and suppose that at times $u\Delta t$, $0 \leqslant u \leqslant m - 1$, there are $\Delta_u f$ particles introduced into compartment 1, where $\Delta_u f = f[(u+1)\Delta t] - f[u\Delta t]$. In this case the resulting probability generating function is

$$\prod_{u=0}^{m-1} G_{\Delta_u f}(z_1, \ldots, z_r; u\Delta t) =$$

$$= \prod_{u=0}^{m-1} \left\{ \sum_{i=0}^{r-1} p_i(t - u\Delta t) z_i + \left[1 - \sum_{i=1}^{r-1} p_i(t - u\Delta t)\right] z_r \right\}^{\Delta_u f}. \tag{5.13}$$

Taking logarithms of both members of (5.13), we have

$$\sum_{i=0}^{m-1} \ln G_{\Delta_u f}(z_1, \ldots z_r; u\,\Delta t) =$$

$$= \sum_{u=0}^{m-1} \Delta_u f \ln \left\{ \sum_{i=1}^{r-1} p_i(t - u\Delta t) z_i + \left[1 - \sum_{i=1}^{r-1} p_i(t - u\Delta t)\right] z_r \right\}. \tag{5.14}$$

The right-hand member of (5.14) is seen to be a Riemann-Stieltjes sum approximating the integral

$$\int_0^t \ln \left\{ \sum_{i=1}^{r-1} p_i(t - \tau) z_i + \left[1 - \sum_{i=1}^{r-1} p_i(t - \tau)\right] z_r \right\} df(\tau). \tag{5.15}$$

Expression (5.15) can be interpreted to be the logarithm of the probability generating function corresponding to the situation in which large numbers of particles are introduced into the biological system over an extended period of time, the so-called chronic feeding case.

Now consider the random vector

$$\xi(t) = (\xi_1(t), \ldots, \xi_r(t)),$$

where $\xi_i(t)$ denotes the number of particles in compartment i at time t. It is easy to see that (5.15) is the logarithm of the probability generating function G of $\xi(t)$. Hence

$$\mathsf{E}\,\xi_i(t) = \left.\frac{\partial G}{\partial z_i}\right|_{\substack{z_j=1\\ 1\leqslant j\leqslant r}} = \int_0^t p_i(t-\tau)\,\mathrm{d}f(\tau),\; 1 \leqslant i \leqslant r. \tag{5.16}$$

Equation (5.12) together with equation (5.16) imply

$$f(t) = \sum_{i=1}^{r} \mathsf{E}\,\xi_i(t).$$

5.1.6.3. WIGGINS (1961) extended the previously presented model by considering an r-dimensional diffusion process, assigning to each compartment an independent feeding source. We shall stress the fact that one open multicompartment system with a single feeding source may be extremely useful for research work in pharmacodynamics.

We shall conclude by quoting the stochastic models proposed by BERMAN (1961), FIRESCU and TAUTU (1966), MATIS and HARTLEY (1971) and SHEPPARD (1962).

5.2. Models in pathology

5.2.1. The process of infection

5.2.1.1. Pathological processes represent a less widely investigated chapter of stochastic models in medicine. This fact is evident since, on the one hand, the modelling of physiological processes has not been systematically carried out, and, on the other, because experimental models of diseases have not yet been thoroughly perfected. The complexity of the elements contributing to the outbreak of the pathological process far exceeds the current capabilities of the mathematician, so that one can hardly build and analyse what has been called by NEYMAN, PARK and SCOTT (1956) "models in the large".

The fundamental concept is that of *response*: the response of a B-object to the pathogenic action of alterexter (or even alteridem, under certain conditions) objects which are either B-objects, or physical or chemical objects (particles, molecules)[11]. We shall denote these objects by the general term D-objects, "damnificer objects" (lat. *damnificus*) and will distinguish biological damnificers as DB-objects. However if it is clear from the nature of the process that we are only dealing with biological damnificers, we shall simply write these as D-objects. Resuming now the problem of response, we add that what was affirmed by SELYE (1969) for the heart is also valid for other physiological systems: it appears that a B-system is capable of responding only with a limited number of reactions to the innumerable agents and combination of agents that can act upon it. There must be a limited number of final common pathways of response.

We shall often refer in pathology to Selye's concept of "pluricausal disease". It is defined as a derangement produced by combined treatment with two or more agents, none of which is effective in this respect by itself. Most of the pluricausal diseases examined to date have been elicited by combined treatment with two agents, one of which (the "conditioner" or "sensitizer") induces a particular type of disease proneness while the other (the "challenger") makes this predisposition manifest and determines the location of the resulting changes.

The most frequent model of "unicausal" disease is the model of infection, that is, of the disease produced by the multiplying of some pathogenic microorganisms (in short, DB-objects) within their hosts. As early as 1935, H. O. HALVORSON distinguished two types of models, the deterministic and the stochastic one (see also ARMITAGE, MEYNELL and WILLIAMS (1965)). The first is derived from the pharmacological work of J. W. TREVAN and postulates an individual effective dose (*IED*) for each host such that it will certainly respond to a dose equalling its *IED*. Inoculated DB-objects can be thought of as co-operating to "saturate" the host defences.

The stochastic model has been influenced by the same pharmacological concept, using the words of *dose* and *threshold*. Inoculated DB-objects are assumed not to co-operate but to act independently — hence the designation "hypothesis of independent action" (MEYNELL and STOCKER (1957)) — and each DB-object is assumed to have an independent probability of multiplying sufficiently to produce a response. The second important assumption is that the population of DB-objects is governed by a birth-and-death process (MEYNELL and MEYNELL (1958), SAATY (1961), YAMAMOTO (1961)). Clearly, the simplest version of this is that for which the process is linear, the parameters b and d being constant from host to host and from DB-population to DB-population

11) Models in pathology involving physical D-objects are known, having been discovered by radiobiology research; BHARUCHA-REID (1952, 1960, p. 223), BHARUCHA-REID and LANDAU (1951). See also HUG and KELLERER (1966).

(B-*homogeneity*). We could agree that DB-objects are characterized by the fact that $b > d$[12]). If there are n_0 DB-objects at time $t = 0$, and one defines response as the first attainment of a given threshold number $\theta > n_0$, and the response time as the value of t when and if this occurs, then the probability of host response will be

$$\begin{cases} \dfrac{1-(d/b)^{n_0}}{1-(d/b)^{\theta}}, & b \neq d, \\[2ex] \dfrac{n_0}{\theta}, & b = d, \end{cases} \tag{5.17}$$

the complementary probability corresponding to absorption at state zero. In the limit ($\theta \to \infty$) we have the well-known probability that the DB-population is not ultimately wiped out

$$\begin{cases} 1-(d/b)^{n_0}, & b > d, \\ 0, & b \leqslant d. \end{cases}$$

The results of WILLIAMS (1965 a, b) regarding the distribution of the number of viable DB-objects, the distribution of incubation periods and the distribution of response times, are important for the mathematical definition of the process of infection.

5.2.1.2. In practice, as WILLIAMS (1965 a) says, the most that can be known about the inoculum is that it is Poisson distributed with mean n_0. Then, the probability that the population is not ultimately wiped out is

$$p_{n_0} = \begin{cases} 1 - \exp(-(1-d/b)\, n_0), & b > d, \\ 0, & b \leqslant d, \end{cases} \tag{5.18}$$

upon taking the appropriate weighted average. The expected population size at time t is

$$\mathsf{E}\,\xi(t) = n_0 \exp((b-d)t). \tag{5.19}$$

---

12) WIGGINS (1957) has built a model where the assumption of pathogeneity has another interpretation. He assumes that the body of the host is divided into three regions R_1, R_2 and R_3. If a DB-object enters R_3, it is rendered noninfectious; if it enters R_1, the DB-object remains therein for a time so short that practically no division occurs ($b = 0$); if it enters R_2, infection occurs (R_2 is a sensitive region which acts as a "trap" in the sense that, once a DB-object reaches R_2 from R_1 it cannot return). It is assumed that when the population in R_2 reaches a certain fixed number θ for this first time, the host dies. The number θ may be called *the lethal threshold. A* DB-object is therefore pathogenic only for a certain region (or state) of the host's body.

One observes that the probability of non-response is an exponential function of the size of the dose: a behaviour typical of infectious but not of physical or chemical D-objects, where the corresponding curve is much steeper at the ED_{50} (50 per cent effective dose). The exponential response is a feature common to all models which obey the independent action hypothesis, viz., that each of the DB-objects has its own probability p_1 of provoking the host response and that they do so without co-operation, the over-all response probability therefore being $p_{n_0} = 1 - e^{-p_1 n_0}$. The above results suggest a suitable scaling of the variables; for, defining

$$\tau \equiv (b - d)\, t,$$

$$\delta_\tau \equiv (1 - d/b)\, \xi(t),$$

we obtain from (5.18) and (5.19)

$$p_{n_0} = 1 - e^{-\delta_0},$$

$$\mathsf{E}\, \delta_\tau = \delta_0 e^{\tau},$$

where δ_0 may be called *the dose in natural units*. Introducing the "unit dose" u through the relation

$$\frac{1}{u} \equiv 1 - \frac{d}{b},$$

we see that $u = 1/p_1$, and so $\delta_\tau = \xi(t) p_1$, which shows that δ_0 is also what MEYNELL and MEYNELL (1958) call *the effective dose*. Seting $p_{n_0} = 1/2$, according to the relations above,

$$ED_{50} = u \log 2.$$

We also want to identify the parameter b; clearly, this is inversely proportional to the mean generation time r per parasite: indeed, comparing deterministic splitting with the expectation for a linear pure birth process, we find $2^{t/r} = e^{bt}$, and so

$$b = \frac{\log 2}{r}.$$

On the basis of the above,

$$\tau = \frac{t\,(\log 2)^2}{r\, ED_{50}}.$$

This completes the identification of the various parameters.

5.2.1.3. The assumption of a fixed threshold has been abandoned in a new class of stochastic models considered by PURI (1967 a). Two reasons brought about this renunciation: a fixed lethal threshold hypothesis is not likely to be strictly correct, and it is mathematically intractable because of the involvement of the first passage problem. P. S. PURI adopts an alternative hypothesis, suggested by L. M. LECAM, namely, that the connection between the number of DB-objects in a host and the host's response is indeterministic in character. The number of these DB-objects determines only the probability of response of the host. Given that at time t the host is alive, the occurrence of its death in time interval $(t, t + \Delta t)$ is treated as a random event, the probability of which can be written as $[a + f(\cdot \mid t)] \Delta t + o(\Delta t)$. The expression within brackets must be interpreted as a risk function: $a \geqslant 0$, and f is a nonnegative and nondecreasing function of the number of DB-objects present in the body of the host at time t and, possibly, of certain other relevant characteristics of the process of infection. The constant a may depend upon the initial average dose injected and/or may possibly be taken as a function of time. One may consider that a represents the constant risk of death due to other causes and may be taken as zero in cases where the response is other than death and/or where no other causes are operating.

An example of a stochastic model belonging to the considered class is the Markov process $(\xi(t), \eta(t), \zeta(t))_{t \geqslant 0}$ where $\xi(t)$ is the number of DB-objects at time t,

$$\eta(t) = \int_0^t \xi(\tau) d\tau,$$

and $\zeta(t)$ is a binary random variable, $\zeta(t) = 1$ if there is no response by time t and 0 otherwise. The function $\eta(t)$ is contemplated as a measure of the amount of toxin produced by the live DB-objects during the interval $(0, t)$, assuming that the rate of toxin excretion is constant per DB-objects per unit time. Taking the above into account, one may write

$$\mathsf{P}\,(\text{host responds during } (t, t + \Delta t) \mid \xi(t) = x, \eta(t) = y, \zeta(t) = 1) =$$

$$= [a + f(x, y \mid t)] \Delta t + o(\Delta t), a \geqslant 0, x \in N.$$

For the sake of simplicity, P. S. PURI assumes that

$$f(x, y \mid t) = bx + cy, \quad b \geqslant 0, \quad c \geqslant 0.$$

Details and developments are to be found in PURI (1967 a).

5.2.1.4. Special attention should be given to the model of infection studied by GANI (1963). The particular character of the model is due to the fact that the infectious process considered in it is the infection of bacteria with bacteriophages and that the process may be repeated in several generations of bacteria. The paper gives expressions for probability generating functions of surviving bacteria conditional on numbers of infections, of probability generating functions for surviving bacteria (summing up all such possible infections), and of the probabilities of extinction (see also GANI (1962 b)). One must also recall GANI's (1967 b) paper on models for antibody attachment to virus and bacteriophage, where one finds applications of geometric probabilities.

5.2.1.5. It is generally considered that all the hosts are uniform in their susceptibility to response causing mechanism and also that DB-odjects are uniform in their action. This general hypothesis has been replaced by MODE (1961) with the following assumptions: (i) the host may be differentiated into *varieties* on the basis of its resistance to the races of DB-objects; (ii) the DB-objects may be differentiated into *races* on the basis of their ability to grow on a set of host varieties; (iii) host resistance to a particular race of the pathogen is genetically controlled. By defining the probability that at time t a member of the host population belonging to variety i and a DB-object belonging to race j are associated, C. J. MODE studied the following four situations:

1) Population number in both the host and pathogen population is constant;
2) Population number in the host population is constant;
3) Population number in the pathogen population is constant;
4) Population number in both the host and pathogen population variable.

In another paper, outlining the genetic terms, MODE (1964) assumes the following situation: the host has two loci for reaction control versus a pathogen population, each locus having two alleles (host resistance or host susceptibility) and a DB-object also has two loci, each with two alleles (virulence and avirulence), in such a way that host resistance occurs only when a gene for resistance and its complementary gene in the pathogen for avirulence are present simultaneously. One supposes that mutations from one race of DB-objects to another are possible. The proposed model is particularly complex and considers a host-pathogen system consisting of m genotypes of the host differing in their resistance to n genotypes of the pathogen. The host reproduces by a mixture of mating systems, namely, selfing and random mating, its generations being discrete; the (asexual) evolution of the pathogen population can be described by a multidimensional birth-and-death process.

Disease as it occurs in nature is indeed the result of an intermixture of many divergent host and pathogen genotypes and their corresponding phenotypes in diverse and varying environments (see GOWEN (1961)). The stratifications of resistance observed for the different strains of

DB-objects may be interpreted as due to a number of genes affecting different morphological, physicochemical and immunological processes within the different hosts in relation to their response to the DB-objects genotypes. The inheritance differences suggest that resistance is probably due to several different characteristics and that not one attribute is all-important. For the distribution of host resistance into many different phenotypic classes, each of which having particular attributes, see e.g. PEARSON (1912) on tuberculosis, heredity and environment.

5.2.2. The clinical process

5.2.2.1. Under the name *clinical process* we shall designate the stochastic process describing the passage of an individual through the various stages of a disease. Disease is a dynamic pathological process, reversible in its early stages. In fact, we have described a model of this type under Subparagraph 3.2.2.6 within the framework of the emigration-immigration process. The construction of a clinical process must start from two components: the stages of the disease and the starting point. ALLING (1958), for example, has considered a Markov model of the evolution of pulmonary tuberculosis, with the following six stages: 1) active tuberculosis that will remain active indefinitely; 2) arrested tuberculosis that will remain arrested indefinitely; 3) tuberculosis that has proved to be fatal; 4) arrested tuberculosis that is certain sometime to become active; 5) active tuberculosis that is certain sometime to become arrested; 6) active tuberculosis that is certain sometime to be fatal without ever becoming arrested. However, the author himself has stated that the clinical status "active tuberculosis" is a markedly heterogeneous classification since it includes both the patient who is nearly dead of tuberculosis and the patient who is nearly well. BERKSON and GAGE (1952) have built a clinical model with four stages, two non-essential (living with uncured cancer, living with cured cancer) and two absorbant ones (dead from cancer, dead from ordinary causes). Since there is no communication between the non-essential states, the process is essentially the sum of two subprocesses, one consisting of a nonessential and an absorbing state, and another of a nonessential state and two absorbing states.

The second specification must be made in connection with the starting point as related to medical action: diagnosis, completion of treatment, or relapse.

5.2.2.2. PESKY and TAUTU (1971) have built a model of the evolution of pulmonary tuberculosis by considering the following nine stages, the majority of which may be defined exactly in terms of roentgenological and bacteriological data: 1) primordial infection; 2) infiltrative tuberculosis; 3) fibro-nodular tuberculosis; 4) fibro-cavitary tuberculosis with ulceration and negative bacilloscopy; 5) fibro-cavitary tuberculosis with ulceration and positive bacilloscopy; 6) fibrosis; 7) recovery; 8) death by tuberculosis; 9) death by other diseases. The model used is an absorbing homogeneous Markov chain. Stages 1 through 6 form the set F of nonessential states while the two states 7 and 8 are the absorbing states specific to the process. The starting point has been the time of diagnosis: 671 patients have been followed for a period of three years, at periodical controls at six months intervals. These patients have been given the classical form of treatment.

Some interesting results have been obtained by applying the theory of finite state Markov chains (see I; 1.1.4). For example, according to the data obtained, for the group of males between 15 to 49 years (249 patients) the fact has been stressed that in tuberculosis there are active states (sensitive to the antibacillary specific treatment) and chronic states with certain anatomo-clinical features (where we obtain high values of p_{ii}). State 3, fibro-nodular tuberculosis, has a special position because it is reached from all the other nonessential states. We have for the group of patients mentioned above the following transition matrix $\mathbf{P}$:

$$\mathbf{P} = \begin{array}{c} \\ 7 \\ 8 \\ 9 \\ 1 \\ 2 \\ 3 \\ 4 \\ 5 \\ 6 \end{array} \begin{array}{c} \begin{array}{ccccccccc} 7 & 8 & 9 & 1 & 2 & 3 & 4 & 5 & 6 \end{array} \\ \begin{pmatrix} 1.000 & 0.000 & 0.000 & 0.000 & 0.000 & 0.000 & 0.000 & 0.000 & 0.000 \\ 0.000 & 1.000 & 0.000 & 0.000 & 0.000 & 0.000 & 0.000 & 0.000 & 0.000 \\ 0.000 & 0.000 & 1.000 & 0.000 & 0.000 & 0.000 & 0.000 & 0.000 & 0.000 \\ 0.033 & 0.000 & 0.010 & 0.895 & 0.000 & 0.031 & 0.000 & 0.000 & 0.031 \\ 0.000 & 0.000 & 0.000 & 0.000 & 0.447 & 0.430 & 0.010 & 0.083 & 0.030 \\ 0.008 & 0.030 & 0.002 & 0.000 & 0.001 & 0.936 & 0.004 & 0.006 & 0.013 \\ 0.000 & 0.047 & 0.000 & 0.000 & 0.000 & 0.328 & 0.401 & 0.113 & 0.111 \\ 0.000 & 0.044 & 0.000 & 0.000 & 0.000 & 0.352 & 0.084 & 0.498 & 0.022 \\ 0.000 & 0.014 & 0.000 & 0.000 & 0.000 & 0.396 & 0.000 & 0.000 & 0.590 \end{pmatrix} \end{array}$$

In Table 4 we have written down the probabilities of entering an absorbing state, starting from one of the six nonessential states, according to data obtained on the same group of patients[13].

Table 4

Initial state	Probability to reach the absorbing state		
	Recovery	Death by tuberculosis	Death by other diseases
Primordial infection	0.19925	0.18155	0.61920
Infiltrative tuberculosis	0.16376	0.31395	0.52228
Fibro-nodular	0.16510	0.30088	0.53402
Fibro-cavitary (BK −)	0.15332	0.37209	0.47459
Fibro-cavitary (BK +)	0.14962	0.37465	0.47572
Fibrosis	0.18721	0.31403	0.49876

[13] The small probabilities of recovery could be explained by the short observation interval (3 years).

5.2.2.3. An interesting paper of Lu (1966) offers the possibility of new applications. This paper deals with the formation process of dental caries considered as a finite absorbing Markov chain. The state space consists of the state of the tooth surfaces as follows: The physical pattern of caries can be represented by a five-digit binary number, because a tooth has five surfaces (occlusal, mesial, distal, lingual and buccal). Thus (00000) denotes a healthy tooth, (10000) denotes a tooth with a carious occlusal surface, and so on. We shall have 32 possible states, state (11111) being the absorbing state and representing the total caries of the tooth.

The process may be interpreted as a random motion on a 5-dimensional lattice. This random motion has nevertheless some restrictions since the process never goes through a state it has already gone through and transition probabilities take equal values for all accessible points (states).

If state i denotes the caries condition of a tooth at some time, the subsequent development of the caries process is assumed to be solely dependent on state i, and not dependent on states the caries process occupied prior to state i. Thus, suppose we observe state (01100), a state with mesial and distal caries. There are three possible routes by which the caries process may arrive to this state:

$$(00000) \to (01100),$$

$$(00000) \to (01000) \to (01100),$$

$$(00000) \to (00100) \to (01100).$$

Provided we can consider that all conditions pertaining to mesial-distal caries are adequately represented by (01100), then the conditions of the resultant state (01100) via the three routes are "similar" under our present definition; the route through which the state (01100) is arrived at does not matter.

By applying the theory of finite Markov chains (see I; 1.1.4), K. H. Lu has calculated the parameters of the considered process according to his data on the maxillary second bicuspid (1081 cases) observed for three years by a single examiner at six months intervals. The transition probabilities were estimated from the method suggested by Zahl (1955). The three most probable pathways from (00000) to (11111) are as follows:

$$(00000) \to (00100) \to (01100) \to (11100) \to (11111), \text{ with probability } 0.341,$$

$$(00000) \to (01000) \to (01100) \to (11100) \to (11111), \text{ with probability } 0.206,$$

$$(00000) \to (10000) \to (10100) \to (11100) \to (11111), \text{ with probability } 0.100.$$

The probability of tooth loss via these three pathways is therefore 0.647. It is also noteworthy that the path to absorption via the two-surface decay (01100) alone is 54.7 per cent. Whenever a bicuspid has any two of the three surfaces (mesial, occlusal and distal) carious, the third one inevitably becomes carious and thus leads to the absorbing state.

5.2.2.4. The model used by K. H. Lu may also be adopted to study other clinical processes by means of an adequate description of clinical states as k-digit binary numbers ($k \in N^*$), or of groups of such numbers. The simplest attempt is to study the clinical evolution of cancer as compared to the state of four elements: the primary tumour (T), regional adenopathy (A), metastases (M) and supplementary criteria of gravity (G): e.g. hormonal ones. Each of these elements may have several stages of expansion, these being expressed through k-digit binary numbers. Let us suppose $k = 4$. The noncancerous state shall be expressed by the group ((0000), (0000), (0000), (0000)). In breast cancer, the group ((1000), (0000), (0000), (0000)) will represent the clinical state when the tumour is single and small (not larger than 2 cm diameter) without adenopathy, without metastases, without pregnancy or breast feeding. Clearly, a knowledge of the probabilities of certain pathways has the character of a prognosis.

In the same way, one can imagine special features of heart disease or collagenoses, for which fixed arrays of binary vectors may also have diagnostic significance.

5.2.2.5. There are several diseases in which possibly recurrent phases may be distinguished. Most papers studying the clinical process consider Markovian models, thus a negative exponential distribution of time spent in a particular phase. Weiss and Zelen (1963, 1965) have noticed that the negative exponential distribution is not sufficient for the study of the statistic relating to leukemia. They have built a semi-Markov model for the clinical process.

The health of a patient can be characterized at any instant of time as being in one of a finite number of states. In clinical terminology, the patient may be in a relapse state, a remissive state, a toxic state, etc. These remissive and relapse states may be further classified according to the degree of remission or relapse and also according to how many and what kinds of relapse or remissions preceded the present state. In addition to these transient states (the set of which will be denoted by F), the patient may have entered a terminal (absorbing) state, such as failure (death), cure, or patient lost. Let us denote by A the set of the absorbing states.

We shall define the following probabilities:

1°. $p_i(t)$, the probability of being in state i at time t;

2°. $r_i(t)\,\Delta t$, the probability of occupying state $i\ (i \in F)$ once during the time interval $(t, t+\Delta t)$, and $R_i(t) = \int_t^{\infty} r_i(x)\mathrm{d}x$;

3°. $u_i(t)\,\Delta t$, the probability of leaving state $i\ (i \in F)$ in the time interval $(t, t+\Delta t)$;

4°. p_{ij}, the probability of passing from state i to state j, conditional on leaving state i ($p_{ii} = 0$ for $i \in F$).

We shall now introduce the column vector analogues of the above quantities. Let us denote

$$\mathbf{p}_F(t) = (p_i(t))_{i\in F}, \quad \mathbf{p}_F(0) = (p_i(0)\,\delta_{ij})_{i\in F},$$

$$\mathbf{p}_A(t) = (p_j(t))_{j\in A},$$

$$\mathbf{r}(t) = (r_i(t)\,\delta_{ij})_{i,j\in F},$$

$$\mathbf{u}(t) = (u_i(t))_{i\in F},$$

$$\mathbf{R}(t) = (R_i(t)\,\delta_{ij})_{i,j\in F}.$$

We deduce the following equations for the dynamics of the process in the set F of nonessential states, using familiar arguments in the theory of semi-Markov processes:

$$\mathbf{p}_F^*(\lambda) = \mathbf{R}^*(\lambda)\,\mathbf{p}_F(0) + \mathbf{R}^*(\lambda)\mathbf{F}'\,\mathbf{u}^*(\lambda), \quad \mathrm{Re}\lambda > 0,$$

$$\mathbf{u}^*(\lambda) = \mathbf{r}^*(\lambda)\,\mathbf{p}_F(0) + \mathbf{r}^*(\lambda)\,\mathbf{F}'\,\mathbf{u}^*(\lambda),$$

where the asterisk and argument λ indicate the Laplace transform of the corresponding vector and $\mathbf{F} = (p_{ij})_{i,j\in F}$. We get the solutions

$$\mathbf{p}_F^*(\lambda) = \mathbf{R}^*(\lambda)\,[\mathbf{I} - \mathbf{F}'\,\mathbf{r}^*(\lambda)]^{-1}\,\mathbf{p}_F(0),$$

$$\mathbf{u}^*(\lambda) = [\mathbf{I} - \mathbf{r}^*(\lambda)\,\mathbf{F}']^{-1}\,\mathbf{r}^*(\lambda)\,\mathbf{p}_F(0).$$

The probability of being in the absorbing state $j \in A$ at time t is $p_A(t) = \sum_{j\in A} p_j(t)$ with the corresponding Laplace transform

$$\mathscr{p}_A(\lambda) = \frac{1}{\lambda} - \sum_{i\in F} \mathscr{p}_i(\lambda).$$

By means of this relation we shall deduce the moments of the time to therapeutical failure (time to reach absorption). The survival function

$$G(t) = \mathsf{P}\,(\text{failure time} > t)$$

has the Laplace transform

$$\mathcal{G}(\lambda) = \sum_{i \in F} \mathcal{P}_i(\lambda) = \mathbf{1}'\,\mathbf{p}_F^*(\lambda),$$

where we denoted by $\mathbf{1}$ the column vector $(\delta_{ij})_{i \in F}$. Let us now set

$$m_i(k) = \int_0^\infty t^k\, r_i(t)\, \mathrm{d}t,$$

$$\mathbf{M}(k) = (m_i(k)\, \delta_{ij})_{i,j \in F},$$

and

$$\mathbf{u}(0) = \lim_{\lambda \to 0+} \mathbf{u}^*(\lambda) = (\mathbf{I} - \mathbf{F}')^{-1}\, \mathbf{p}_F(0).$$

Note that $u_i(0)$ is the expected number of times state i is visited. Then, the mean of the time to reach absorptions is

$$\lim_{\lambda \to 0+} \mathcal{G}(\lambda) = \mathbf{1}'\mathbf{M}(1)\,[\mathbf{I} - \mathbf{F}']^{-1}\,\mathbf{p}_F(0) = \mathbf{1}'\,\mathbf{M}(1)\,\mathbf{u}(0).$$

The total time spent in a given transient state is one of the important parameters of the process. Let us consider a single remissive state $i \in F$ and partition the transition probability matrix $\mathbf{P} = (p_{ij})_{i,j \in A \cup F}$ according to

$$\begin{array}{cc} & \begin{array}{ccc} A & i & F - \{i\} \end{array} \\ P = & \begin{pmatrix} \mathbf{I} & \mathbf{0} & \mathbf{0} \\ \boldsymbol{\delta} & \mathbf{0} & \boldsymbol{\beta} \\ \boldsymbol{\gamma} & \boldsymbol{\alpha} & \boldsymbol{\varepsilon} \end{pmatrix} \begin{array}{l} A \\ i \\ F - \{i\} \end{array} . \end{array}$$

Then the probability of entering state i at least once starting from state j is

$$h_{ji} = [(\mathbf{I} - \boldsymbol{\varepsilon})^{-1}\,\boldsymbol{\alpha}]_j, \quad j \neq i$$

$$h_{ii} = \boldsymbol{\beta}\,[\mathbf{I} - \boldsymbol{\varepsilon}]^{-1}\,\boldsymbol{\alpha}.$$

The first two moments of the total sojourn time in state i conditional on starting from state j are

$$\frac{m_i(1)}{1-h_{ii}}\,\delta_{ij}+\frac{m_i(1)\,h_{ji}}{1-h_{ii}}\,(1-\delta_{ij})$$

and

$$\left(\frac{m_i(2)}{1-h_{ii}}+\frac{2m_i^2(1)\,h_{ii}}{(1-h_{ii})^2}\right)[\delta_{ij}+h_{ji}(1-\delta_{ij})],$$

respectively.

5.2.2.6. Weiss and Zelen (1963) illustrate the application of their model to data from 25 patients with acute leukemia. The process considered has six states. Considering the effects of the specific chemotherapy of these states, 2 and 3 are first and second partial remission states (including subsequent relapse), states 4 and 5 are first and second complete remission states (including subsequent relapse). State 1 is the initial relapse state, and state 0 is the final (death) state. In Fig. 8 one can see the relations between states 1 through 5, state 0 not being illustrated; it can be reached out of any of the nonessential states of the process.

An investigation of the distribution of sojourn times within the various nonessential states showed that with the exception of the initial state, these distributions can be well approximated by the gamma distribution. A series of numerical results is also given regarding the matrix **P** and the distribution of time to reach the absorbing state.

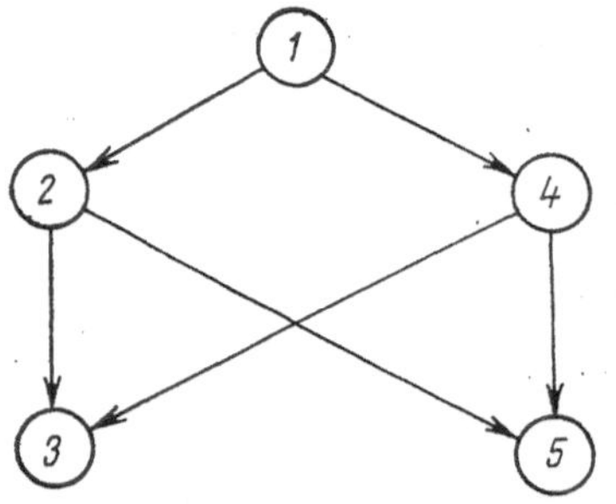

Fig. 8. — *1*: Initial relapse state; *2*: First partial remission; *3*: Second partial remission; *4*: First complete remission; *5*: Second complete remission.

5.2.2.7. The states presented here in the models of the clinical process generally describe the clinical evolution of a disease. However, if we refer to states preceding the attack of disease on an individual, it is easy to understand that there sometimes existed a disease-free state and that, even before the appearance of the symptoms of the disease, especially if it is a chronic disease, there was a pre-clinical state (or several pre-clinical states). A pre-clinical state can be regarded as a state in which clinical symptoms have not been exhibited and the individual is unaware of the disease. The idealized history of a chronic disease will be taken as a transition from no disease to pre-clinical disease and then to clinical disease.

This process is very important in modern medicine where a series of special diagnostic procedures have been created precisely to detect individuals in the pre-clinical state of an illness. The sooner this discovery takes place, the greater the chances of reconversion. The time by which the diagnosis is advanced by using the special diagnostic procedure will be called *the lead time*; it is synonymous in our language with *the forward recurrence time*. For a mathematical treatment of these questions, the reader is referred to FEINLEIB (1967) and ZELEN and FEINLEIB (1969 a).

5.2.2.8. A simple discrete parameter stochastic model for the age distribution of a chronic disease has been built by BERMAN (1963 a). Time intervals are measured in such a way that in the population considered at each time unit one new individual is born and one individual dies. Each individual lives a fixed number $m \in N^*$ of time units.

At each time unit each living individual, including one just born, has probability p of contracting a certain incurable disease. If an individual contracts the disease, then he has the disease during the remainder of his life time. If he does not contract the disease, then he again has probability p of contracting it at the next time unit, and so on until he dies. Contractions of the disease are stochastically independent for the different individuals.

Let $\delta_k^{(n)}$, $1 \leqslant k \leqslant m$, $n \in N^*$, be the indicator of the event that an individual of age k at time n has the disease. The sequence of random variables $\eta_n = \sum_{k=1}^{m} \delta_k^{(n)} 2^{-k}$, $n \in N^*$, is a finite state Markov chain.

5.2.2.9. Readers interested in the problem of periodic control of individuals for the purpose of discovering incipient states of chronic diseases will find in LINCOLN and WEISS (1964) criteria for random and periodic policies, and solutions for optimal examination schedules.

5.2.3. Stochastic models for tumour growth

5.2.3.1. HEIDELBERGER (1964) has stated that the mechanism of carcinogenesis is a mirror into which every man looks and sees himself (alluding to the multitude of points of view in this question). NIELS ARLEY has also stressed the extremely complicated mechanism of tumour growth: this process is of a multistage type governed by continued genotype variation, adaptation, and selection within the genetic mosaic of cell clones constituting a tumour. A viral mechanism, a biochemical mechanism and a genetic mechanism need not be mutually exclusive possibilities. According to this conception, studies of carcinogenesis (as well as studies of other biologic processes) can be made on a variety of levels, or scales, among which are the (i) somatic scale, (ii) clone and cell scale and (iii) molecular biology (NEYMAN and SCOTT (1967)). The three levels of the phenomenon of carcinogenesis just enumerated are obviously interconnected.

Most stochastic models are limited to the second level, the cell-clone scale, and, as could be observed in other models in this subsection, they represent a combination of several stochastic processes. The list of these models starts with that of ARLEY and IVERSEN (1952) and includes in 1970 the model of IYER and SAKSENA. When analysing these models, one can also observe that in some of them, carcinogenesis is a multistage process, while in others the appearance of the cancerous cell is due to a "multi-hit" mechanism. Let us develop this subtle distinction made by NEYMAN and SCOTT (1967). The term multi-hit is used to describe the mechanism in which a cancer initiating mutation requires not just one hit but a certain minimum number $k > 1$ of hits on the same cell. The term multistage mutation mechanism is used to describe a mechanism involving several successive mutations, each generating a clone of mutant cells. Thus, the first mutation (whether induced by a single hit or by $k > 1$ successive hits on the same cell) leads to a benign clone of cells described as first order mutants. Each of the first order mutant cells is subject to the risk of a second mutation which can produce (or cannot yet produce) a cancer cell. Thus, we contemplate a double hierarchy of hypothetical mechanism of carcinogenesis, determined by the number of stages and by the number of hits of the carcinogen required to produce each particular mutation. We may therefore characterize a carcinogenesis model by a notation similar to that used in queueing theory, according to the following order: the level (1, 2, 3), the hit number k, and the stages number r, ($k, r \in \mathbb{N}^*$). Thus, we shall denote the original ARLEY and IVERSEN (1952) model as a 2/1/1 and NORDLING's (1953) model as $2/k/1$ carcinogenesis model.

For an understanding of the problem as a whole, see the two excellent papers of ARMITAGE and DOLL (1961) and NEYMAN and SCOTT (1967).

5.2.3.2. We shall now present KENDALL's (1960) 2/1/2 model. Suppose we have a large (say infinite) population of normal cells subject to carcinogenetic action (perhaps by irradiation), usually of two kinds: a "background", which is always present, and an experimental enhancement of this of known intensity acting over a known period. Suppose that the joint effect of these agents is to transform some normal cells into first-order mutants, which it will be convenient to call *grey* cells.

Quantitatively, we assume that in a time interval (t_1, t_2), the number of grey cells produced in this way is a Poisson variable with mean

$$\int_{t_1}^{t_2} f(t)\, \mathrm{d}t,$$

where $f(t)$ represents the intensity of the carcinogenic action (of both kinds)[14]. In most experimental situations $f(t)$ will be a step-function.

It is next supposed that as a result of the same carcinogenic influences, the grey cells are capable of being transformed into second-order mutants, which we shall call *black* cells.

Two alternative assumptions can be made:

(a) the effect of a second-order mutation is to transform a single grey cell into a black cell; such mutations occur independently of the process controlling the growth of the clone to which the grey cell belongs;

(b) second-order mutations occur (if at all) at epochs of cell division, and their effect is to convert one of the two fission products into a black cell, so that we have

$$\text{grey} \rightarrow \text{grey} + \text{black}.$$

We shall now introduce the assumptions of the growth and mutation process:

1°. Each grey cell will be supposed to generate a clone; these benign growths are governed by a subcritical linear birth-and-death process ($b_1 < d_1$);

2°. Each black cell will be supposed to generate a clone; these malignant growths are governed by a supercritical linear birth-and-death process ($b_2 > d_2$);

3°. Mutations grey → black or grey → grey + black occur in the time interval $(t, t + \Delta t)$ with probability $\alpha\Delta t$. If α is time-dependent, we put $\alpha = \alpha_0 f(t)$, where f is the function already used to measure the intensity of the carcinogenic action at various times[15].

The consequence of these multiplying processes is that at time t there will be $\xi(t)$ grey clones and $\eta(t)$ black clones of different sizes. Small size clones will not be discovered during experimental research: we cannot suppose that the grey and black clones of equal size will be detected with equal frequency. Out of the whole number of clones only $\xi^*(t)$ and $\eta^*(t)$, say, will have the chance to be observed. This chance is

---

[14] The function f is also called alternatively the "time pattern function" or the "feeding function" by NEYMAN and SCOTT (1967) interpreting f as determining the time pattern of the application of the carcinogen.

[15] While revising the model of the two-stage mutation mechanism, NEYMAN and SCOTT (1967) set $\alpha = \alpha_0 + \alpha_1 f(t)$ where α_0 and α_1 are nonnegative constants not both equal to zero.

associated with the clone size and "colour". We must, therefore introduce a function $c_1(n)$ giving the probability that a grey clone of size n will be counted by the experimenter, and a similarly defined function $c_2(n)$ for the black clones. J. NEYMAN has suggested that one might take

$$c_1(n) = 1 - \gamma^n, \text{ for grey clones,}$$

$$c_2(n) = 1 - \beta^n, \text{ for black clones,}$$

where β and γ are positive numbers less than unity. This is a plausible assumption, D. G. KENDALL says, because we might argue that if β is the chance that a single black cell will be missed, then β^n will be the chance that a clone of size n will be missed, and $1 - \beta^n$ will be the chance that it will be counted. (Since the experimenter has in front of him only a two-dimensional section in a clone, D. G. KENDALL proposes the replacement of n by $n^{2/3}$.)

Consider first the history of a single grey clone descended from a single grey cell formed at $t = 0$. At a subsequent time t we will have $x(t)$ grey cells and $\eta(t)$ black clones (of various sizes). Consider one of these black clones, formed, say, at time u $(0 < u < t)$; it will have been growing for a period $(t - u)$ and if its size is y, then we shall have (see I; 2.3.4.7.5),

$$\mathsf{E}\, z^y = \frac{d_2(a-1) - (a\,d_2 - b_2)\,z}{(ab_2 - d_2) - b_2\,(a-1)\,z},$$

where $a = \exp((b_2 - d_2)(t - u))$. The chance that it will be counted is $1 - \beta^y$ and so (averaging over the y-distribution) the final chance that it will be counted is

$$q\,(t - u) = \frac{(b_2 - d_2)\,(1 - \beta)\,a}{(ab_2 - d_2) - b_2(a-1)\beta}.$$

When $(t - u) \to \infty$, then this chance tends to the limit $q(\infty) = 1 - \frac{d_2}{b_2}$, which is just the chance that the black clone will become infinitely large instead of becoming extinct.

Let us introduce the probability generating function for the joint distribution of the number of grey cells and the number of detected black clones:

$$F(z_1, z_2; t) = \mathsf{E}\,[z_1^{x(t)}\ z_2^{\eta^*(t)}].$$

We shall use assumption (a) and consider that α is constant. The function F satisfies the equation

$$F(z_1, z_2; t) = z_1 \exp(-(b_1 + d_1 + \alpha)\,t) +$$

$$+ \int_0^t \{b_1 [F(z_1, z_2; t-u)]^2 + d_1 + \alpha [1 - (1 - z_2)\, q\,(t - u)]\} \times$$

$$\times \exp(-(b_1 + d_1 + \alpha)\; u)\,\mathrm{d}u. \qquad (5.20)$$

By putting $\tau = t - u$, multiplying through by $\exp\,(b_1 + d_1 + \alpha)\,t$, and carrying out a differentiation with regard to t, we obtain from (5.20) the following differential equation of the Riccati type for F:

$$\frac{\mathrm{d}F}{\mathrm{d}t} = b_1 F^2 - (b_1 + d_1 + \alpha)\, F + d_1 + \alpha\, [1 - (1 - z_2) q(t)], \qquad (5.21)$$

with the initial condition $F(z_1, z_2; 0) = z_1$.

We can consider an approximation to (5.21) replacing $q(t)$ by its limit $q(\infty)$. We then have

$$\frac{\mathrm{d}F}{\mathrm{d}t} = b_1 F^2 - (b_1 + d_1 + \alpha)\, F + (d_1 + \alpha \bar{q}), \qquad (5.22)$$

where

$$\bar{q} = 1 - (1 - z_2)\, q(\infty) = 1 - (1 - z_2)\left(1 - \frac{d_2}{b_2}\right).$$

We can see that when $t \to \infty$, F will ultimately become independent of t and of z_1, and will converge to a function of z_2 alone which will generate the distribution of the total number of black clones which ultimately escape extinction. When $t \to \infty$, the solutions to (5.21) and (5.22), in each case with the initial condition $F(z_1, z_2;\ 0) = z_1$, will converge to the same limit. Thus, as an exact result, we have the formula

$$\lim_{t \to \infty} F(z_1, z_2; t) = c(z_2), \quad 0 \leqslant z_1 \leqslant 1, 0 \leqslant z_2 \leqslant 1,$$

where

$$c\,(z_2) = \frac{1}{2b_1}\,(b_1 + d_1 + \alpha - \sqrt{[(b_1 + d_1 + \alpha)^2 - 4b_1(d_1 + \alpha\bar{q})]}).$$

Next, let us consider the formation of the grey clones and the ensemble of black clones which are derived from them. One starts by considering the events which follow from the possibility of a grey cell being formed in the time interval $(u, u + \Delta u)$, where $0 < u < t$, and $(0, t)$ is the interval of exposure. A grey cell will be formed then with a probability $f(u)\,\Delta u$. Suppose that one is formed, and at the end of the experiment it yields x grey cells and η^* detected black clones. Suppose that it yields ξ^* detected grey clones, where $\xi^* = 0$ or 1. Then,

$$\mathsf{E}\,[z_1^{\xi^*} z_2^{\eta^*} \mid x, \eta^*] = (1 - \gamma^x)\, z_1 z_2^{\eta^*} + \gamma^x z_2^{\eta^*},$$

and so

$$\mathsf{E}\,[z_1^{\xi^*} z_2^{\eta^*}] = F(\gamma, z_2; t - u) + z_1\,[F(1, z_2; t - u) - F(\gamma, z_2; t - u)].$$

From these relations it is possible to deduce the probability generating function for the joint distribution of $\xi^*(t)$ and $\eta^*(t)$:

$$G(z_1, z_2; t) = \mathsf{E}\,[z_1^{\xi^*(t)} z_2^{\eta^*(t)}] = \exp\,(-R(z_2, t) - (1 - z_1)\,S(z_2, t)),$$

where

$$R(z_2, t) = \int_0^t [1 - F(1, z_2; t - u)]\,f(u)\,du,$$

$$S(z_2, t) = \int_0^t [F(1, z_2; t - u) - F(\gamma, z_2; t - u)]\,f(u)\,du.$$

Hence,

$$\mathsf{E}\,\xi^*(t) = S(1, t),$$

$$\mathsf{E}\,\eta^*(t) = -\left.\frac{\partial R(z_2, t)}{\partial z_2}\right|_{z_2=1},$$

and

$$\mathsf{D}\,\xi^*(t) = S(1, t),$$

$$\mathsf{D}\,\eta^*(t) = -\left.\left(\frac{\partial^2 R(z_2, t)}{\partial z_2^2} + \frac{\partial R(z_2, t)}{\partial z_2}\right)\right|_{z_2=1},$$

$$\operatorname{Cov}\,[\xi^*(t), \eta^*(t)] = \left.\frac{\partial S(z_2, t)}{\partial z_2}\right|_{z_2=1}.$$

Assuming the function f to be constant, the following asymptotic formulae will then hold:

$$\lim_{t \to \infty} \mathsf{E}\, \xi^*(t) = \frac{f}{b_1} \log\left(\frac{d_1 + \alpha - b_1\gamma}{d_1 + \alpha - b_1}\right), \tag{5.23}$$

$$\lim_{t \to \infty} \frac{\mathsf{E}\, \eta^*(t)}{t} = \frac{f\alpha(1 - d_2/b_2)}{d_1 + \alpha - b_1}. \tag{5.24}$$

5.2.3.3. A non-Markovian variant was given by WAUGH (1961) introducing two age-dependent birth-and-death processes. In the same year, H. G. TUCKER considered a vector process $(\xi(t), \eta(t), \zeta(t))_{t \geqslant 0}$ where $\xi(t)$ denotes the number of observable "hyperplastic loci" (distinct pretumoral clones), $\eta(t)$ is the number of tumours that can be counted and $\zeta(t)$ denotes the total number of cells in all the tumours, at time t. The model is based on SHIMKIN and POLLISSAR's (1955) two-stage theory of induced pulmonary carcinogenesis (adult strain A mouse exposed to urethane). See also SHIMKIN et al. (1967).

5.2.4. Some stochastic aspects of chemotherapy

5.2.4.1. In this brief paragraph we shall present in a quite general way the few papers dealing with the stochastic modelling of chemotherapy. Only very recently a general stochastic model (PADGETT and TSÓKOS (1971)) has been considered which is, in fact, the stochastic alternative of the deterministic model developed by BELLMAN, JACQUEZ and KALABA (1961). However, even the original model was a model of the kinetics of distribution of an injected compound in some tissue compartments rather than a model of chemotherapy.

HORN and GRIMM (1969) have dealt with the effects of cancer chemotherapy on tumour growth. They considered two linear birth-and-death processes: one with parameters b_1 and d_1 under conditions of tumour growth, and a second process with parameters b_2 and d_2 under conditions of anti-cancer treatment. The two actions of the specific pharmacological agent have been distinguished, namely the antimitotic action ($b_2 > b_1$) and the cytotoxic action ($d_2 < d_1$). The value of the mitotic index and the ratio of the number of dead and yet visible tumour cells and that of living tumour cells have been calculated.

5.2.4.2. We think that a stochastic model in chemotherapy must take into account the following schema:

(a) There is a population of BD-objects developing within another B-object population, each of these populations having different growth laws;

(b) The two antagonistic populations are or are not taxonomically different;

(c) Under the action of a drug C, the population of DB-objects may react either by death, or by the appearance of the property R of resistance, resulting from a mutation process. Therefore, in the presence of C, there are DB-objects with property S (sensitivity) and DB-objects with property R (resistance);

(d) It is supposed that at least one part of the B-objects considered has only property S versus drug C;

(e) The strategy of chemotherapy is as follows: to use compounds with a DB-selective toxicity, in such a way that they destroy the population of DB-objects with both properties S and R with the smallest possible damage of B-objects.

A suggestion for a model of DB-objects population growth submitted to the action of a drug C could be given by a *limit process* (case (V), considered by KARLIN and MCGREGOR (1964)).

5.2.4.3. A model which takes these conditions into account has been developed by NISSEN-MEYER (1966) for the analysis of effects of antibiotics on bacteria. He considers a two-dimensional birth-and-death process with mutation in a single direction, $S \to R$. The resistant bacteria will be subject to a birth-and-death process with a time-dependent feeding function, and formulas derived for the probability of their extinction may be used as a measure of the effectiveness of the treatment. As in chronic diseases (for example, tuberculosis), a lengthy, periodical treatment is necessary, the effects of the periodical treatment are studied as well as the optimal conditions of application. The important result of this analysis is that there may be some merit in starting with relatively high doses at the beginning of each period and then reducing them gradually.

5.2.5. Competing risks of illness

5.2.5.1. First studied by actuaries (see e.g. CHIANG (1961)), the problem of competing risks of illness and competing risks of deaths has been given a stochastic treatment with the building of an illness-death process by FIX and NEYMAN (1951) and the subsequent works of CORNFIELD (1957) and KIMBALL (1958). A detailed treatment of the problem has been given by CHIANG (1964); we shall give it here.

Suppose that there are r diseases operating in a human population. Due to their competition, an individual may be afflicted with any one disease or with any combination of them. Consequently, there are 2^r possible states of illness and they can be grouped into $r + 1$ mutually exclusive sets of states according to the concurrent presence of $0, 1, \ldots,$

or r diseases. C. L. CHIANG introduces the following Table for enumerating the states of disease:

Table 5

Types of states according to multiplicity of illness	States	Number of states of the given type
State of no illness	1	$\binom{r}{0}$
States with single disease	$2, \ldots, r+1$	$\binom{r}{1}$
States with two diseases	$r+2, \ldots, \frac{r(r+1)}{2}+1$	$\binom{r}{2}$
States with c diseases	$\left[\binom{r}{0}+\cdots+\binom{r}{c-1}+1\right], \ldots, \left[\binom{r}{0}+\cdots+\binom{r}{c}\right]$	$\binom{r}{c}$
States with r diseases	2^r	$\binom{r}{r}$

An individual is said to be in state α when afflicted with the corresponding diseases ($2 \leqslant \alpha \leqslant 2^r$); a healthy person is in state 1. An individual may leave a state at any given time either through recovery or by being afflicted with new diseases; all the 2^r states are thus communicating states.

In a similar way, one considers 2^r states of death, likewise divided into $r+1$ types, according to the multiplicity of causes of death. An individual dead from unknown cause was in state 1 of death. The 2^r states of death constitute a closed set, in that no state outside the set may be reached from any state belonging to the set. In fact, each state itself is absorbing since an individual arriving at state δ ($1 \leqslant \delta \leqslant 2^r$) of death from state α of illness will stay in δ forever.

At time $t = 0$ let there be n_α individuals in state α ($1 \leqslant \alpha \leqslant 2^r$). These n_α individuals will travel independently from one state to another. At the end of the interval $(0, t)$ after having left the original state α of illness m times, $m \in N$, a number $n_{\alpha\beta}^{(m)}(t)$ of individuals will be found in state β of illness and a number $d_{\alpha\delta}^{(m)}(t)$ in state δ of death ($1 \leqslant \delta \leqslant 2^r$). The sum

$$n_{\alpha\beta}(t) = \sum_{m=0}^{\infty} n_{\alpha\beta}^{(m)}(t), \quad 1 \leqslant \alpha, \beta \leqslant 2^r,$$

is the number of individuals in state α of illness at time 0 and in state β of illness at time t; we shall correspondingly denote

$$d_{\alpha\delta}(t) = \sum_{m=0}^{\infty} d_{\alpha\beta}^{(m)}(t), \quad 1 \leqslant \alpha, \delta \leqslant 2^r,$$

the total number of individuals transferred from state α of illness to state δ of death in time interval $(0, t)$. For each α we have

$$n_\alpha = \sum_{\beta=1}^{2^r} n_{\alpha\beta}(t) + \sum_{\delta=1}^{2^r} d_{\alpha\delta}(t), \quad 1 \leqslant \alpha \leqslant 2^r.$$

Transitions among the 2^{r+1} states are governed by "risks of illness", $s_{\alpha\beta}$, and "risks of death", $v_{\alpha\delta}$. We shall therefore consider the following probabilities:

1°. The probability that an individual in state α of illness at time t will be in state β of illness at time $t + \Delta t$, is $s_{\alpha\beta}(t)\,\Delta t + o(\Delta t)$;

2°. The probability that an individual in state α of illness at time t will be in state δ of death at time $t + \Delta t$, is $v_{\alpha\delta}(t)\,\Delta t + o(\Delta t)$. Transition intensities $s_{\alpha\beta}$, $1 \leqslant \alpha, \beta \leqslant 2^r$ will be the components of the morbidity risk matrix, and intensities $v_{\alpha\delta}$, $1 \leqslant \alpha, \delta \leqslant 2^r$, will be the components of the mortality risk matrix. The strict formulation of an illness-death process requires the assumption that no death takes place without a cause and that an individual is afflicted at the time of death with the diseases corresponding to the cause of his death. Many of the $v_{\alpha\delta}$ risks would then be zero. Similarly, if an individual is unlikely to be afflicted with more than one disease within an infinitesimal interval Δt, many of the $s_{\alpha\beta}$ would also be zero.

But the fundamental quantities underlying the probability distributions of $n_{\alpha\beta}^{(m)}(t)$, $n_{\alpha\beta}(t)$, $d_{\alpha\delta}^{(m)}(t)$, and $d_{\alpha\delta}(t)$ are the four transition probabilities defined as follows;

3°. The probability that an individual in state α of illness at time 0 will leave state α $(1 \leqslant \alpha \leqslant 2^r)$ m times $(m \in N)$ and be in state β of illness at time t: $p_{\alpha\beta}^{(m)}(t)$;

4°. The probability that an individual in state α of illness at time 0 will be in state β of illness $(1 \leqslant \alpha, \beta \leqslant 2^r)$ at time t: $p_{\alpha\beta}(t)$;

5°. The probability that an individual in state α of illness at time 0 will leave state α m times and be in absorbing state δ at time $t = q_{\alpha\delta}^{(m)}(t)$;

6°. The probability that an individual in state α of illness at time 0 will be in absorbing state δ at time t: $q_{\alpha\delta}(t)$.

In the mentioned paper, C. L. CHIANG deals with expressions for the probability generating functions for the above probabilities, of the probability distributions of $n_{\alpha\beta}(t)$ and $d_{\alpha\delta}(t)$, as well as the expected length of stay in state β of illness within time interval $(0, t)$ for an individual in state α of illness at time 0.

5.2.5.2. We remind the reader that in a previous paper C. L. CHIANG (1961) has studied the probabilities of death with respect to a specific risk. Clearly, this is an actuarial aspect of the problem of competing risks. See also BERMAN (1963 b) for a semi-Markov treatment.

5.2.6. Control of biological processes

5.2.6.1. The applications of control theories to biological processes are relatively recent. As it has been conceived until now, the control is directed toward a population of DB-objects which grows according to a certain law of growth. This is why BECKER (1970) calls these DB-objects: "pests".

The control involves cost due to labour, materials, risk, etc. and it is therefore necessary to weigh the damage done by the DB-objects against the cost of the control action and so choose a suitable level at which to apply the control. For example, in an epidemic spread by the infectives or carriers in a large population, the "damnificers" are the infectives or carriers, and the control might be to increase the rate of removal of infectives or to inoculate part of the population. The damage due to an epidemic involves a cost — which must include not only human losses, but also damages inflicted on production, increase of medical activity in hospitals, increased production of drugs, etc. TAYLOR (1968), on the other hand, considers an epidemic control problem in which he regards the susceptibles as "damnificers", since a large number of susceptibles creates a threat of a major epidemic outbreak. He finds the optimal moment of time to remove all susceptibles by vaccination when the population of susceptibles grows according to a pure birth process.

5.2.6.2. Let us consider (following BECKER (1970 b)) the growth of a damnificer population (shortly D-population), given by a Kendall process $(\xi(t))_{t\geqslant 0}$ with rates $b_n = bn + \nu$ and $d_n = dn$ (see Subparagraph 2.1.2.7).

Consider a control action that is applied at discrete points of time and has the effect of "removing" from the habitat, either by death or emigration, each D-object present at that time with probability q and not affecting a particular D-object with probability $p = 1 - q$. Then, if there are n D-objects present just before the control is applied, there are r D-objects present in the habitat just after the control is applied with probability $\binom{n}{r} p^r q^{n-r}$.

Let us suppose that control is exerted k times in a time interval $(0, t')$. Each of the k actions (e.g., treatments) is assumed to be equally effective so that survivors of previous actions and newly arrived D-objects are assumed to be equally susceptible to a given action. Having thus specified the nature and total amount of control we aim to find the time points at which to apply the action in order that this control be most effective. If the cost of applying the k actions is independent of the times at which they are applied, the cost of the control action is fixed, since apart from these times the control action is completely specified. Taking the cost of the control action as fixed, the requirement is to minimize the damage due to the D-objects and, therefore, we aim to find the time points at which to apply the treatment in order that the mean number of D-objects in the interval $(0, t')$ is a minimum.

As shown in Subparagraph 3.2.1.2 the mean size of our D-population at time t, when the initial number of D-objects is n_0, is given by

$$m(t) = \begin{cases} \nu t + n_0 & \text{, when } b = d, \\ (\nu/(b-d))(e^{(b-d)t} - 1) + n_0 e^{(b-d)t} & \text{, when } b \neq d. \end{cases}$$

Denoting by n_i the expected number remaining just after the control action at time $\tau_1 + \ldots + \tau_i$, $1 \leqslant i \leqslant k$, is applied, we may write the objective function as

$$\sum_{i=1}^{k+1} \int_0^{\tau_i} m_{n_{i-1}}(t)\,dt, \tag{5.25}$$

where

$$\tau_{k+1} = t' - \sum_{i=1}^{k} \tau_i .$$

It can be shown that the n_i satisfy the recurrence relation

$$n_i = m_{n_{i-1}}(\tau_i)\, p, \qquad 1 \leqslant i \leqslant k.$$

The problem is now to find the positive τ_i such that the objective function (5.25) is minimized.

The results that may be obtained are obviously determined by the character of the growth process. Thus:

(a) When $\nu = 0$, we find the objective function to be an increasing function of $\tau_1, \ldots, \tau_{k-1}$ and τ_k, so that their common optimal value is zero, i.e., the action is applied k times at $t = 0$.

(b) When $\nu > 0$ and $b = d$, the objective function has an optimal value when the control action is applied at time intervals given by

$$\tau_1 = L' - \frac{n_0}{\nu}, \qquad \tau_i = L'q, \qquad 2 \leqslant i \leqslant k, \qquad \tau_{k+1} = L',$$

where

$$L' = \frac{t' + n_0/\nu}{1 + p + qk},$$

provided only that $\tau_1 \geqslant 0$, i.e.,

$$\frac{n_0}{\nu} \leqslant \frac{t'}{p + qk}.$$

(c) When $\nu > 0$ and $b \neq d$ and the action is totally effective, i.e., $p = 0$, the damage due to this D-population, as measured by the objective function (5.25), is minimized when the treatment is applied at time intervals

$$\tau_1 = L' - \frac{1}{b-d}\log\left(1 + \frac{\nu n_0}{b-d}\right), \qquad \tau_i = L', \qquad 2 \leqslant i \leqslant k+1,$$

where

$$L' = \left[t' + \frac{1}{b-d}\log\left(1 + \frac{\nu n_0}{p}\right)\right][k+1]^{-1},$$

provided only that $\tau_1 \geqslant 0$. If τ_1, as defined, is negative, the optimal times are given by $\tau_1 = 0$, $\tau_i = t'/k$, $2 \leqslant i \leqslant k+1$.

5.2.6.3. A similar control model has been proposed by NEUTS (1968). The D-population grows according to a Yule-Furry process (see Subparagraph 2.1.2.2) and the survival probability of j D-objects is as above

$$p_j = \binom{n}{j} p^{n-j} q^j, \qquad 0 \leqslant j \leqslant n.$$

Let us consider the control action as the application of a treatment to these D-objects (bacteria, parasites, cancer cells, etc.). The probability p will depend on the dosage and on the type of D-objects. One assumes that the therapeutist can choose the value of p, but that a high value of p is likely to destroy not only the D-objects, but the host as well (see the strategy of chemotherapy in Subparagraph 5.2.4.2). If we denote by $\psi(p)$ the probability that the host survives a treatment with parameter p, then in general ψ will be a decreasing function on the interval [0,1] such that $\psi(0) = 1$. We will assume that ψ is known.

One postulates that if the host is alive at time t, the probability that he dies during $(t, t + \Delta t)$ depends only on the size of D-population at time t, and on the number, dosage, and times of treatments already received. Specifically, one assumes that an organism that has n D-objects in it and has already received k treatments has a probability

$$(d + d^*n + f(p_1, \ldots, p_k;\ t_1, \ldots, t_k))\,\Delta t,$$

of dying in the interval $(t, t + \Delta t)$.

The quantity d is, as usual, the death rate, but for the host in the absence of D-objects. The quantity d^*n expresses the death rate due to the disease; it is proportional to the number of D-objects. The last term corresponds to the long-range aftereffects of the earlier treatments. In further calculations we will make the following simplifying assumption: Since in practice the treatments will be roughly of equal dosage and will last a more or less equal length of time τ, we will assume that the last term reduces to

$$kf(p, \tau)\,\Delta t,$$

where $f(p, \tau)$ is some function of the (average) dosage and of the length of time between treatments.

Suppose that at time $t = k\tau +$ [immediately after the kth treatment], there are n_k D-objects present and the host is still alive, $k \in N^*$. The conditional probability $P^{(k)}(n; t)$ that at time $t(k\tau < t < (k + 1)\tau)$ the organism is still alive and that there are then n D-objects present, satisfies the following system of differential equations,

$$\begin{aligned}\frac{\mathrm{d}}{\mathrm{d}t} P^{(k)}(n; t) &= -(d + (b + d^*)n + kf)\,P^{(k)}(n; t) + \\ &\quad + b(n - 1)P^{(k)}(n - 1; t), \qquad n \geqslant n_k\end{aligned} \tag{5.26}$$

with

$$P^{(k)}(n; t) = 0,\ n < n_k, \qquad P^{(k)}(n_k; k\tau +) = 1.$$

The system of equations can be solved by using the generating function

$$G^{(k)}(z, t) = \sum_{n \in N} P^{(k)}(n; t) z^n,$$

by means of which one obtains

$$\frac{\partial G^{(k)}}{\partial t} = (bz^2 - (b + d^*)\,z)\,\frac{\partial G^{(k)}}{\partial z} - (d + kf)\,G^{(k)}, \tag{5.27}$$

with $G^{(k)}(z, k\tau+) = z^{n_k}$. Equation (5.27) with initial conditions can be solved by classical methods, with which the reader is already acquainted. Finally, the following result is obtained:

$$G^{(k)}(z, t) = \exp(-(t - k\tau)(d + kf + bn_k + d^*n_k)) \times$$
$$\times z^{n_k}\left[1 - \frac{bz}{b + d^*}(1 - \exp(-(b + d)(t - k\tau)))\right]^{-n_k}.$$

Let $P(m, t)$ be the conditional probability that at time t the host is alive and is infected by m D-objects, given that there were n D-objects at $t = 0$. The $P(m, t)$ are continuous functions of t except at the points $k\tau$, $k \in N^*$, where they have jumps. We define

$$H(z, t) = \sum_{m \in N} P(m, t)\, z^m,$$

for all t except $t = k\tau$, $k \in N^*$. We have $H(z, 0) = z^n$, and for $k\tau < t < (k + 1)\tau$,

$$H(z, t) = \sum_{n_k = 0}^{\infty} P(n_k, k\tau +)\, G^{(k)}(z, t) =$$
$$= \exp(-(t - k\tau)(d + kf)) \times \tag{5.28}$$
$$\times H\left(\frac{\exp - (t - k)(b + d)z}{1 - \dfrac{bz}{b + d}(1 - \exp(-(b + d^*)(t - k\tau)))},\ k\tau+\right).$$

Formula (5.28) shows that it suffices to know $H(z, k\tau +)$ for $k \in N$ in order to express $H(z, t)$ at all other points t. It can be shown that $H(z, k\tau +)$ is defined recursively by

$$H(z, 0 +) = z^n,$$

$$H(z, (k + 1)\tau +) = \psi(p) \exp(-\tau(d + kf)) \times$$
$$\times H\left(\frac{(p + qz)\exp(-\tau(b + d^*))}{1 - \dfrac{b(p + qz)}{b + d^*}(1 - \exp(-\tau(b + d^*))},\ k\tau +\right), \quad k \in N.$$

The most important problems will be to find τ and p as to maximize the probability of survival in a given interval $[0, t]$ or to maximize the expected lifetime of organisms under treatment. Here we shall only deal with the first problem.

If no treatment at all is given ($k = 0$), the probability $H_0(t)$ that the host is alive at time t is given by

$$H_0(t) = H(1, t) = \exp(-dt - nt(b + d^*)) \times$$
$$\times \left[1 - \frac{b}{b + d^*}(1 - \exp(-(b + d^*)t))\right]^{-n},$$

using equation (5.28). If k treatments are given in $(0, t]$ then we have necessarily

$$\frac{t}{k+1} < \tau \leqslant \frac{t}{k}, \qquad k \geqslant 1.$$

The probability that the host is then alive at time t, which we will denote by $H_k(t)$, is then given by

$$H_k(t) = \exp(-(t - k\tau)(d + kf)) \times$$
$$\times H\left(\frac{\exp(-(t - k\tau)(b + d^*))}{1 - \dfrac{b}{b + d^*}(1 - \exp(-(b + d^*)(t - k\tau))}, k\tau +\right), \quad (5.29)$$

using again (5.28). The values of τ and p for which the probability $H_k(t)$ is maximum can of course be found by numerical methods only. It is possible that the maximum will be attained for many pairs (τ, p). If this is the case, then we have a further degree of freedom in choosing a policy (τ, p). Whether this is so will depend on the actual choice of the functions $\psi(p)$ and $f(\tau, p)$.

5.3. Epidemic processes

5.3.1. The classical models

5.3.1.1. The enormous number of papers on stochastic models of epidemics makes a comprehensive to say nothing of an analytical presentation impossible in the space available here. The reader is invited to consult, in addition to the books of BAILEY (1957) and BARTLETT (1960 b), the excellent critical report given by KENDALL (1956) and the interesting survey of DIETZ (1967), along with GLINEUR's (1967) and MÉLARD's (1968) theses. One may also consult the chapters dealing with this problem in the books of BAILEY (1964 a, 1967 b), BARTLETT (1960 b) and BHARUCHA-REID (1960).

5.3.1.2. We shall first try to point out the fundamental elements of an epidemic model. The population generating an epidemic process is a population of alterexter B-objects (which we usually call the "focus population" or epidemic population). The size of this population is fixed. The number of types or categories of B-objects forming the epidemic population varies depending on the degree of complexity of the model, but to trigger an epidemic, two components are necessary: the infected individuals (infectives) and the uninfected susceptibles. Precisely the relations between these two categories of B-objects, which are supposed to be mixed homogeneously together, determine the dynamics of the process.

The third category of B-objects distinguished in an epidemic population represents in some way the result of the epidemics onset: these B-objects have been called "removed cases". They are in fact a non-homogeneous category comprising isolated, immune or dead individuals. As we immediately see, the occurrence of an epidemic disturbs the apparent homogeneity of the population. The first distinctions that can be made are the introduction besides the category of infected individuals, of the category of healthy carriers; one may also suppose that there are two types of susceptibles. A classification of categories of B-objects on the basis of epidemiologic and clinical criteria can be found in FIRESCU and TAUTU (1967 a).

The time component of epidemic models is not well (or not at all) specified. The evolution of a contagious disease has been divided into several periods, among which *the latent period* (when the disease develops without clinical signs) and *the infectious period*. In the known models it is supposed that the latent period is zero, so that an infected individual becomes infectious to others immediately after contagion. It is also quite convenient to assume that the length of the infectious period has a negative exponential distribution. The estimation of these periods has been the object of thorough studies (BAILEY (1957), BAILEY and ALFF-STEINBERGER (1970)). For a Bayesian approach, see HILL (1963).

According to BAILEY (1967 b) epidemic theory falls into two distinct, though complementary, parts. On the one hand, there is the study of small groups such as individual families. From these it is possible, although this is not often done, to collect detailed data to which relatively realistic models can be fitted, yielding information about such biological or clinical entities as contact rate, length of latent and infectious periods and so forth. However, little can be deduced from this about the spread of infections through a community.

The latter requires the special theory of large groups. This provides some insight into the behaviour of population outbreaks with regard to such features as threshold phenomena or the general graphical appearance of "epidemic curves". Unfortunately, even the simplest mathematical models, especially if they are stochastic, present formidable problems of analysis.

5.3.1.3. The stochastic epidemic models are of two types: (a) "classical" birth-and-death models, and (b) Markov chain models, e.g. binomial chains, Galton-Watson chains, etc. In both cases it is a question of processes with a finite or countable state space.

Before presenting the two types of models in their essential aspects, we shall have to distinguish, in the case of the so-called "classical" models, between the two fundamental models: the *simple epidemic* and the *general epidemic.*

5.3.1.4. The simple epidemic describes the infection of a closed group of n susceptibles into which a infectives are introduced. This might be a reasonable approximation to the early stages of some upper respiratory (mild) infections such that there is no isolation of infectives.

Suppose that we have a homogeneous mixing group of $n + a$ individuals. For simplicity, consider that we have one initial infective, $a = 1$. Let there be $\xi(t) = x$ susceptibles and $\eta(t) = y$ infectives at time t, so that $x + y = n + 1$.

The basic assumption is that the chance of any susceptible's becoming infected in a short interval of time is jointly proportional to the number of infectives in circulation and the length of the interval. This means that the chance of one new infection in the whole group in a short interval of time will be proportional to the product of the number of infectives and the number of susceptibles, as well as the length of the interval. We thus have a transition intensity $\beta(n - x + 1)x$ which is a non-linear function of the group size: this is the chief source of the subsequent difficulties in the appropriate mathematical analysis. Here, β is called the *contact rate.*

If we change the time scale to $\tau = \beta t$, the chance of one new infection in the whole group in time interval $(t, t + \Delta t)$ becomes $x(n-x+1)\Delta\tau$. For the probabilities $p_x(t)$ that there are still x uninfected susceptibles at time t we have the following differential equations:

$$\left.\begin{aligned} \frac{dp_x}{d\tau} &= (x + 1)(n - x)p_{x+1} - x(n - x + 1)p_x, \quad 0 \leqslant x \leqslant n-1 \\ \frac{dp_n}{d\tau} &= -np_n, \end{aligned}\right\} \tag{5.30}$$

with the initial condition $p_x(0) = \delta_{xn}$, $0 \leqslant x \leqslant n$. Here δ is Kronecker's. BAILEY (1963) gave an explicit solution in terms of the hypergeometric function for the probability generating function $\sum_{j=0}^{n} p_j(t)z^j$.

5.3.1.5. The general stochastic epidemic involves both infection and removal. The model considered is that for which at time $t \geqslant 0$, there are in circulation in a closed population of size $n + a$, with $n, a \geqslant 1$,

$$0 \leqslant x \leqslant n \qquad \text{uninfected susceptibles,}$$

$$0 \leqslant y \leqslant n + a - x \qquad \text{infectives,}$$

the remaining $n + a - x - y \geqslant 0$ individuals having been removed. At time $t = 0$ the population is known to consist of n susceptibles and a infectives. During an interval of time Δt the chance of one new infection is taken to be $\beta xy\Delta t$ and the chance of one removal is $\gamma y \Delta t$, where β is the contact rate and γ is the *removal rate*. It has also been customary to make use of the *relative removal rate* $\rho \equiv \gamma/\beta$ (BAILEY (1955)), which will be called the threshold.

The general stochastic epidemic is a two-dimensional Markov jump process $(\xi(t), \eta(t))_{t \geqslant 0}$ where, as before, $\xi(t)$ is the number of uninfected susceptibles at time t, and $\eta(t)$ is the number of infectives at time t. The transition intensities are as follows:

$(x, y) \to (x - 1, y + 1)$	xy
$(x, y) \to (x, y - 1)$	ρy
$(x, y) \to (x, y)$	$-(x + \rho)y$,

where the time scale is chosen such that $\beta = 1$. The transition probabilities $p_{xy}(t) = \mathrm{P}(\xi(t) = x,\ \eta(t) = y \mid \xi(0) = n,\ \eta(0) = a)$ satisfy the equations

$$p'_{x0} = \rho p_{x1}, \qquad 0 \leqslant x \leqslant n,$$

$$p'_{xy} = (x + 1)(y - 1)p_{x+1,\,y-1} - y(x + \rho)p_{xy} + \rho(y + 1)p_{x,\,y+1},$$
$$0 \leqslant x \leqslant n, \qquad 0 < y < n + a - x,$$

$$p'_{ny} = -y(n + \rho)p_{ny} + \rho(y + 1)p_{n,\,y+1}, \qquad 0 \leqslant y < a,$$

$$p'_{na} = -a(n + \rho)p_{na},$$

$$p'_{0,\,n+a} = (n + a - 1)p_{1,\,n+a-1} - \rho(n + a)p_{0,\,n+a},$$

with the initial condition $p_{xy}(0) = \delta_{xn}\delta_{ya}$.

It is easily deduced that the associated probability generating function

$$G(z_1, z_2; t) = \sum_{x, y} p_{xy}(t) z_1^x z_2^y, \qquad |z_1|, |z_2| \leqslant 1,$$

satisfies the partial differential equation

$$\frac{\partial G}{\partial t} = z_2(z_2 - z_1)\frac{\partial^2 G}{\partial z_1 \partial z_2} + \rho(1 - z_2)\frac{\partial G}{\partial z_2}, \tag{5.31}$$

with the initial condition $G(z_1, z_2; 0) = z_1^n z_2^a$. The essence of GANI's (1965 c) and SISKIND's (1965) methods of solution consists of noting that if we write

$$G(z_1, z_2; t) = \sum_{x=0}^{n} z_1^x f_x(z_2, t), \tag{5.32}$$

where

$$f_x(z_2, t) = \sum_{y=0}^{n+a-x} z_2^y p_{xy}(t),$$

then the order of equation (5.31) may be reduced to the first. Substituting (5.32) in (5.31) and equating coefficients of z_1^x on right and left-hand sides, we obtain

$$\frac{\partial f_n}{\partial t} = -[(n + \rho) z_2 + \rho]\frac{\partial f_n}{\partial z_2},$$

$$\frac{\partial f_x}{\partial t} = z_2^2(x + 1)\frac{\partial f_{x+1}}{\partial z_2} - [(x + \rho) z_2 - \rho]\frac{\partial f_x}{\partial z_2}, \qquad 0 \leqslant x \leqslant n - 1.$$

At this stage, V. SISKIND proceeds by direct recursive integration of the $f_x(z; t)$ and J. GANI makes use of Laplace transforms.

5.3.1.6. One of the related problems raised by classical models is that of determining the total size of epidemic. In fact, for general epidemics, the total number of cases, not counting the initial ones, is the limit of $n - \xi(t)$ as $t \to \infty$. It is easy to see that this limit must equal a constant (possibly zero). If this problem is treated deterministically, there results an asymptotic relation between the total number of cases and the ratio of the initial number of susceptibles to the relative removal rate. For the number of removed cases $n - x - y$, a further approximation is possible, and we can write

$$\lim_{a \to 0} \lim_{t \to \infty} (n - x - y) = \begin{cases} 2(n - \rho), & \text{for } n > \rho \\ 0 \quad , & \text{for } n \leqslant \rho. \end{cases}$$

This equation asserts (i) that there will be no epidemic if n is less than the threshold ρ, and (ii) that in all other cases the number of susceptibles will fall as far below the threshold as it was initially above it. This is the celebrated threshold theorem of KERMACK and MCKENDRICK (1927). [See also BARTLETT (1956, 1966).] BAILEY (1957) obtained a stochastic analogue of this theorem, also finding the expression of the corresponding distribution of the total size of an epidemic. His calculations revealed a gradual transition from J-shaped distributions, containing only small epidemics for population sizes below the threshold, to U-shaped distributions containing either large or small epidemics but practically no epidemics of intermediate size when the threshold is exceeded. There is also an interesting transitional form of distribution near the threshold value. The interested reader may consult FOSTER (1955) and WHITTLE (1955). An algebraic proof of the threshold theorem was given by WILLIAMS (1971).

KENDALL (1956) interpretes BAILEY's (1953) findings in the following way:

(i) there will be a minor epidemic if $n \leqslant \rho$, and (ii) there will be either a minor or a major epidemic if $n > \rho$, epidemics of intermediate magnitude being very unlikely.

A moment's reflexion shows that this mode of behaviour might well have been expected. At $t = 0$ the population of infectious individuals is subject to a "tangential" birth-and-death process with a constant death parameter equal to γ and a stochastic birth parameter equal to βn. Thus, if n is reasonably large, we shall for a short time have in effect a birth-and-death process in which the ratio death parameter / birth parameter has the value $\gamma/\beta n = \rho/n$.

The chance of ultimate extinction in such a birth-and-death process is $(\rho/n)^a$ if $n > \rho$, and is unity if $n \leqslant \rho$. Thus, if $n \leqslant \rho$, we may expect a small number of secondary infection cases and then quiescence. If $n > \rho$, however, the tangential birth-and-death process will either die out quickly or grow indefinitely. Thus, we may then expect with complementary probabilities that the number of infectives will either fall speedily to zero, producing a small number of secondary infection cases or build up to large values, so that a major epidemic results. D. G. KENDALL thus proposes the following approximating stochastic system $(\tilde{\eta}(t))_{t\geqslant 0}$:

(a) If $n \leqslant \rho$, then $\tilde{\eta}(t)$ is the number of individuals at time t in a population controlled by a linear birth-and-death process with birth parameter $n\beta$ and death parameter γ, satisfying the initial condition $\tilde{\eta}(0) = a$;

(b) If $n > \rho$, then the system has two modes of behaviour, A and B, with

$$\mathsf{P}(A) = (\rho/n)^a,$$
$$\mathsf{P}(B) = 1 - (\rho/n)^a.$$

In mode A, $\tilde{\eta}(t)$ behaves as if it were the number of individuals at time t in a linear birth-and-death process with birth parameter $n\beta$ and death parameter γ, satisfying the initial condition $\tilde{\eta}(0) = a$, and further conditioned by the requirement that $\lim_{t\to\infty} \tilde{\eta}(t) = 0$. In mode B, $\tilde{\eta}(t)$ behaves as if it were the function $y(t)$ describing the evolution of the corresponding deterministic epidemic.

An interesting approach to this problem has been made by DANIELS (1967). As we are concerned only with the final distribution of the number of susceptibles, it is simpler to work with the random walk of transitions in the x, y plane. An alternative formulation of the random walk in terms of the numbers of new cases and removals is possible and the problem can then be described in terms of a game involving a mixture of sampling with and without replacement. One of the important results obtained by H. E. DANIELS is the following: When the threshold is large but the population size is much larger, the distribution of the number remaining uninfected in a large epidemic has approximately the Poisson form with mean $ne^{-n/\rho}$.

5.3.1.7. It was pointed out that the spread of a rumour in a closed community could hardly be more different from the spread of an epidemic. Indeed, the salient feature of the rumour's spread is the absence of any threshold effect; the fraction of the population which ultimately learns the rumour is approximately independent of the population size (DALEY and KENDALL (1965)).

5.3.1.8. By definition, the epidemic curve is the probability density function $f(t)$ of the time of occurrence of new cases; therefore, if k out of n cases are observed to fall in the interval $(t, t + \Delta t)$, we must have

$$f(t)\,\Delta t = \mathrm{E}\left[\frac{k}{n}\right] = \frac{1}{n}\,[m(t) - m(t + \Delta t)],$$

hence

$$f(t) = -\frac{1}{n}\,\frac{\mathrm{d}m(t)}{\mathrm{d}t},$$

where $m(t)$ is the mean number of susceptibles still uninfected at time t. For the simple epidemic,

$$m(t) = \sum_{x=1}^{n} x p_x(t),$$

and WILLIAMS (1965) determined the moment-generating function

$$\int_0^\infty e^{\theta t} f(t)\,\mathrm{d}t = 1 + \frac{\theta}{n}\int_0^\infty e^{\theta t} m(t)\,\mathrm{d}t = 1 + \frac{\theta}{n}\sum_{r=a}^{n+a-1}\frac{1}{r}\times$$

$$\times\frac{1}{\left(1 - \dfrac{\theta}{an}\right)\left[1 - \dfrac{\theta}{(a+1)(n-1)}\right]\cdots\left[1 - \dfrac{\theta}{r(n-r+a)}\right]},$$

so that the mean of the epidemic curve is

$$\frac{1}{n} \sum_{r=0}^{n+a-1} \frac{1}{r}.$$

See also BAILEY (1963) and HASKEY (1954).

5.3.2. A general approach to epidemics

5.3.2.1. We shall now present, according to SEVERO (1967, 1969 a, b) the general structure of epidemic models using Markov jump processes. We shall define an epidemic population as a set Ω of B-objects ω. A B-object considered as an element ω is in Ω iff, for some time $t \geqslant 0$, ω is an uninfected susceptible or an infective. For each $\omega \in \Omega$ and for each $t \geqslant 0$ we define

$$W_1(\omega, t) = \begin{cases} 1, & \text{if } \omega \text{ is an uninfected susceptible at time } t \\ 0, & \text{otherwise} \end{cases}$$

$$W_2(\omega, t) = \begin{cases} 1, & \text{if } \omega \text{ is an infective at time } t \\ 0, & \text{otherwise} \end{cases}$$

$$W_3(\omega, t) = \begin{cases} 1, & \text{if } \omega \text{ is neither an uninfected susceptible nor an infective at time } t \\ 0, & \text{otherwise} \end{cases}$$

We shall assume that the number of elements in Ω is n, a finite positive integer. Let

$$\xi(t) = \Sigma\, W_1(\omega, t),$$

$$\eta(t) = \Sigma\, W_2(\omega, t),$$

$$\zeta(t) = \Sigma\, W_3(\omega, t),$$

where the summation is over all $\omega \in \Omega$. Then $\xi(t) + \eta(t) + \zeta(t) = n$, and we interpret the random vector $(\xi(t), \eta(t), \zeta(t))_{t \geqslant 0}$ as an epidemic model.

We shall denote the size of an epidemic population at time $t \geqslant 0$ by $m(t)$, i.e.,

$$m(t) = \xi(t) + \eta(t).$$

Let a_{ij} be the probability that at $t = 0$ we have $\xi(0) = i$, $\eta(0) = j$, so that $\zeta(0) = n - i - j$, and assume $\Sigma a_{ij} = 1$, where the sum is over A_n', the set of nonnegative integer pairs (i, j) such that $i + j \leqslant n$. Under the usual assumptions, one can obtain a system of differential equations for the probabilities $p_{xy}(t)$ that $\xi(t) = x$ and $\eta(t) = y$, i.e.,:

$$p_{xy}(t) = \sum_{(i,j) \in A_n'} \mathrm{P}(\xi(t) = x, \eta(t) = y \mid \xi(0) = i, \eta(0) = j)\, a_{ij}.$$

The equations are

$$p_{xy}'(t) = f_{1,0}(x, y; \boldsymbol{\theta}, n)\, p_{x+1,y}(t) + f_{1,-1}(x, y; \boldsymbol{\theta}, n) p_{x+1,y-1}(t) +$$
$$+ f_{0,1}(x, y; \boldsymbol{\theta}, n)\, p_{x,y+1}(t) + f_{0,0}(x, y; \boldsymbol{\theta}, n)\, p_{xy}(t), \qquad (5.33)$$

where $\boldsymbol{\theta}$ is a vector of parameters, $(x, y) \in A_n'$, and $p_{x+v,y+w}(t) \equiv 0$ whenever $(x + v, y + w) \notin A_n'$. The coefficients f are assumed to exist for all $(x, y) \in A_n'$ and for all $\boldsymbol{\theta}$ in the appropriate domain, and to be independent of t.

Let us confine attention to the special sub-family of models defined by

$$f_{1,0} = \lambda (x + 1)^{1-c_2} y^{c_1} \delta(n - x - y - 1),$$
$$f_{1,-1} = \pi (x + 1)^{1-c_2} (y - 1)^{c_1} \delta(y - 1),$$
$$f_{0,1} = \rho (y + 1)^{1-c_3} \delta(n - x - y - 1),$$
$$f_{0,0} = -(x^{1-c_2} y^{c_1} + \rho\, y^{1+c_3}),$$

in which the parameter vector $\boldsymbol{\theta}$ can be defined as the 6-tuple $(\lambda, \pi, \rho, c_1, c_2, c_3)$ and where we assume $c_1 > 0$, $c_2 < 1$, and $c_3 > -1$, and

$$\delta(x) = \begin{cases} 1, & \text{for } x \geqslant 0 \\ 0, & \text{for } x < 0. \end{cases}$$

Then the following vectors are special epidemic cases:

$\boldsymbol{\theta} = (0, 1, 0, 1, 0, 0)$	simple stochastic epidemic (BAILEY (1957))
$\boldsymbol{\theta} = (0, 1, \rho, 1, 0, 0)$	general stochastic epidemic (BAILEY (1957))
$\boldsymbol{\theta} = (1, 0, \rho, 1, 0, 0)$	WEISS's (1965) carrier model
$\boldsymbol{\theta} = (\lambda, \pi, \rho, 1, 0, 0)$	DOWNTON's (1968) carrier model
$\boldsymbol{\theta} = (0, 1, \rho, c_1, c_2, c_3)$	SEVERO's (1969 a) model
$\boldsymbol{\theta} = (0, 1, 0, c_1, c_2, 0)$	KRYSCIO and SEVERO's (1969) model.

Here λ is the direct removal rate of susceptibles, π a constant representing the proportion of those infected who become carriers rather than full-fledged and detectable infectives, and ρ is the (relative) removal rate.

5.3.2.2. We effect a change in notation by letting, for each ordered pair of integers (x, y),

$$p_{xy}(t) = \delta(x)\,\delta(y)\,\delta(n - x - y)\,r_k(t),$$

$$a_{xy} = a_k,$$

where

$$k \equiv k\,(x, y : n) = \frac{1}{2}(n + 1)(n + 2) - (n + 1)\,x - y + \frac{1}{2}(x - 1)\,x.$$

We let A_n denote the set of ordered triplets (k, x, y) in which $(x, y) \in A_n'$ and $k = k\,(x, y; n)$. Then the system (5.33) becomes the elegant triangular system

$$r_k'(t) = b_{k,k-n+x-1}\,r_{k-n+\ \,-1}(t) + b_{k,k-n+x}\,r_{k-n+x}(t) +$$
$$+ b_{k,k-1}\,r_{k-1}(t) + b_{k,k}\,r_k(t), \text{ for } (k, x, y) \in A_n$$

where

$$b_{k,k-n+x-1} = f_{1,0}(x, y; \boldsymbol{\theta}, n)\;\delta\,(n - x + y - 1),$$
$$b_{k,k-n+x} = f_{1,-1}(x, y; \boldsymbol{\theta}, n)\;\delta\,(y - 1),$$
$$b_{k,k-1} = f_{0,1}\,(x, y; \boldsymbol{\theta}, n)\;\delta\,(n - x - y - 1),$$
$$b_{k,k} = f_{0,0}\,(x, y; \boldsymbol{\theta}, n),$$

and for which the initial conditions are $r_k(0) = a_k$.

5.3.2.3. A recursive procedure for solving equations (5.33) may be described as follows. We define $\boldsymbol{\Theta}$ as the set of all possible values of the vector parameter $\boldsymbol{\theta}$. For fixed n let $\boldsymbol{\Theta}_n$ denote that subset of $\boldsymbol{\Theta}$ such that whenever there exist distinct pairs (x, y) and $(x', y') \in A_n'$ such that

$$b_{k,k}(x, y; \boldsymbol{\theta}, n) = b_{k',k'}(x', y'; \boldsymbol{\theta}, n),$$

where $k = k\,(x, y; n)$ and $k' = k\,(x', y'; n)$, then, assuming $k < k'$, either

$$b_{k+\gamma,k} = 0 \qquad \text{for } 1 \leqslant \gamma \leqslant k' - k,$$

or

$$b_{k',\gamma} = 0 \qquad \text{for } k \leqslant \gamma \leqslant k' - 1.$$

Then $\boldsymbol{\theta} \in \boldsymbol{\Theta}_n$ is sufficient to enable us to obtain the solution recursively. For $\boldsymbol{\theta} \in \boldsymbol{\Theta}_n$, and for $1 \leqslant i \leqslant s \equiv 1/2\,(n+1)\,(n+2)$, we get

$$r_i(t) = \sum_{v=1}^{s} c(i, v)\, \exp\,(b_{vv} t). \tag{5.34}$$

Here, for $(k, x, y) \in A_n$ and $1 \leqslant j \leqslant s$,

$$c(k,j) = \begin{cases} 0, \text{ for } k < j, \\ a_1, \text{ for } k = j = 1, \\ (b_{k,k-n+x-1} c^*(k-n+x-1, j)\, \delta(k-n+x-1-j) + \\ \qquad + b_{k,k-n-x}\, c^*(k-n-x, j)\, \delta(k-n-x, j) + \\ \qquad + b_{k,k-1}\, c^*(k-1, j))\, \alpha\,(b_{jj} - b_{kk}), \text{ for } k > j, \\ a_j - \sum_{v=1}^{j-1} c(j, v), \text{ for } k = j > 1, \end{cases}$$

in which

$$c^*(\mu, \nu) = \begin{cases} c(\mu, \nu), \text{ for } 1 \leqslant \mu, \nu \leqslant s \\ 0, \text{ otherwise} \end{cases}$$

and

$$\alpha\,(r) = \begin{cases} 1/r, & r \neq 0 \\ 0, & r = 0 \end{cases}$$

In this treatment are also included the special cases considered by GANI (1965 c) and SISKIND (1965).

5.3.3. Epidemic Markov chains

5.3.3.1. According to BAILEY (e.g., 1964 a, p. 182), discrete-time epidemic models have been used quite successfully in the statistical fitting of certain epidemic theories to data relating to small groups such as families. This statement refers to the so-called binomial chains. A transcription of the epidemic process into the language of Galton-Watson chains will be also given.

In the case of binomial chains, the main idea is that we have a fixed latent period, which may be used as the unit of time, and an infectious period which is contracted to a single point. It follows, starting with one or more initial cases all infectious simultaneously, that successive cases will occur in batches in discrete generations separated by the incubation period.

Let ξ_m, $m \in N$, denote the number of susceptibles in the group just before time m, who becomes infectious at m, and η_m the remaining number of uninfected susceptibles at time m. Clearly, $\eta_m = \xi_{m+1} + \eta_{m+1}$. Consider that the total initial population consists of n susceptibles of which x_0 $(1 \leqslant x_0 < n)$ become infective at time $m = 0$, leaving $y_0 = n - x_0$ still susceptible.

We further define the chance of *adequate contact* $r = (1 - q)$ which is the probability of contact at any time between any two individuals sufficient to produce a new infection if one of them is susceptible and one infective. The probability of new infections during $(m, m + 1)$ may be considered under two mutually exclusive assumptions: (I) the probability is independent of the number x $(\geqslant 1)$ of infectives in circulation at time m, or (II) it is dependent on the number x. The conditional probability that $\xi_{m+1}=x'$ and $\eta_{m+1} = y'$, given that $\xi_m=x$ and $\eta_m = y$, will be, under assumption I,

$$\frac{y!}{y'!(y-y')!} r^{y-y'} q^{y'}. \tag{5.35}$$

This is the Greenwood model. Under assumption II, we have

$$\frac{y!}{x'!(y-x')!} (1 - q^{x})^{x'} (q^{x})^{y-x'}, \tag{5.36}$$

which is the Reed-Frost model. The above formulation is that used by GANI and JERWOOD (1971); it differs from the "original" version (see BAILEY (1957, Ch. 6; 1964 a, p. 183; 1967 b, p. 197)) bringing out the fact that the Greenwood model is a finite Markov chain $(\eta_m)_{m \in N}$ while the Reed-Frost model is a bivariate Markov chain $(\xi_m, \eta_m)_{m \in N}$.

Let us consider the Greenwood model (5.35). The corresponding transition matrix $\mathbf{P}$, for an initial number $\eta_0 = n$ of susceptibles has entries

$$p_{ij} = \begin{cases} \binom{i}{j} r^{i-j} q^{j}, & 0 \leqslant j \leqslant i, \\ 0, & i < j \leqslant n, \end{cases}$$

and takes the lower triangular form.

$$\begin{array}{c} \\ 0 \\ 1 \\ 2 \\ \vdots \\ n \end{array} \begin{array}{c} \begin{array}{cccccc} 0 & 1 & 2 & & & n \end{array} \\ \begin{pmatrix} 1 & 0 & 0 & \dots & 0 \\ r & q & 0 & \dots & 0 \\ r^2 & 2rq & q^2 & \dots & 0 \\ \vdots & \vdots & \vdots & & \vdots \\ r^n & nr^{n-1}q & \binom{n}{2} r^{n-2} q^2 & \dots & q^n \end{pmatrix} \end{array}.$$

One can observe that **P** is the transition matrix of the Markov chain imbedded at times $j\tau(\tau =$ the latent period) in a pure death process.

If m is the smallest natural number such that $\eta_m = \eta_{m-1}$ we can say that the epidemic terminates at time $M = m$. The probability of terminating in state j, starting in state i, is

$$\mathsf{P}(M = m,\ \eta_m = \eta_{m-1} = j \mid \eta_0 = i) = r_{ij}^{(m-1)} p_{jj},$$

where $\mathbf{R} = (r_{ij}) = \mathbf{P} - \mathbf{Q}$, that is the matrix of remaining probabilities with diagonal elements zero ($\mathbf{Q} = \operatorname{diag}(p_{jj})$). Summing over all j, $0 \leqslant j \leqslant n$, we obtain

$$\mathsf{P}(M = m \mid \eta_0 = i) = \mathbf{A}_i \mathbf{R}^{m-1} \mathbf{Q}\,\mathbf{1}, \tag{5.37}$$

where $\mathbf{A}_i = (0, 0, \ldots, 1, \ldots)$ is the $1 \times (n+1)$ row vector with 1 in the $(i+1)$th column and zeros elsewhere, and $\mathbf{1}$ is the $(n+1) \times 1$ column vector of unit elements.

The probability generating function of the time to termination can be expresses as

$$\sum_{m \in N} \mathsf{P}(M = m \mid \eta_0 = i) z^m = \mathbf{A}_i (\mathbf{I} - z\mathbf{R})^{-1} z\mathbf{Q}\mathbf{1},\ 0 \leqslant z \leqslant 1,$$

where **I** is the identity matrix. J. Gani and D. Jerwood refer to the distribution (5.37) as a Markov geometric distribution because of its similarity of structure with the ordinary geometric distribution.

If we suppose that after an initial outbreak of infectiousness in the population considered, the remaining susceptibles are inoculated, the effect of this action could be regarded as an increase in the probability q of no contact with the infection. This probability will now vary from $q = q_0$ at $m = 0$ to some maximum value $q_1 > q_0$ after one latent period, say, after which it will gradually decrease to $q_\infty = q_0$ again. In this case $(\eta_m)_{m \in N}$ will be a non-homogeneous Markov chain and the probability of duration M of the epidemic starting with i susceptibles will now be given by

$$\mathbf{A}_i \mathbf{R}_0 \mathbf{R}_1 \ldots \mathbf{R}_{M-2} \mathbf{Q}_{M-1} \mathbf{1}.$$

A similar treatment of the bivariate Markov chain $(\xi_m, \eta_m)_{m \in N}$ may also be found in Gani and Jerwood (1971).

5.3.3.2. Another chain-binomial model was devised by Sugiyama (1961) who abandoned the assumption of the impossibility of an infection from without once the primary cases have occurred.

Very interesting are the papers of Elveback et al (1964, 1968, 1971) on extensions of the Reed-Frost model for the study of competition between viral agents in the presence of interference.

5.3.3.3. We shall close our account with a transcription of the epidemic process into the language of the theory of Galton-Watson chains. Within this theory one can assume that all infectives present in the population at a given time infect the susceptibles independently of each other. More precisely, if there are k infectives in the mth generation of the epidemic, then the distribution of the next $(m + 1)$ st generation can be represented as the sum of k independent identically distributed random variables with a specified distribution. These random variables represent the "progeny" of the k infectives of nth generation.

The epidemic process will thus be described on the basis of the following assumptions (BARTOSZINSKI (1967)):

I. Every individual who becomes infected passes first through an incubation period of length τ followed by a period of infectiousness of length T. During the period of incubation he is harmless to others, whereas during the period T he may infect those with whom he comes in contact. We shall measure time in appropriate discrete units (calling them days, for simplicity), and we assume that τ and T are random variables with the joint probability distribution

$$p_{ij} = \mathsf{P}(\tau = i,\ T = j).$$

Let $F(z_1, z_2) = \sum_{i,j \in N^*} p_{ij} z_1^i z_2^j$ be the probability generating function of the pair (τ, T).

II. At each day during the period τ every individual has a probability $1 - \alpha$, $0 < \alpha \leqslant 1$, of being discovered and isolated.

III. During the period T the probability of being discovered and isolated is $1 - \beta$, $(0 < \beta \leqslant 1)$, each day. Both these probabilities do not depend on the number of days during which the individual in question remains undiscovered.

IV. Every infective who has not been previously discovered makes a certain number of contacts with uninfected members of the population during each day of his period of infectiousness. The number of contacts made on different days are assumed to be independent and identically distributed with the distribution $(r_k)_{k \in N}$. Let $R(z) = \sum_{k \in N} r_k z^k$ be the probability generating function of the number of daily contacts.

V. We assume that each contact with an infective results in an infection with probability γ, $0 < \gamma \leqslant 1$, independently or the results of other contacts.

VI. The events occurring to an individual at a given time unit (day) are independent of the events that occurred to him or other members of the population at previous time units and they are independent of the events that occur to other members of the population at the same time unit.

VII. The expected number of daily contacts is finite.

Obviously, the probabilities introduced in hypotheses II, III and V have a medical interpretation. Thus, α may represent the efficiency of health service in its attempts to trace the individuals who may have had contact with the disease in order to isolate them, while β may measure the efficiency of periodic checkups and health control. Also, γ may represent the degree of contagiousness of the disease, or the resistance of the population due to vaccination, climate and so on.

Let us suppose that we have a single infective at a certain moment and let φ_j be the probability that he will be "effectively infective" (infective and not isolated) during exactly j time units ($j \in N^*$). He must therefore remain undiscovered during his entire period of incubation and during the first j days of his infectiousness period. If $\tau = i$, $T = j$, this probability is $\alpha^i\beta^j$; in case $\tau = i$, $T = h$ $(h > j)$, this probability equals $\alpha^i\beta^j(1 - \beta)$.

Denoting for simplicity

$$\mathsf{P}(\tau = i, T > j) = p_{i,j+1} + p_{i,j+2} + \ldots = q_{ij},$$

we get

$$\varphi_j = \beta^j \sum_{i \in N^*} p_{ij}\alpha^i + (1 - \beta)\beta^j \sum_{i \in N^*} q_{ij}\alpha^i.$$

The last formula holds for all $j \in N^*$; for $j = 0$ we must add a term which accounts for the possibility of being discovered during the period of incubation. Adding all φ_j, one could easily obtain $\sum_{j \in N} \varphi_j = F(\alpha, 1)$.

Now let ψ_k be the probability of exactly k contacts leading to the disease during one time unit of the period T. Taking into account hypotheses IV and V,

$$\psi_k = \sum_{l \geqslant k} \binom{l}{k} \gamma^k (1 - \gamma)^{l-k} r_l.$$

For the probability generating function of the sequence $(\psi_k)_{k \in N}$ we deduce

$$\sum_{k \in N} \psi_k w^k = \sum_{k \in N} \frac{(\gamma w)^k}{k!} R^{(k)}(1 - \gamma) = R(\gamma w + 1 - \gamma);$$

the inner series in the antepenultimate expression is the kth derivative of the function R at the point $1 - \gamma$, and the penultimate expression is the expansion of $R(\gamma w + 1 - \gamma)$ into a Taylor series in the neighbourhood of the point $1 - \gamma$. By the independence assumption, the probability generating function of the joint number of infectives during j time units of effective infectiousness is $[R(\gamma w + 1 - \gamma)]^j$.

Thus, for the probability generating function of the number of direct "descendants" of an infective we get

$$G(z) = 1 - F(\alpha, 1) + \sum_{j \in N^*} \varphi_j [R(\gamma z + 1 - \gamma)]^j =$$

$$= 1 - F(\alpha, 1) + \sum_{j \in N^*} \sum_{i \in N^*} \beta^j \alpha^i p_{ij} [R(\gamma z + 1 - \gamma)]^j +$$

$$+ (1 - \beta) \sum_{j \in N^*} \sum_{i \in N^*} \beta^j \alpha^i q_{ij} [R(\gamma z + 1 - \gamma)]^j =$$

$$= 1 - F(\alpha, 1) + F[\alpha, \beta R(\gamma z + 1 - \gamma)] +$$

$$+ (1 - \beta) H[\alpha, \beta R(\gamma z + 1 - \gamma)], \tag{5.38}$$

where

$$H(z_1, z_2) = \sum_{i, j \in N^*} q_{ij} z_1^i z_2^j = [1 - z_2]^{-1} [F(z_1, 1) - F(z_1, z_2)].$$

Hence one can obtain the equivalent of I; Theorem 1.2.9 for the considered epidemics. *Putting* $\partial F(z_1, z_2)/\partial z_2|_{z_1 = \alpha, z_2 = 1} = D(\alpha)$, *a single individual has a positive probability of causing an epidemic iff* $D(\alpha) > 1/\gamma R'(1)$ *and* $\beta_0 < \beta < 1$, *where* β_0 *is the smallest root of the equation*

$$R'(1)\gamma x F(\alpha, x) - R'(1)\gamma x F(\alpha, 1) + 1 - x = 0.$$

If the above conditions are satisfied, then the probability of an epidemic equals $1 - \rho$, *where* ρ *is the smallest positive root of the equation* $x = G(x)$ *with* G *given by* (5.38).

This proposition can be considered as the branching version of the threshold theorem.

One can see that in order to prevent an epidemic one must try to make either $\gamma D(\alpha) \leqslant 1/R'(1)$ for arbitrary β, or, if for some reason the above inequality is unattainable, one should try to make $\beta \leqslant [1 + \gamma R'(1) F(\alpha, 1)]^{-1}$, $(< \beta_0)$.

The case in which the incubation period is of constant length c gives $D(\alpha) = D\alpha^c$, and $F(\alpha, 1) = \alpha^c$, D being the average length of the period of infectiousness. Then the "safe" domain would be at least either $\gamma\alpha^c \leqslant 1/R'(1)D$, β arbitrary, or $\beta \leqslant (1 + \gamma R'(1)\alpha^c)$.

References

ABBEY, H.

(1952) An examination of the Reed-Frost theory of epidemics. Human Biology **24**, 201–233.

ABERCROMBIE, M., and J. E. M. HEAYSMAN

(1953) Social behaviour of cells in tissue culture. Exp. Cell Res. **5**, 111–131.

ADKE, S. R.

(1964 a) A multi-dimensional birth and death process. Biometrics, **20**, 212–216.

(1969) A birth, death and migration process. J. Appl. Probability **6**, 687–691.

ADKE, S. R., and J. E. MOYAL

(1963) A birth, death, and diffusion process. J. Math. Anal. Appl. **7**, 209–224.

AHMED, M. S.

(1963) A stochastic model for the tunnelling and retunnelling of the flour beetle. Biometrics, **19**, 341–351.

ALLEN, S. L.

(1968) Genetic and epigenetic control of several isozymic systems in Tetrahymena. Ann. N. Y. Acad. Sci. **151**/1, 190–207.

ALLING, D. W.

(1958) The after-history of pulmonary tuberculosis.: A stochastic model. Biometrics. **14**, 527–548.

ALLISON, A. C.

(1954) Notes on sick-cell polymorphism. Ann. Human Genetics **19**, 39–57.

ALMOND, J.

(1954) A note on the χ^2 test applied to epidemic chains. Biometrics **10**, 459–477.

ANDERSEN, F. S.

(1960) Competition in population consisting of one age group. Biometrics **16**, 19–27.

ANDREWARTHA, H. G., and L. C. BIRCH

(1954) The distribution and abundance of animals. Chicago: Univ. Chicago Press.

ANSCOMBE, F. Y.

(1949) The statistical analysis of insect counts based on the negative binomial distribution. Biometrics **5**, 165–173.

(1950) Sampling theory of the negative binomial and logarithmic series distributions. Biometrika **37**, 358–382.

ARLEY, N.
(1943) On the theory of stochastic processes and their application to the theory of cosmic radiations. Copenhagen; G. F. C. Gads.
(1964) Applications of stochastic models for the analysis of the mechanisms of carcinogenesis. In GURLAND (Ed.), (1964), pp. 3–39.
ARLEY, N., and S. IVERSEN
(1952) On the mechanism of experimental carcinogenesis. Acta Path. Microbiol. Scand. **31**, 164–171.
ARMITAGE, P.
(1952) The statistical theory of bacterial populations subject to mutation. J. Roy. Statist. Soc. Ser. B **14**, 1–40.
ARMITAGE, P., and R. DOLL
(1961) Stochastic models for carcinogenesis Proc. 4th Berkeley Symp. Math. Statist. Prob. Vol. IV, pp. 19–38. Berkeley; Univ. California Press.
ARMITAGE, P., G. G. MEYNELL, and T. WILLIAMS
(1965) Birth-death and other models for microbial infection. Nature **207**, 570–572.
ARNOLD, B. C.
(1968) A modification of a result due to Moran. J. Appl. Probability **5**, 220–223.
ASHBY, W. R., H. VON FOERSTER, and C. C. WALKER
(1962) Instability of pulse activity in a net with threshold. Nature **196**, 561–562.
ATHREYA, K. B., and S. KARLIN
(1970) Branching processes with random environments. Bull. Amer. Math. Soc. **76**, 865–870.
BAILEY, N. T. J.
(1952) A study of queues and appointment systems in hospital out-patient departments, with special reference to waiting-times. J. Roy. Statist. Soc. Ser. B **14**, 185–199.
(1955) A note on equalising the mean waiting times of successive customers in a finite queue. J. Roy. Statist. Soc. Ser. B **17**, 262–263.
(1957) The mathematical theory of epidemics. London; Griffin.
(1961) Introduction to the mathematical theory of genetic linkage. Oxford; Clarendon Press.
(1963) The simple stochastic epidemic; a complete solution in terms of known functions. Biometrika **50**, 235–240.
(1964a) The elements of stochastic processes with applications to the natural sciences. New York: Wiley.
(1964b) Some stochastic models for small epidemics in large populations. Appl. Statist. **13**, 9–19.
(1967a) The stimulation of stochastic epidemics in two dimensions. Proc. 5th Berkeley Symp. Math. Statist. Prob. Vol. IV, pp. 237–257. Berkeley: Univ. of California Press.
(1967b) The mathematical approach to biology and medicine. New York: Wiley.
(1968a) Stochastic birth, death and migration processes for spatially distributed populations. Biometrika **55**, 189–198.
(1968b) A perturbation approximation to the simple stochastic epidemic in a large population. Biometrika **55**, 199–209.

BAILEY, N. T. J., and C. ALFF-STEINBERGER
(1970) Improvements in the estimation of the latent and infectious periods of a contagious disease. Biometrika **57**, 141–153.
BAILEY, V. A., A. J. NICHOLSON, and E. J. WILLIAMS
(1962) Interactions between hosts and parasites when some host individuals are more difficult to find than others. J. Theoret. Biol. **3**, 1–18.
BARAKAT, R.
(1959) A note on the transient stage of the random dispersal of logistic populations. Bull. Math. Biophys. **21**, 141–151.
BARNETT, V. D.
(1962) The Monte Carlo solution of a competing species problem. Biometrics **18**, 76–103.
(1964) The joint distribution of occupation totals for a simple random walk. J. Austral. Math. Soc. **4**, 518–528.
BARRETT, J. C.
(1966) A mathematical model of the mitotic cycle and its application to the interpretation of percentage labelled mitoses data. J. Nat. Cancer Inst. **37**, 443–450.
BARTHOLOMAY, A. F.
(1962) Enzymatic reaction-rate theory: a stochastic approach. Ann. N. Y. Acad. Sci. **96**/4, 897–912.
BARTHOLOMEW, D. J.
(1963) A multi-stage renewal process. J. R. Statist. Soc. Ser. B. **25**, 150–168.
BARTKO, J. J., and G. A. WATTERSON
(1963) Inference on a genetic model of the Markov chain type. Biometrika **50**, 251–264.
BARTLETT, M. S.
(1949) Some evolutionary stochastic processes. J. Roy. Statist. Soc. Ser. B **11**. 211–229.
(1956) Deterministic and stochastic models for recurrent epidemics. Proc. 3rd Berkeley Symp. Math. Statist. Prob. Vol. IV, pp. 81–109. Berkeley: Univ, of California Press.
(1957a) On theoretical models for competitive and predatory biological systems. Biometrika **44**, 27–42.
(1957b) Measles periodicity and community size. J. Roy. Statist. Soc. Ser. A **120**, 48–70.
(1960a) Some stochastic models in ecology and epidemiology. In *Contributions to probability and statistics: Essays in honor of Harold Hotelling*, pp. 89–96. Stanford: Stanford Univ. Press.
(1960b) Stochastic population models in ecology and epidemiology. London: Methuen.
(1961) Monte Carlo studies in ecology and epidemiology. Proc. 4th Berkeley Symp. Math. Statist. Prob. Vol. IV, pp. 39–56. Berkeley: Univ. California Press.
(1964a) The relevance of stochastic models for large-scale epidemiological phenomena. Appl. Statist. **13**, 2–8.
(1964b) The spectral analysis of two-dimensional point processes. Biometrika **51**, 299–311.

(1966) An introduction to stochastic processes, with special reference to methods and applications. 2nd edition. London: Cambridge Univ. Press.

(1967) Inference and stochastic processes. J. Roy. Statist. Soc. Ser. A **130**, 457–474.

(1969) Distributions associated with cell populations. Biometrika **56**, 315–324.

(1970) Age distributions. Biometrics **26**, 377–385.

BARTLETT, M. S., J. C. GOWER, and P. H. LESLIE

(1960) A comparison of theoretical and empirical results for some stochastic population models. Biometrika **47**, 1–12.

BARTLETT, M. S., and D. G. KENDALL

(1951) On the use of the characteristic functional in the analysis of some stochastic processes occurring in physics and biology. Proc. Cambridge Phil. Soc. **46**, 65–76.

BARTOSZYŃSKI, R

(1967) Branching processes and the theory of epidemics. Proc. 5th Berkeley Symp. Math. Statist. Prob. Vol. IV, pp. 259–269. Berkeley: Univ. of California Press.

(1969) Branching processes and models of epidemics. Dissertationes Math. Rozprawy Mat. **61**, 1–51.

BARTOSZYŃSKI, R., J. ŁOŚ, and M. WYCECH-ŁOŚ

(1965) Contribution to the theory of epidemics. *Bernoulli*, *Bayes*, *Laplace* (Berkeley Seminar), pp. 1–8. Berlin: Springer.

BARTOSZYŃSKI, R., L. LUBIŃSKA, and S. NIEMIERKO

(1962) A stochastic model of AChE transportation in the peripheral nerve trunks. Biometrika **49**, 447–454.

BATHER, J. A.

(1963) Two non-linear birth and death processes. J. Austral. Math. Soc. **3**, 104–116.

BECKER, N. G.

(1968) The spread of an epidemic to fixed groups within the population. Biometrics **24**, 1007–1014.

(1970a) A stochastic model for two interacting populations. J. Appl. Probability **7**, 544–564.

(1970b) Control of a pest population. Biometrics **26**, 365–375.

BELLMAN, E., and T. E. HARRIS

(1948) On the theory of age-dependent stochastic branching processes. Proc. Nat. Acad. Sci. U.S.A. **34**, 601–604.

BELLMAN, R. E., J. A. JACQUEZ, and R. KALABA

(1961) Mathematical models of chemotherapy. Proc. 4th Berkeley Symp. Math. Statist. Prob. Vol. IV, pp. 57–66. Berkeley: Univ. California Press.

BENAYOUN, R.

(1964) Sur un modèle stochastique utilisé dans la théorie mathématique des épidémies. C. R. Acad. Sci. Paris **258**, 5789–5791.

BENTLEY, D. L.

(1963a) A contribution to counter theory. J. Roy. Statist. Soc. Ser. B **25**, 169–178.

(1963b) A mathematical representation for Wald's compartment theory. J. Opt. Soc. Amer. **53**, 287–292.

BERGER, A., and R. Z. GOLD

(1961) On comparing survival times. Proc. 4th Berkeley Symp. Math. Statist. Prob. Vol. IV, pp. 67–76. Berkeley: Univ. California Press.

BERGNER, P. E. E.

(1961) Tracer dynamics. I. A tentative approach and definition of fundamental concepts. J. Theoret. Biol. **1**, 120–140.

(1965a) Stochastic concepts in clinical red cell survival studies. Nature **205**, 975–977.

(1965b) On the ergodic properties of the tracer system. Biometrics **21**, 761–762 (Abstract).

BERKSON, J., and R. P. GAGE

(1952) Survival cure for cancer patients following treatment. J. Amer. Statist. Assoc. **47**, 501–515.

BERMAN, M.

(1961) Application of differential equations to the study of the thyroid system. Proc. 4th Berkeley Symp. Math. Statist. Prob. Vol. IV, pp. 87–99. Berkeley: Univ. California Press.

(1963) The formulation and testing of models. Ann. N. Y. Acad. Sci. **108**/1, 182–194.

BERMAN, S. M.

(1963a) A Markov process on binary numbers. Ann. Math. Statist. **34**, 416–423.

(1963b) Note on extreme values, competing risks and semi-Markov processes. Ann Math. Statist. **34**, 1104–1106.

(1965) A renewal theoretic model for chronic disease statistics. J. Appl. Probability **2**, 119–128. Correction ibid. (1967), **4**, 214.

BERNARD, S. R., L. R. SHENTON, and V. R. RAO UPPULURI

(1967) Stochastic models for the distribution of radioactive material in a connected system of compartments. Proc. 5th Berkeley Symp. Math. Statist. Prob. Vol. IV, pp. 481–510. Berkeley: Univ. California Press.

BERNARDELLI, H.

(1941) Population waves. J. Burma Res. Soc. **31**, 1–18.

BERTALANFFY, L. VON

(1960) Principles and theory of growth. In W. W. NOWINSKI (Ed.) *Fundamenta aspects of normal and malignant growth.* Amsterdam; Elsevier.

BEURLE, R. L.

(1956) Properties of a mass of cells capable of regenerating pulses. Philos. Trans. Roy. Soc. London Ser. B **240**, 55–94.

BEUTLER, E., M. YEH, and V. F. FAIRBANKS

(1962) The normal human female as a mosaic of X-chromosome activity studies using the gene for G–6–PD deficiency as a marker. Proc. Nat. Acad. Sci. U.S.A. **48**, 9–16.

BHARUCHA-REID, A. T.

(1952) A probability model of radiation damage. Nature **169**, 369–370.

(1956) On the stochastic theory of epidemics. Proc. 3rd Berkeley Symp. Math. Statist. Prob. Vol. IV, pp. 111–119. Berkeley: Univ. California Press.

(1960) Elements of the theory of Markov processes and their applications. New York; McGraw-Hill.

BHARUCHA-REID, A. T., and H. G. LANDAU

(1951) A suggested chain process for radiation damage. Bull. Math. Biophys. **13**, 153–163.

BHAT, B. R.

(1968) On an extension of Gani's model for attachment of phages to bacteria. J. Appl. Probability **5**, 572–578.

BIRTA, L.

(1965) A formal approach to concepts of interaction. Ph. D. Thesis, Case Inst. of Technology.

BISHIR, J.

(1962) Maximum population size in a branching process. Biometrics **18**, 394–403.

BLACKWELL, D., and D. G. KENDALL

(1964) The Martin boundary for Pólya's urn scheme, and an application to stochastic population growth. J. Appl. Probability **1**, 284–296.

BLATT, J. M.

(1967) Enzymatic break-up of polypeptides as a stochastic process. J. Theoret. Biol. **17**, 282–303.

BODMER, W. F.

(1960) Discrete stochastic processes in population genetics. J. Roy. Statist. Soc. Ser. B **22**, 218–244.

(1967) Models for DNA mediated bacterial transformation. Proc. 5th Berkeley Symp. Math. Statist. Prob. Vol. IV, pp. 377–401. Berkeley: Univ. California Press.

(1968) Demographic approaches to the measurement of differential selection in human populations. Proc. Nat. Acad. Sci. U.S.A. **59**, 690–699.

BODMER, W. F., and P. A. PARSONS

(1962) Linkage and recombination in evolution. Advances in Genetics, **11**, 1–100.

BOEN, J. R., and D. SYLVESTER

(1966) A quantitative discussion of the effectiveness of voiding as a defense against bladder infection. Biometrics **22**, 53–57.

BOL'ŠEV, L. N., and JU. I. KRUOPIS

(1969) On the question of modelling epidemic processes. Litovsk. Mat. Sb. **9**, 243–253. (Russian)

BOSSO, J. A., O. M. SORARRAIN, and E. E. A. FAVRET

(1969) Application of finite absorbent Markov chains to sib mating populations with selection. Biometrics **25**, 17–26.

BOSWELL, M. T., and G. P. PATIL

(1970a) Chance mechanisms generating the logarithmic series distributions used in the analysis of individuals and species. *Statistical Ecology*, Vol. I. University Park: Pennsylvania State Univ. Press.

(1970b) Chance mechanisms generating the negative binomial distribution. *Random counts in scientific work*, Vol. I. University Park: Pennsylvania State Univ. Press.

BOYD, J. P.
(1968) Algebra and consanguineal kinship. *Calcul et formalisation dans les sciences de l'homme*, pp. 47–58. Paris: CNRS.
BRAZEE, R. D., G. E. HALL, and O. K. HEDDEN
(1969) Image processing problems in research on aerosols and biological rheology. Ann. N. Y. Acad. Sci. **157**/1, 285–297.
BRAZIER, M. A. B.
(1967) The challenge of biological organization to mathematical description. Proc. 5th Berkeley Symp. Math. Statist. Prob. Vol. IV, pp. 1–10. Berkeley: Univ. California Press.
BROADBENT, S. R., and D. G. KENDALL
(1953) The random walk of *Trichostrongylus retortaeformis*. Biometrics **9**, 460–466.
BROCKWELL, P. J., and J. GANI
(1970) A population process with Markovian progenies. J. Math. Anal. Appl. **32**, 264–273.
BRONK, B. V., G. J. DIENES, and A. PASKIN
(1968) The stochastic theory of cell proliferation. Biophys. J. **8**, 1353–1398.
BURNETT-HALL, D. G., and W. A. O'N. WAUGH
(1967a) Sensitivity of a birth process to changes in the generation time distribution. Proc. 5th. Berkeley Symp. Math. Statist. Prob. Vol. IV, pp. 609–623. Berkeley: Univ. California Press.
(1967b) Indices of synchrony in cellular cultures. Biometrics **23**, 693–716.
BUHLER, W. J.
(1966) A theorem concerning the extinction of epidemics. Biometrische Z. **8**, 10–14.
(1967a) Single cell against multicell hypotheses of tumor formation. Proc. 5th Berkeley Symp. Math. Statist. Prob. Vol. IV, pp. 635–637. Berkeley: Univ. California Press.
(1967b) Quasi-competition of two birth and death processes. Biometrische Z. **9**, 76–83.
BÜNNING, E.
(1967) Known and unknown principles of biological chronometry. Ann. N. Y. Acad. Sci. **138**/2, 515–524.
BUSH, R. S., and W. R. BRUCE
(1964) The radiation sensitivity of transplanted lymphoma cells as determined by the spleen colony method. Radiation Res. **21**, 612–621.
CAIANIELLO, E. R.
(1961) Outline of a theory of thought processes and thinking machines. J. Theoret. Biol. **1**, 204–235.
CAIANIELLO, E. R. (Ed.)
(1968) Neural networks. Berlin: Springer.
CANE, V. R.
(1966) A note on the size of epidemics and the number of people hearing a rumour. J. Roy. Statist. Soc. Ser. B **28**, 487–490.
(1967) Mathematical models for neural networks. Proc. 5th Berkeley Symp. Math. Statist. Prob. Vol. IV, pp. 21–36. Berkeley: Univ. California Press.

CARR, R. N., and R. F. NASSAR
(1970) Effects of selection and drift on the dynamics of finite populations. I. Ultimate probability of fixation of a favorable allele. Biometrics **26**, 41—49.

CASSIE, R. M.
(1962) Frequency distribution models in the ecology of plankton and other organisms. J. Animal Ecol. **31**, 65—92.

CASTILLO, J. DEL, and B. KATZ
(1956) Biophysical aspects of neuro-muscular transmission. Progress in Biophys. and Biophys. Chem. **6**, 122—170.

CHANDRASEKHAR, S.
(1943) Stochastic problems in physics and astronomy. Rev. Modern Phys. **15**, 1—89.

CHANG, M. L., and T. S. CHANG
(1969) Direct solution of Markovian phage attachment to bacteria in suspension. Math. Biosci. **5**, 9—18.

CHANG, P. C.
(1970) Statistical models for animal survival time in mouse lymphoma. Biometrics **26**, 749—766.

CHAPMAN, D. G.
(1967) Stochastic models in animal population ecology. Proc. 5th Berkeley Symp. Math. Statist. Prob. Vol. IV, pp. 147—162. Berkeley: Univ. California Press.

CHIA, A. B.
(1968) Random mating in a population of cyclic size. J. Appl. Probability **5**, 21—30.

CHIA, A. B., and G. A. WATTERSON
(1969) Demographic effects on the rate of genetic evolution. I. Constant size populations with two genotypes. J. Appl. Probability **6**, 231—248.

CHIANG, C. L.
(1954) Competition and other interactions between species. *Statistics and Mathematics in Biology*, pp. 197—215. Ames: Iowa State Univ. Press.
(1957) An application of stochastic processes to experimental studies of flour beetles. Biometrics **13**, 79—97.
(1961) On the probability of death from specific causes in the presence of competing risks. Proc. 4th Berkeley Symp. Math. Statist. Prob. Vol. IV, pp. 169—180. Berkeley: Univ. California Press.
(1964) A stochastic model of competing risks of illness and competing risks of death. In GURLAND (Ed.) (1964), pp. 323—351.
(1968) Introduction to stochastic processes in biostatistics. New York: Wiley.
(1970) Competing risks and conditional probabilities. Biometrics **26**, 767—776.
(1971) A stochastic model of human fertility. I. Multiple transition probabilities. Biometrics **27**, 345—356.

CHOI, K., and W. G. BULGREN
(1968) An estimation procedure for mixtures of distributions. J. Roy. Statist. Soc. Ser. B **30**, 444—460.

CLEAVER, J. E.
(1967) Thymidine metabolism and cell kinetics. *Frontiers of Biology*, **6**. Amsterdam: North Holland.

Coggshall, J. C., and G. A. Bekey
(1970) A stochastic model of skeletal muscle based on motor unit properties. Math. Biosci. **7**, 405–419.

Coleman, R., and J. L. Gastwirth
(1969) Some models for interaction of renewal processes relating to neuron firing. J. Appl. Probability **6**, 38–58.

Conner, H. E.
(1964) Extinction probabilities for age- and position-dependent branching processes. SIAM J. Appl. Math. **12**, 899–909.
(1966) Limiting behaviour for age- and position-dependent branching processes. J. Math. Anal. Appl. **13**, 265–295.

Consaei., R.
(1949) Sur quelques processus stochastiques discontinus à deux variables aléatoires. I, II. Acad. Roy. Belg. Bull. Cl. Sci. **35**, 399–416; 743–755.
(1952) Sur les processus de Poisson du type composé. Acad. Roy. Belg. Bull. Cl. Sci. **38**, 442–461.

Cornfield, J., J. Steinfeld, and S. W. Greenhouse
(1960) Models for the interpretation of experiments using tracer compounds. Biometrics **16**, 212–234.

Cox, D. R., and P. A. W. Lewis
(1966) The statistical analysis of series of events. London: Methuen.

Cox, D. R., and H. D. Miller
(1965) The theory of stochastic processes. London: Methuen.

Cox, D. R., and W. L. Smith
(1953) The superposition of several strictly periodic sequences of events. Biometrika **40**, 1–11.
(1954) On the superposition of renewal processes. Biometrika **41**, 91–99.
(1957) On the distribution of *Tribolium confusum* in a container. Biometrika **44**, 393–404.

Crow, J. F.
(1968) Rates of genetic change under selection. Proc. Nat. Acad. Sci. U.S.A. **59**, 655–661.
(1969) Molecular genetics and population genetics. Proc. 12th Intern. Congress Genetics. Vol. 3, pp. 105–113. London: Pergamon Press.

Crow, J. F., and M. Kimura
(1955) Some genetic problems in natural populations. Proc. 3rd Berkeley Symp. Math. Statist. Prob. Vol. IV, pp. 1–22. Berkeley: Univ. California Press.

Crump, K. S.
(1970) Migratory populations in branching processes. J. Appl. Probability **7**, 565–572.

Crump, K. S., and C. J. Mode
(1968, 1969) A general age-dependent branching process I, II. J. Math. Anal. Appl. **24**, 494–508; **25**, 8–17.
(1969) An age-dependent branching process with correlation among sister cells. J. Appl. Probability **6**, 205–210.

CUISENIER, J.

(1968) L'utilisation des calculatrices électroniques dans l'étude des systèmes de parenté. *Calcul et formalisation dans les sciences de l'homme*, pp. 31–46. Paris: CNRS.

DALEY, D. J.

(1967) Concerning the spread of news in a population of individuals who never forget. Bull. Math. Biophys. **29**, 373–376.

(1968) Extinction conditions for certain bisexual Galton-Watson branching processes. Z. Wahrscheinlichkeitstheorie **9**, 315–322.

DALEY, D. J., and D. G. KENDALL

(1965) Stochastic rumours. J. Inst. Math. Appl. **1**, 42–55.

DANIELS, H. E.

(1967) The distribution of the total size of an epidemic. Proc. 5th Berkeley Symp. Math. Statist. Prob. Vol. IV, pp. 281–293. Berkeley: Univ. of California Press.

DARVEY, I. G., and P. J. STAFF

(1967) The application of the theory of Markov processes to the reversible one substrate – one intermediate – one product enzymic mechanism. J. Theoret. Biol. **14**, 157–172.

DARWIN, CH.

(1869) On the origin of species by means of natural selection or the preservation of favoured races in the struggle for life. Fifth edition. London: J. Murray.

DARWIN, J. H.

(1960) An ecological distribution akin to Fisher's logarithmic distribution. Biometrics **16**, 51–60.

DAVIDSON, R. G.

(1968) The Lyon hypothesis. Ann. N. Y. Acad. Sci. **151**/1, 157–158.

DAVIDSON, R. G., H. M. NITOWSKI, and B. CHILDS

(1963) Demonstration of two populations of cells in the human female heterozygous for glucose-6-phosphate dehydrogenase variants. Proc. Nat. Acad. Sci. U.S.A. **50**, 481–485.

DAVIS, A. W.

(1964) A note on the characteristic functional for a replacement process of Gani. J. Appl. Probability **1**, 157–160.

(1970) Some generalizations of Bailey's birth, death and migration model. Advances in Appl. Probability **2**, 83–109.

DAWE, C. J.

(1960) Cell sensitivity and specificity of response to polyoma virus. National Cancer Institute Monograph No. 4, pp. 67–92. New York: V. S. Dept. of Health, Education & Welfare.

DE LUCA, A., and L. M. RICCIARDI

(1968) Probabilistic description of neurons. In CAIANIELLO (Ed.) (1968), pp. 100–109.

DIETZ, K.

(1966) On the model of Weiss for the spread of epidemics by carriers. J. Appl. Probability **3**, 375–382.

(1967) Epidemics and rumours; A survey. J. Roy. Statist. Soc. Ser. A **130**, 505–528.

(1968) Erzeugung multimodeller Intervallverteilungen durch Ausdünnung von Erneuerungsprozess. Kybernetik **4**, 131–136.

(1971) Mathematical models for malaria in different ecological zones. Biometrics **27**, 249 (Abstract).

DIETZ, K., and F. DOWNTON

(1968) Carrier-borne epidemics with immigration. I. Immigration of both susceptibles and carriers. J. Appl. Probability **5**, 31–42.

DOBZHANSKY, T.

(1965) Evolutionary and population genetics. Proc. 11th Intern. Congress Genetics. Vol. 2, pp. 81–89. London: Pergamon Press.

(1967) Genetic diversity and diversity of environments. Proc. 5th Berkeley Symp. Math. Statist. Prob. Vol. IV, pp. 295–304. Berkeley; Univ. of California Press.

DOWNTON, F.

(1967 a) A note on the ultimate size of a stochastic epidemic. Biometrika **54**, 314–316.

(1967b) Epidemics with carriers: A note on a paper of Dietz. J. Appl. Probability **4**, 264–270.

(1968) The ultimate size of carrier-borne epidemics. Biometrika **55**, 277–289.

DWASS, J. T.

(1969) The total progeny in a branching process and a related random walk. J. Appl. Probability **6**, 682–686.

EDSALL, J. T.

(1964) Recent advances in protein research. In M. SELA (Ed.) *New perspectives in biology*, pp. 1–17. Amsterdam; Elsevier.

ELANDT-JOHNSON, R.

(1969) Survey of histocompatibility testing: biological background, probabilistic and statistical models and problems. Biometrics **25**, 207–283.

ELLISON, B. E.

(1966) Limit theorems for random mating in infinite populations. J. Appl. Probability **3**, 94–114.

ELTON, C. S.

(1958) The ecology of invasion by animals and plants. London: Methuen.

ELVEBACK, L., E. ACKERMAN, L. GATEWOOD, and J. P. FOX

(1971) Stochastic two-agent epidemic simulation models for a community of families. Amer. J. Epidemiology **93**, 267–280.

ELVEBACK, L., J. P. FOX, and E. ACKERMAN

(1968) A probabilistic model for interference between viral agents; Enteroviruses and the live poliovirus vaccine. Proc. 8th Intern. Congress Tropical Medicine and Malaria.

ELVEBACK, L., J. P. FOX, and A. VARMA
(1964) An extension of the Reed-Frost epidemic model for the study of competition between viral agents in the presence of interference. Amer. J. Hyg. **80**, 356–364.

ENGELBERG, J.
(1961) A method of measuring the degree of synchronization of cell populations. Exp. Cell Res. **23**, 218.

EPSTEIN, B.
(1967) Bacterial extinction time as an extreme value phenomenon. Biometrics **23**, 835–839.

EWENS, W. J.
(1963 a) Numerical results and diffusion approximations in a genetic process. Biometrika **50**, 241–249.
(1963 b) Diploid populations with selection depending on gene frequency. J. Austral. Math. Soc. **3**, 359–374.
(1963 c) The mean time for absorption in a process of genetic type. J. Austral. Math. Soc. **3**, 375–383.
(1963 d) The diffusion equation and a pseudodistribution in genetics. J. Roy. Statist. Soc. Ser. B **25**, 405–412.
(1965) The adequacy of the diffusion approximation to certain distributions in genetics. Biometrics **21**, 386–394.
(1968) Some applications of multiple-type branching processes in population genetics. J. Roy. Statist. Soc. Ser. B **30**, 164–175.
(1969) Population genetics. London: Methuen.

FALCONER, D. S.
(1967) Introduction to quantitative genetics. Edinburgh-London; Oliver & Boyd.

FATT, P., and B. KATZ
(1952) Spontaneous sub-threshold activity at motor nerve endings. J. Physiol. **117**, 109–128.

FEICHTINGER, G.
(1971) Stochastische Modelle demographischer Prozesse. Berlin: Springer.

FEINLEIB, M.
(1967) The stable disease model. Biometrics **23**, 1299 (Abstract).

FELDMAN, M. W
(1966) On the offspring number distribution in a genetic population. J. Appl. Probability **3**, 129–141.

FELLER, W.
(1939) Die Grundlagen der Volterraschen Theorie des Kampfes ums Dasein in wahrscheinlichkeitstheoretischer Behandlung. Acta Biotheoretica **5**, 11–40.
(1941) On the integral equation of renewal theory. Ann. Math. Statist. **12**, 243–267.
(1949) On the theory of stochastic processes, with particular reference to applications. Proc. Berkeley Symp. Math. Statist. Prob., pp. 403–432. Berkeley: Univ. of California Press.
(1951) Diffusion processes in genetics. Proc. 2nd Berkeley Symp. Math. Statist., pp. 227–246. Berkeley: Univ. of California Press.

(1959) The birth and death processes as diffusion processes. J. Math. Pures Appl. **38**, 301 – 345.

(1966) An introduction to probability theory and its applications. Vol. II. New York: Wiley.

(1968) An introduction to probability theory and its applications. Vol. I, 3rd edition. New York: Wiley.

FIENBERG, S. E.

(1970) A note on the diffusion approximation for single neuron firing problems. Kybernetika **7**, 227 – 229.

FINNEY, D. J., and L. MARTIN

(1951) A re-examination of Rahn's data on the number of genes in bacteria. Biometrics **7**, 133 – 144.

FIRESCU, D., and P. TAUTU

(1965 a) On a stochastic model of haematopoesis. Stud. Cerc. Mat. **17**, 1345 – 1359 (Romanian).

(1965 b) A stochastic model of embryonic endoderm formation at inferior metazoans. Rev. Roumaine Math. Pures Appl. **11**, 753 – 761.

(1966) A stochastic theory of multicompartment systems. An. Univ. Bucureşti Ser. Sti. Natur. Mat.-Mec. **15**, 75 – 87 (Romanian).

(1967 a) A stochastic model of a focal epidemic. Rev. Roumaine Math. Pures Appl. **12**, 653 – 664.

(1967 b) A simple stochastic model of population growth, heterogeneous by marker characters. Bull. Math. Soc. Sci. Math. R. S. Roumanie **11**, 133 – 142.

(1969) Stochastic considerations of all-or-none biological systems. Biometrische Z. **11**, 145 – 154.

FIRESCU, D., S. SAVU, and P. TAUTU

(1970) On the distribution function of pharmacoreceptors response-time. Rev. Roumaine Math. Pures Appl. **15**, 35 – 44.

FISHER, R. A.

(1922) On the dominance ratio. Proc. Roy. Soc. Edinburgh **42**, 321 – 341.

(1930 a) The genetical theory of natural selection. Oxford: Clarendon Press.

(1930 b) The distribution of gene ratios for rare mutations. Proc. Roy. Soc. Edinburgh **50**, 205 – 220.

(1953) Population genetics. Proc. Roy. Soc. Ser. B **141**, 510 – 523.

FISHER, R. A., A. S. CORBET, and C. B. WILLIAMS

(1943) The relation between the number of species and the number of individuals in a random sample of an animal population. J. Animal Ecol. **12**, 42 – 58.

FIX, E., and J. NEYMAN

(1951) A single stochastic model of recovery, relapse, death, and loss of patients Human Biology, **23**, 205 – 241.

FLOR, H. H.

(1956) The complementary genic systems in flax and flax rust. Advances in Genetics **8**, 29 – 54.

FOERSTER, H. VON

(1959) Some remarks on changing populations. In F. STOHLMAN (Ed.) *The kinetics of cellular proliferation*, pp. 382 – 407. New York: Grune & Stratton.

(1962) Bio-logic. Proc. 2nd Ann. Bionics Symp., pp. 1–15. New York: Plenum Press.

FOSTER, F. G.

(1955) A note on Bailey's and Whittle's treatment of a general stochastic epidemic. Biometrika **42**, 123–125.

FRAENKEL, G. S., and D. L. GUNN

(1940) The orientation of animals. Oxford: Oxford Univ. Press.

FRANCK, U. F.

(1956) Models for biological excitation processes. Progress in Biophys. and Biophys. Chem. **6**, 171–206.

FREESE, E., and A. YOSHIDA

(1965) The role of mutations in evolution. In V. BRYSON, H. J. VOGEL (Eds.) *Evolving genes and proteins*, pp. 341–355. New York: Academic Press.

FUCHS, G

(1965) Mathematische Theorie der Entwicklung einer in vitro isolierten Knochenmark-population und deren Anwendungsmöglichkeiten. Blut **11**, 1–17.

(1966) Theorie und Praxis stochastischer Modelle für die Entwicklung von in vitro isolierten Knochenmarkpopulationen. Methods of Information in Medicine **5**, 86–93.

(1969) Stochastische Modelle in der Nierenphysiologie. Biometrische Z. **11**, 25–49.

FURRY, W. H.

(1937) On fluctuation phenomena in the passage of high energy electrons through lead. Phys. Rev. **52**, 569–581.

GALTON, F., and H. W. WATSON

(1874) On the probability of the extinction of families. J. Roy. Anthropol. Inst. **4**, 138–144.

GANI, J.

(1957) Problems in the probability theory of storage systems. J. R. Statist. Soc. Ser. B **19**, 181–206.

(1961) On the stochastic matrix in a genetic model of Moran. Biometrika **48**, 203–206.

(1962 a) An approximate stochastic model for phage reproduction in a bacterium. J. Austral. Math. Soc. **2**, 478–483.

(1962 b) The extinction of a bacterial colony by phages: a branching process with deterministic removals. Biometrika **49**, 272–276.

(1963) Models for a bacterial growth process with removals. J. Roy. Statist. Soc. Ser. B **25**, 140–149.

(1965 a) On the age distribution of replaceable ranked elements. J. Math. Anal. Appl. **10**, 587–597.

(1965 b) Stochastic models for bacteriophage. J. Appl. Probability **2**, 225–268.

(1965 c) On a partial differential equation of epidemic theory. Biometrika **52**, 617–622.

(1967 a) On the general stochastic epidemic. Proc. 5th Berkeley Symp. Math. Statist. Prob. Vol. IV, pp. 271–279. Berkeley: Univ. of California Press.

(1967 b) Models for antibody attachment to virus and bacteriophage. Proc. 5th Berkeley Symp. Math. Statist. Prob. Vol. IV, pp. 537–547. Berkeley: Univ. of California Press.

(1971) Models in applied probability. J. Roy. Statist. Soc. Ser. A **134**, 26–38.

GANI, J., and D. JERWOOD

(1971) Markov chain methods in chain binomial epidemic models. Biometrics, **27**, 591–603.

GANI, J., and R. PYKE

(1960) The content of a dam as the supremum of an infinitely divisible process. J. Math. Mech. **9**, 639–652.

GANI, J., and R. C. SRIVASTAVA

(1968) A stochastic model for the attachment and detachment of antibodies to virus. Math. Biosci. **3**, 307–321.

GANI, J., and G. F. YEO

(1962) On the age distribution of n ranked elements after several replacements. Austral. J. Statist. **4**, 55–60.

(1965) Some birth-death and mutation models for phage reproduction. J. Appl. Probability **2**, 150–161.

GARFINKEL, D., R. H. MACARTHUR, and R. SACK

(1964) Computer simulation and analysis of simple ecological systems. Ann. N. Y. Acad. Sci. **115**/2, 943–951.

GART, J. J.

(1965) Some stochastic models relating time and dosage in response curves. Biometrics **21**, 583–599.

(1968) The mathematical analysis of an epidemic with two kinds of susceptibles. Biometrics **24**, 557–566.

GART, J. J., and J. L. DE VRIES

(1966) The mathematical analysis of concurrent epidemics of yaws and chickenpox. J. Hyg. **64**, 431–439.

GEISLER, C. D., and J. M. GOLDBERG

(1966) A stochastic model of the repetitive activity of neurons. Biophys. J. **6**, 53–69.

GELFANT, S.

(1963) A new theory on the mechanism of cell division. Intern. Soc. Cell Biology **2**, pp. 229–259. New York: Academic Press.

GERSTEIN, G. L., and D. MANDELBROT

(1964) Random walk models for the spike activity of a single neuron. Biophys. J. **4**, 41–68.

GILLOIS, M.

(1965/66) Relation d'identité en génétique. Ann. Inst. H. Poincaré Sect. B **2**, 1–94.

GLEASON, A. M.

(1966) Fundamentals of abstract analysis. Reading: Addison-Wesley.

GLINEUR, L.

(1967) Études de modèles épidémiques. Ph. D. Thesis, Bruxelles.

GLUSS, B.

(1967) A model for neuron firing with exponential decay of potential resulting in diffusion equations for probability density. Bull. Math. Biophys. **29**, 233–243.

GOFFMAN, W., and V. A. NEWILL

(1964) Generalization of epidemic theory. An application to the transmission of ideas. Nature **204**, 225–228.

GOOD, I. J.

(1953) The population frequencies of species and the estimation of population parameters. Biometrika **40**, 237–264.

(1955) The joint distribution for the sizes of the generations in a cascade process. Proc. Cambridge Philos. Soc. **51**, 240–242.

GOODMAN, L. A.

(1953) Population growth of the sexes. Biometrics **9**, 212–225.

(1967 a) On the reconciliation of mathematical theories of population growth. J. Roy. Statist. Soc. Ser. A **130**, 541–553.

(1967 b) The probabilities of extinction for birth-and-death processes that are age-dependent or phase-dependent. Biometrika **54**, 579–596.

(1968) Stochastic models for the population growth of the sexes. Biometrika **55**, 469–487.

(1969) The analysis of population growth when the birth and death rates depend upon several factors. Biometrics **25**, 659–681.

GORDON, M.

(1962) Good's theory of cascade processes applied to the statistics of polymer distributions. Proc. Roy. Soc. Ser. A **268**, 240–256.

GORDON, P.

(1965) Théorie des chaînes de Markov finies et ses applications. Paris: Dunod.

GOWEN, J. W.

(1961) Experimental analysis ot genetic determinants in resistance to infectious disease. Ann. N. Y. Acad. Sci. **91**/3, 689–709.

GREEN, M.

(1970) Effect of oncogenic DNA viruses on regulatory mechanisms of cells. Fed. Proc. **29**, 1265–1275.

GREENHOUSE, S. W.

(1961) A stochastic process arising in the study of muscular contraction. Proc. 4th Berkeley Symp. Math. Statist. Prob. Vol. IV, pp. 257–265. Berkeley; Univ. of California Press.

GREENWOOD, M.

(1946) The statistical study of infectious diseases. J. Roy. Statist. Soc. Ser. A **109**, 85–110.

GREENWOOD, M., and G. U. YULE
(1920) An inquiry into the nature of frequency distributions representative of multiple happenings with particular reference to the occurrence of multiple attacks of disease or of repeated accidents. J. Roy. Statist. Soc. **82**, 255–279.
GRIFFITH J. S.
(1963) On the stability of brain-like structures. Biophys. J. **3**, 187–195.
GRUNDFEST, H.
(1961) General physiology and pharmacology of junctional transmission. In A. M. SHANES (Ed.) *Biophysics of physiological and pharmacological actions*, pp. 329–389. Washington: Amer. Assoc. for Adv. Sci.
GUPTA, S. K.
(1965) Queues with Poisson input and mixed Erlangian service time distribution with finite waiting space. J. Operations Res. Soc. Japan **8**, 24–31.
GURLAND, J.
(1957) Some interrelations among compound and generalized distributions. Biometrika **44**, 265–268.
GURLAND, J. (Ed.)
(1964) Stochastic models in medicine and biology. Madison: Univ. Wisconsin Press.
HAIGHT, F. A.
(1967) Handbook of the Poisson distribution. New York: Wiley.
HAJNAL, J.
(1957) Mathematical models in demography. Cold Spring Harbor Symp. Quant. Biol. **22**, 97–103.
HALDANE, J. B. S.
(1949) Some statistical problems arising in genetics. J. Roy. Statist. Soc. Ser. B **11**, 1–14.
HARARY, F., and B. LIPSTEIN
(1962) The dynamics of brand loyalty: A Markovian approach. Operations Res. **10**, 19–40.
HARARY, F., R. Z. NORMAN, and D. CARTWRIGHT
(1965) Structural models: An introduction to the theory of directed graphs. New York: Wiley.
HARLAMOV, B. P.
(1968) Some properties of branching processes with an arbitrary set of particle types. Teor. Verojatnost. i Primenen. **13**, 82–95 (Russian).
HARRIS, H.
(1970 a) Genetical theory and the "inborn errors of metabolism". Brit. Med. J. **1**, 321–327.
(1970 b) The principles of human biochemical genetics. Amsterdam: North-Holland.
HARRIS, T. E.
(1951) Some mathematical models for branching processes. Proc. 2nd Berkeley Symp. Math. Statist. Prob., pp. 305–328. Berkeley: Univ. California Press.
(1959) A mathematical model for multiplication by binary fission. In F. STOHLMAN (Ed.) *The kinetics of cellular proliferation*, pp. 368–381. New York: Grune & Stratton.
(1963) The theory of branching processes. Berlin: Springer.

HARRIS, T. E., P. MEIER, and J. TUKEY
(1950) Timing of the distribution of events between observations: a contribution to the theory of follow-up studies. Human Biology **22**, 249—270.
HASKEY, H. W.
(1954) A general expression for the mean in a simple stochastic epidemic. Biometrika **41**, 272—275.
(1957) Stochastic cross-infection between two otherwise isolated groups. Biometrika **44**, 193—204.
HAUSCHKA, T. S., S. T. GRINNELL, L. RÉVÉSZ, and G. KLEIN
(1957) Quantitative studies on the multiplication of neoplastic cells in vivo. IV. J. Nat. Cancer Inst. **19**, 13—30.
HEARON, J. Z.
(1963) Theorems on linear systems. Ann. N. Y. Acad. Sci. **108**/1, 36—68.
HEIDELBERGER, C.
(1964) Discussion to N. Arley's paper. In GURLAND (Ed.) (1964), p. 41.
HILL, B. M.
(1963) The three-parameter lognormal distribution and Bayesian analysis of a point-source epidemie. J. Amer. Statist. Assoc. **58**, 72—84.
HILL, R. T., and N. C. SEVERO
(1969) The simple stochastic epidemic for small populations with one or more initial infectives. Biometrika **56**, 183—196.
HILL, W. G., and A, ROBERTSON
(1968) Linkage disequilibrium in finite populations. Theoret. Appl. Genet. **38**. 226—231.
HIRSCH, H. R., and J. ENGELBERG
(1966) Decay of cell synchronization: solutions of the cell growth equation. Bull Math. Biophys. **28**, 391—409.
HOFFMAN, J. G.
(1949) Theory of the mitotic index and its application to tissue growth measurement. Bull. Math. Biophys. **11**, 139—144.
HOFFMAN, J. G., N. METROPOLIS, and V. GARDINER
(1956) Digital computer studies of cell multiplication by Monte Carlo methods. J. Nat. Cancer Inst. **17**, 175—188.
HOLGATE, P.
(1965 a) The distance from a random point to the nearest point of a closely packed lattice. Biometrika **52**, 261—263.
(1965 b) Tests of randomness based on distance methods. Biometrika **52**, 345—353.
(1966) A mathematical study of the founder principle of evolutionary genetics. J. Appl. Probability **3**, 115—128.
(1967 a) The size of elephant herds. Math. Gaz. **51**, 302—304.
(1967 b) Divergent population processes and mammal outbreaks. J. Appl. Probability **4**, 1—8.
(1969) Species frequency distributions. Biometrika **56**, 651—660.
HOLLANDER, J. L., D. J. MCCARTHY, G. ASTORGA, and E. CASTRO-MURILLO
(1965) Studies on pathogenesis of rheumatoid joint inflammation. I. "R. A. Cell' and working hypothesis. Ann. Intern. Med. **62** 271—284.

HORN, M., and H. GRIMM
(1969) An application of birth and death processes in cancer chemotherapy. 37th Session Inst. Internat. Statist, Contributed papers, 222–224.
HORRIDGE, G. A.
(1956) The co-ordination of the protective retraction of coral polyps. Philos. Trans. Roy. Soc. London Ser. B **240**, 495–528.
HOUSEHOLDER, A. S., and A. W. KIMBALL
(1954) A stochastic model for the selection of macronuclear units in *Paramecium* growth. Biometrics **10**, 361–374.
HUG, O., and A. M. KELLERER
(1966) Stochastic der Strahlenwirkung. Berlin: Springer.
HYRENIUS, H.
(1965) New technique for studying demographic-economic-social interrelations. Rep. No. 3, Demographic Institute, Göteborg.
IGLEHART, D. L.
(1964 a) Multivariate competition processes. Ann. Math. Statist. **35**, 350–361.
(1964 b) Reversible competition processes. Z. Wahrscheinlichkeitstheorie **2**, 314–331.
IOSIFESCU, M., and P. TAUTU
(1968) Applications of the theory of random systems with complete connections in evolutionary biology. 3rd Coll. Prob. Theory, Braşov. (unpublished)
IRWIN, J. O.
(1963) The place of mathematics in medical and biological statistics. J. Roy. Statist. Soc. Ser. A **126**, 1–41.
(1964) The contribution of G. U. Yule and A. G. McKendrick to stochastic process methods in biology and medicine. In GURLAND (Ed.) (1964), pp. 147–163.
(1967) Discussion of Bartlett's (1967) paper. J. Roy. Statist. Soc. Ser. A **130**, 476–477.
IYER, K. S. S., and V. N. SAKSENA
(1970) A stochastic model for the growth of cells in cancer. Biometrics **26**, 401–410.
JAGERS, P.
(1969 a) A general stochastic model for population development. Skand. Aktuarietidskr. **52**, 84–103.
(1969 b) The proportion of individuals of different kinds in two-type populations. A branching process arising in biology. J. Appl. Probability **6**, 249–260.
(1970) The composition of branching populations. Math. Biosci. **8**, 227–238.
JAQUETTE, D. L.
(1970) A stochastic model for the optimal control of epidemics and pest populations. Math. Biosci. **8**, 343–354.
JAY, G. E.
(1955) Variation in response of various mouse strains to hexobarbital (Evipal). Proc. Soc. Exptl. Biol. Med. **90**, 378–385.
JAYKAR, S. D.
(1970) A mathematical model for interaction of gene frequencies in a parasite and its host. Theoret. Population Biol. **1**, 140–164.

JENSEN, L,. and E. POLLAK
(1969) Random selective advantages of a gene in a finite population. J. Appl. Probability **6**, 19–37.

JOHANNESMA, P. I. M.
(1968) Diffusion models for the stochastic activity of neurons. In CAIANIELLO (Ed.) (1968), pp. 116–144.

JOSHI, D. D.
(1954) Les processus stochastiques en démographie. Publ. Inst. Statist. Univ. Paris **3**, 153–177.

KARLIN, S.
(1964) Discussion of Karlin and McGregor's (1964) paper. In GURLAND (Ed.) (1964), p. 273.
(1966) A first course in stochastic processes. New York: Academic Press.
(1968) Equilibrium behavior of population genetic models with non-random mating. I, II. J. Appl. Probability 5, 231–313; 487–566.

KARLIN, S., and J. L. MCGREGOR
(1957) The differential equations of birth and death processes and the Stieltjes moment problem. Trans. Amer. Math. Soc. **85**, 489–546.
1962) On a genetics model of Moran. Proc. Cambridge Philos. Soc. **58**, 299–311.
1964) On some stochastic models in genetics. In GURLAND (Ed.) (1964), pp. 245–271.
(1965) Direct product branching processes and related induced Markoff chains. I. Calculation of rates of approach to homozygosity. *Bernoulli*, *Bayes*, *Laplace* (Berkeley Seminar), pp. 111–145. Berlin: Springer.
(1967 a) Uniqueness of stationary measures for branching processes and applications. Proc. 5th Berkeley Symp. Math. Statist. Prob. Vol. II, Part 2, pp. 243–254. Berkeley: Univ. California Press.
(1967 b) The number of mutant forms maintained in a population. Proc. 5th Berkeley Symp. Math. Statist. Prob. Vol. IV, pp. 415–438. Berkeley: Univ. California Press.
(1968) The role of the Poisson progeny distribution in population genetic models. Math. Biosci. **2**, 11–17.

KARLIN, S., J. MCGREGOR, and W. F. BODMER
(1967) The rate of production of recombinants between linked genes in finite populations. Proc. 5th Berkeley Symp. Math. Statist. Prob. Vol. IV, pp. 403–414. Berkeley; Univ. California Press.

KATTI, S. K.
(1966) Interrelations among generalized distributions and their components. Biometrics **22**, 44–52.

KATTI, S. K., and J. GURLAND
(1961) The Poisson Pascal distribution. Biometrics **17**, 527–538.

KELLY, C. D., and O. RAHN
(1932) The growth rate of individual bacterial cells. J. Bact. **23**, 147–153.

KEMENY, J. G., and J. L. SNELL
(1960) Finite Markov chains. Princeton: Van Nostrand.

KENDALL, D. G.
(1948 a) On the generalized birth-and-death process. Ann. Math. Statist. **19**, 1–15.
(1948 b) On some modes of population growth leading to R. A. Fisher's logarithmic series distribution. Biometrika **35**, 6–15.
(1948 c) On the role of a variable generation time in the development of a stochastic birth process. Biometrika **35**, 316–330.
(1949) Stochastic processes and population growth. J. Roy. Statist. Soc. Ser. B **11**, 230–264.
(1950) Random fluctuations in the age-distribution of a population whose development is controlled by the simple "birth-and-death" process. J. Roy. Statist. Soc. Ser. B **12**, 278–285.
(1952) Les processus stochastiques de croissance en biologie. Ann. Inst. H. Poincaré **13**, 43–108.
(1953) Stochastic processes and the growth of bacterial colonies. Symp. Soc. Exper. Biol. **7**, 55–65.
(1956) Deterministic and stochastic epidemics in closed populations. Proc. 3rd Berkeley Symp. Math. Statist. Prob. Vol. IV, pp. 149–165. Berkeley: Univ. California Press.
(1957) Some problems in the theory of dams. J. Roy. Statist. Soc. Ser. B **19**, 207–212.
(1960) Birth-and-death processes and the theory of carcinogenesis. Biometrika **47**, 13–21.
(1964) Some recent work and further problems in the theory of queues. Teor. Vero jatn. i Primenen. **9**, 3–15.
(1966) Branching processes since 1873. J. London Math. Soc. **41**, 385–405.

KERMACK, W. O., and A. G. MCKENDRICK
(1927) Contributions to the mathematical theory of epidemics. Proc. Roy. Soc. Ser. A **115**, 700–721.
(1932) Contributions to the mathematical theory of epidemics. II. The problem of endemicity. Proc. Roy. Soc. Ser. A **138**, 55–83.

KERNER, E. H.
(1962) Gibbs ensemble and biological ensemble. Ann. N. Y. Acad. Sci. **96**/4, 975–984.

KESTEN, H.
(1970) Quadratic transformations: A model for population growth. I, II. Advances in Appl. Probability **2**, 1–82; 179–228.

KEYFITZ, N.
(1967 a) Reconciliation of population models: matrix, integral equation and partial fraction. J. Roy. Statist. Soc. Ser. A **130**, 61–83.
(1967 b) Estimating the trajectory of a population. Proc. 5th Berkeley Symp. Math. Statist. Prob. Vol. IV, pp. 81–113. Berkeley: Univ. California Press.
(1968) Introduction to the mathematics of population. Reading: Addison-Wesley.

KEYFITZ, N., and E. M. MURPHY
(1967) Matrix and multiple decrement in population analysis. Biometrics **23**, 485–503.

KHATRI, C. G., and J. R. PATEL
(1961) Three classes of univariate discrete distributions. Biometrics **17**, 567–575.

KHAZANIE, R. G.
(1968) An indication of the asymptotic nature of the Mendelian Markov process. J. Appl. Probability **5**, 350–356.

KHAZANIE, R. G., and H. E. MCKEAN
(1966 a) A Mendelian Markov process with binomial transition probabilities. Biometrika **53**, 37–48.
(1966 b) A Mendelian Markov process with multinomial transition probabilities. J. Appl. Probability **3**, 353–364.

KILLMANN, S. A., E. P. CRONKITE, T. M. FLIEDNER, V. P. BOND, and G. BRECHER
(1963) Mitotic indices of human bone marrow cells. II. The use of mitotic indices for estimation of time parameters of proliferation in serially connected multiplicative cellular compartments. Blood **21**, 141–162.

KIMBALL, A. W.
(1965) A model for chemical mutagenesis in bacteriophage. Biometrics **21**, 875–889.

KIMURA, M.
(1955) Solution of a process of random genetic drift with a continuous model. Proc. Nat. Acad. Sci. U.S.A. **41**, 144–149.
(1957) Some problems of stochastic processes in genetics. Ann. Math. Statist. **28**, 882–901.
(1964) Diffusion models in population genetics. J. Appl. Probability **1**, 177–232.
(1965) A stochastic model concerning the maintenance of genetic variability in quantitative characters. Proc. Nat. Acad. Sci. U.S.A. **54**, 731–736.
(1969 a) The rate of molecular evolution considered from the standpoint of population genetics. Proc. Nat. Acad. Sci. U.S.A. **63**, 1181–1188.
(1969 b) The number of heterozygous nucleotide sites maintained in a finite population due to steady flux of mutations. Genetics **61**, 893–903.
(1970 a) The length of time required for a selectively neutral mutant to reach fixation through random frequency drift in a finite population. Genet. Res. **15**, 131–133.
(1970 b) Stochastic processes in population genetics, with special reference to distribution of gene frequencies and probability of gene fixation. In KOJIMA (Ed.) (1970), pp. 178–209.

KIMURA, M., and F. CROW
(1964) The number of alleles that can be maintained in a finite population. Genetics **49**, 725–738.

KIMURA, M., and T. OHTA
(1969 a) The average number of generations until fixation of a mutant gene in a finite population. Genetics **61**, 763–771.
(1969 b) The average number of generations until extinction of an individual mutant gene in a finite population. Genetics **63**, 701–709.
(1971) Theoretical aspects of population genetics. Princeton, N. J.: Princeton Univ. Press.

KING, J. L., and T. H. JUKES
(1969) Non-Darwinian evolution. Science **164**, 788–798.

KINGMAN, J. F. C.
(1963) Poisson counts for random sequences of events. Ann. Math. Statist. **34**, 1217–1232.
(1969) Markov population processes. J. Appl. Probability **6**, 1–18.
KIRK, D.
(1968) Patterns of survival and reproduction in the United States: Implications for selection. Proc. Nat. Acad. Sci. U.S.A. **59**, 662–670.
KIRKMAN, H. N.
(1968) Glucose-6-phosphate dehydrogenase variants and drug-induced hemolysis. Ann. N. Y. Acad. Sci. **151**/2, 753–764.
KLEIN, G., and E. KLEIN
(1956) Detection of an allelic difference at a single gene locus in a small fraction of a large tumour-cell population. Nature **178**, 1389–1391.
KLONECKI, W.
(1965) A method for derivation of probabilities in a stochastic model of population growth for carcinogenesis. Colloq. Math. **13**, 273–288.
(1967) On the distribution of the number of survivors and deaths in a birth and death process. Colloq. Math. **16**, 269–279.
(1970) Identifiability questions for chance mechanisms underlying stochastic models for carcinogenesis. Math. Biosci. **7**, 365–377.
KOCH, A. L., and M. SCHAECHTER
(1962) A model for statistics of the cell division process. J. Gen. Microbiol. **29**, 435–454.
KODLIN, D.
(1967) A new response time distribution. Biometrics **23**, 227–239.
KOJIMA, K. (Ed.)
(1970) Mathematical topics in population genetics. Berlin: Springer.
KOJIMA, K., and H. E. SCHAFFER
(1964) Accumulation of epistatic gene complexes. Evolution **18**, 127–129.
(1967) Survival process of linked mutant genes. Evolution **21**, 518–531.
KOLMOGOROV, A. M.
(1959) The transition of branching processes into diffusion processes and associated problems of genetics. Teor. Verojatnost. i Primenen. **4**, 233–236 (Russian).
KÖRNER, S.
(1960) The philosophy of mathematics. An introductory essay. London: Hutchinson.
KOSTITZIN, V. A.
(1939) Mathematical biology. London: Harrap.
KOŽEŠNÍK, J.
(1965 a) Stochastic theory of configurations. I, II. Bull. Acad. Polon. Sci. Ser. Sci. Tech. **13**, 497–503; 505–512.
(1965 b) Stochastic processes in linear configurations in which the transition probability densities do not depend on states of particular components. I, II. Bull. Acad. Polon. Sci. Ser. Sci. Tech. **13**, 559–567; 569–574.
(1965 c) Stochastic theory of biological and economic configuration. Sankhyā Ser. A **27**, 213–230.

KRYSCIO, R. J., and N. C. SEVERO

(1969) Some properties of an extended simple stochastic epidemic model involving two additional parameters. Math. Biosci. **5**, 1–8.

KUBITSCHEK, H. E.

(1961) Normal distribution of cell generation rate. Exper. Cell Res. **26**, 439–450.

(1967) Cell generation times: ancestral and internal controls. Proc. 5th Berkeley Symp. Math. Statist. Prob. Vol. IV, pp. 549–572. Berkeley: Univ. California Press.

LAJTHA, L. G., and R. OLIVER

(1963) Cell population kinetics following different regimes of irradiation. Brit. J. Radiol. **35**, 131–140.

LAJTHA, L. G., R. OLIVER, and C. W. GURNEY

(1962) Kinetic model of a bone-marrow stem-cell population. Brit. J. Haematol. **8**, 442–460.

LAMENS, A.

(1957) Sur le processus non homogène de naissance et de mort à deux variables aléatoires. Acad. Roy. Belg. Bull. Cl. Sci. **43**, 711–719.

LAMENS, A., and R. CONSAEL

(1957) Sur le processus non homogène de naissance et de mort. Acad. Roy. Belg. Bull. Cl. Sci. **43**, 597–605.

LEA, D. A., and C. A. COULSON

(1949) The distribution of the number of mutants in bacterial populations. J. Genetics **49**, 264–285.

LEITNER, F.

(1965) The use of two radiotracers for transfer-rate determinations in closed two compartment systems. Bull. Math. Biophys. **27**, 431–434.

LESLIE, P. H.

(1945) On the use of matrices in certain population mathematics. Biometrika **33**, 183–212.

(1948) Some further notes on the use of matrices in population mathematics. Biometrika **35**, 213–245.

(1957) An analysis of the data for some experiments carried-out by Gause with populations of the protozoa, *Paramecium aurelia* and *Paramecium caudatum*. Biometrika **44**, 314–327.

(1958) A stochastic model for studying the properties of certain biological systems by numerical methods. Biometrika **45**, 16–31.

(1962) A stochastic model for two competing species of *Tribolium* and its application to some experimental data. Biometrika **49**, 1–26.

LESLIE, P. H., and J. C. GOWER

(1958) The properties of a stochastic model for two competing species. Biometrika **45**, 316–330.

LESLIE, R. T.

(1967) Recurrent composite events. J. Appl. Probability **4**, 34–61.

LEVENE, H.

(1967) Genetic diversity and diversity of environment: mathematical aspects. Proc. 5th Berkeley Symp. Math. Statist. Prob. Vol. IV, pp. 305—316. Berkeley: Univ. California Press.

LEVINS, R.

(1965) Genetic consequences of natural selection. In WATERMAN and MOROWITZ (Eds.) (1965), pp. 371—387.

(1969) The effect of random variations of different types on population growth Proc. Nat. Acad. Sci. U.S.A. **62**, 1061—1065.

LEWIS, E. G.

(1942) On the generation and growth of a population. Sankhyā **6**, 93—96.

LEWIS, P. A. W.

(1965) Some results on tests for Poisson processes. Biometrika **52**, 67—77.

LEWONTIN, R. C.

(1957) The adaptation of population to varying environments. Cold Spring Harbor Symp. Quant. Biol. **22**, 395—408.

(1963) Models, mathematics and metaphors. Synthese **15**, 222—244.

(1967) The genetics of complex systems. Proc. 5th Berkeley Symp. Math. Statist. Prob. Vol. IV, pp. 439—455. Berkeley: Univ. California Press.

(1970) On the irrelevance of genes. *Towards a theoretical biology* **3**, pp. 63—72. Edinburgh: Edinburgh University Press.

LEWONTIN, R. C., and D. COHEN

(1969) On population growth in a randomly varying environment. Proc. Nat. Acad. Sci. U.S.A. **62**, 1056—1060.

LI, C. C.

(1967) Genetic equilibrium under selection. Biometrika **23**, 397—484.

LIČKO, V.

(1965) On compartmentalization. Bull. Math. Biophys. **27**, Suppl. 15—19.

LILLESTØL, J.

(1968) Another approach to some Markov chain models in population genetics. J. Appl. Probability **5**, 9—20.

LINCOLN, T. L., and G. H. WEISS

(1964) A statistical evaluation of recurrent medical examinations. Operations Res. **12**, 187—205.

LINDER, D., and S. M. GARTLER

(1965) Glucose-6-phosphate dehydrogenase mosaicism: utilization as a cell marker in the study of leiomyomas. Science **150**, 67—68.

(1967) Problem of single cell versus multicell origin of a tumor. Proc. 5th Berkeley Symp. Math. Statist. Prob. Vol. IV, pp. 625—633. Berkeley: Univ. California Press.

LIPKIN, M.

(1965) Cell proliferation in the gastrointestinal tract of man. Fed. Proc. **24**, 10—15.

LLOYD, E. H.

(1963) The epochs of emptiness of a semi-infinite discrete reservoir. J. Roy. Statist. Soc. Ser. B **25**, 131—136.

Lock, R. H.
(1906) Recent progress in the study of variation, heredity and evolution. London: Murray.

Loève, M.
(1963) Probability theory. 3rd edition. Princeton: Van Nostrand.

Lotka, A. J.
(1922) The stability of the normal age distribution. Proc. Nat. Acad. Sci. U.S.A. **8**, 339–345.
(1939) Théorie analytique des associations biologiques. Part II: Analyse démographique avec application particulière à l'espèce humaine. Actualités Sci. **780**, 1–149. Paris: Hermann.

Lounsbury, F. G.
(1964) A formal account of the Crow-and Omaha-type kinship terminologies. In W. H. Goodenough (Ed.) *Explorations in cultural anthropology*, pp. 351–393. New York: McGraw-Hill.

Lu, K. H.
(1966) A path-probability approach to irreversible Markov chains with an application in studying the dental caries process. Biometrics **22**, 791–809.

Lucas, H. L.
(1964) Stochastic elements in biological models: their sources and significance. In Gurland (Ed.) (1964), pp. 355–383.

Luchak, G.
(1958) The continuous time solution of the equations of the single channel queue with a general class of service time distributions by the method of generating functions. J. Roy. Statist. Soc. Ser. B **20**, 176–181.

Luria, S. E.
(1959) The reproduction of viruses: a comparative survey. *The Viruses*, Vol. 1, pp. 549–568. New York: Academic Press.

Luria, S. E., and M. Delbrück
(1943) Mutations of bacteria from virus sensitivity to virus resistance. Genetics **28**, 491–511.

Lyon, M. F.
(1961) Gene action in the X-chromosomes of the mouse. Nature **190**, 372–373.

MacArthur, R. H.
(1962) Some generalized theorems of natural selection. Proc. Nat. Acad. Sci. U.S.A. **48**, 1893–1897.

Macdonald, P. D. M.
(1970) Statistical inference from the fraction labelled mitoses curve. Biometrika **57**, 489–503.

Malécot, G.
(1944) Sur un problème de probabilités en chaîne que pose la génétique. C. R. Acad. Sci. Paris **219**, 379–381.
(1948) Les mathématiques de l'hérédité. Paris: Masson.
(1955) La génétique de population. Principes et applications. Population **10**, 239–262.

MARCHAND, H.
(1956) Essai d'étude mathématique d'une forme d'épidémie. Ann. Univ. Lyon Sect. A **19**, 13–46.

MARKERT, C. L.
(1958) Chemical concepts of cellular differentiation. In W. D. MCELROY and B. GLASS (Eds.) *The chemical basis of development*, pp. 3–16. Baltimore: John Hopkins Press.
(1964) Cellular differentiation – an expression of differential gene function. Proc. 2nd Intern. Conf. on Congenital Malformations, pp. 163–174. NewYork: Intern. Medical Congress.
(1966) Biological limits on population growth. Bio Science **16**, 859–862.

MARSHALL, A. W., and H. GOLDHAMER
(1955) An application of Markov processes to the study of the epidemiology of mental diseases. J. Amer. Statist. Assoc. **50**, 99–129.

MARTIN, L.
(1949) Evolution de la biométrie de Quetelet au Congrès International de Biométrie, Genève, 1949. Bull. Inst. Agronom. Stations Recherches Gembloux **17**, 43–66.
(1961) Stochastic processes in physiology. Proc. 4th Berkeley Symp. Math. Statist. Prob. Vol. IV, pp. 307–320. Berkeley: Univ. California Press.

MARTINEZ, H. M.
(1966) On the derivation of a mean growth equation for cell cultures. Bull. Math. Biophys. **28**, 411–416.

MARUYAMA, T.
(1970) Rate of decrease of genetic variability in a subdivided population. Biometrika. **57**, 299–311.

MATIS, J. H., and H. O. HARTLEY
(1971) Stochastic compartmental analysis: model and least squares estimation from time series data. Biometrics **27**, 77–102. (Abstract)

MATTHEWS, J. P.
(1970) A central limit theorem for absorbing Markov chains. Biometrika **57**,129–139.

MAYR, E.
(1964) The evolution of living systems. Proc. Nat. Acad. Sci. U.S.A. **51**, 934–941.

MCCULLOCH, W. S.
(1948) The statistical organization of nervous activity. Biometrics **4**, 91–99.

MCCULLOCH, E. A., and J. E. TILL
(1962) The sensitivity of cells from normal bone marrow to gamma radiation in vitro and in vivo. Radiation Res. **16**, 822–832.

MCKENDRICK, A. G.
(1914) Studies on the theory of continuous probabilities, with special reference to its bearing on natural phenomena of a progressive nature. Proc. London Math. Soc. Ser. II **13**, 401–416.
(1926) Applications of mathematics to medical problems. Proc. Edinburgh Math. Soc. **44**, 98–130.

MELARD, G.
(1968) Epidémie stochastique générale. Ph. D. Thesis, Bruxelles.

MELLORS, R. C., T. AOKI, and R. J. HUEBNER
(1963) Further implication of murine leukemia-like virus in the disorders of NZB mice. J. Exp. Med. **129**, 1045–1062.

MERTZ, D. B., and R. B. DAVIES
(1968) Cannibalism of the pupal stage by adult flour beetles: An experiment and a stochastic model. Biometrics **24**, 247–275.

MERTZ, D. B., T. PARK, and W. J. YOUDEN
(1965) Mortality patterns in eight strains of flour beetles. Biometrics **21**, 99–114.

MEYNELL, G. G., and E. W. MEYNELL
(1958) The growth of micro-organisms in vivo with particular reference to the relation between dose and latent period. J. Hyg. **56**, 323–346.

MEYNELL, G. G., and B. A. D. STOCKER
(1957) Some hypotheses on the aetiology of fatal infections in partially resistant hosts and their application to mice challenged with *Salmonella paratyphi* B or *Salmonella typhimurium* by intraperitoneal injection. J. Gen. Microbiol. **16**, 38–58.

MILCH, P. R.
(1968) A multi-dimensional linear growth birth-and-death process. Ann. Math. Statist. **30**, 727–754.

MILLER, G. F.
(1962) The evaluation of eigenvalues of a differential equation arising in a problem in genetics. Proc. Cambridge Philos. Soc. **58**, 588–593.

MIRASOL, N. M.
(1963) The output of an $M/G/\infty$ queueing system is Poisson. Operations Res. **11**, 282–284.

MODE, C. J.
(1961) A generalized model of a host-pathogen system. Biometrics **17**, 386–404.
(1964) A stochastic model of the dynamics of host-pathogen system with mutation. Bull. Math. Biophys. **26**, 205–233.
(1966 a) Some multi-dimensional branching processes as motivated by a class of problems in mathematical genetics. I, II. Bull. Math. Biophys. **28**, 25–50; 181–190.
(1966 b) Restricted transition probabilities and their applications to some problems in the dynamics of biological populations. Bull. Math. Biophys. **28**, 315–331.
(1966 c) A multi-dimensional birth process and its applications to some problems in the dynamics of biological populations. Bull. Math. Biophys. **28**, 333–345.
(1968) A multi-dimensional age-dependent branching process with applications to natural selection. I, II. Math. Biosci. **3**, 1–18; 231–247.
(1971 a) Multitype age-dependent branching processes and cell cycle analysis. Math. Biosci. **10**, 177–190.
(1971 b) Multitype branching processes. Theory and applications. New York: Elsevier.

MODE, C. J., and J. J. BIRCHER
(1970) On the foundations of age-dependent branching processes with arbitrary state space. J. Math. Anal. Appl. **32**, 435–444.

MOORE, G. P., D. H. PERKEL, and J. P. SEGUNDO
(1966) Statistical analysis and functional interpretation of neuronal spike data. Ann. Rev. Physiol. **28**, 493–522.

MORAN, P. A. P.
(1956) The theory of storage. London: Methuen.
(1958) Random processes in genetics. Proc. Cambridge Philos. Soc. **54**, 60–71.
(1959) The theory of some genetical effects of population subdivision. Austral. J. Biol. Sci. **12**, 109–116.
(1961) The survival of a mutant under general conditions. Proc. Cambridge Philos. Soc. **57**, 304–314.
(1962) The statistical processes of evolutionary theory. Oxford: Clarendon Press.
(1963) Some general results on random walks, with genetic applications. J. Austral. Math. Soc. **3**, 468–479.
(1967) Unsolved problems in evolutionary theory. Proc. 5th Berkeley Symp. Math. Statist. Prob. Vol. IV, pp. 457–480. Berkeley: Univ. California Press.

MORGAN, R. W.
(1965) The estimation of parameters from the spread of a disease by considering households of two. Biometrika **52**, 271–274.

MORGAN, R. W., and D. J. A. WELSH
(1965) A two-dimensional Poisson growth process. J. Roy. Statist. Soc. Ser. B **27**, 497–504.

MOROWITZ, H. J.
(1967) Biological self-replicating systems. *Progress in Theoret. Biol.* **1**, pp. 35–58. New York: Academic Press.

MOSCONA, A. A.
(1963) Studies on cell aggregation: demonstration of material with selective cell-binding activity. Proc. Nat. Acad. Sci. U.S.A. **49**, 742–747.

MOTULSKY, A. G., and G. STAMATOYANNOPOULOS
(1968) Drugs, anesthesia and abnormal hemoglobins. Ann. N.Y. Acad. Sci. **151**/2, 807–820.

MOYAL, J. E.
(1957) Discontinuous Markov processes. Acta Math. **98**, 221–264.
(1962 a) The general theory of stochastic population processes. Acta Math **108**, 1–31.
(1962 b) Multiplicative population chains. Proc. Roy. Soc. Ser. A **266**, 518–526.
(1964) Multiplicative population processes. J. Appl. Probability **1**. 267–283.

MYCIELSKI, J., and S. M. ULAM
(1969) On the pairing process and the notion of genealogical distance. J. Combinatorial Theory **6**, 227–234.

NAGAEV, A. V., and A. N. STARCEV
(1968) A threshold theorem for a model of an epidemic. Mat. Zametki **3**, 179–185. (Russian)

(1970) Asymptotic analysis of a certain stochastic model of an epidemic. Teor. Verojatn. i Primenen. **15**, 97 – 105 (Russian).

NEEL, J. V., and N. A. CHAGNON

(1968) The demography of two tribes of primitive relatively unacculturated American Indians. Proc. Nat. Acad. Sci. U.S.A. **59**, 680 – 689.

NEUTS, M. F.

(1968) Controlling a lethal growth process. Math. Biosc. **2**, 41 – 55.

NEYMAN, J.

(1961) A two-step mutation theory of carcinogenesis. Bull. Inst. Internat. Statist. **38**, 123 – 135.

(1965) Certain chance mechanisms involving discrete distributions. Sankhyā Ser. A **27**, 249 – 258.

NEYMAN, J., T. PARK, and E. L. SCOTT

(1956) Struggle for existence. The Tribolium model: biological and statistical aspects. Proc. 3rd Berkeley Symp. Math. Statist. Prob. Vol. IV, pp. 41 – 79. Berkeley: Univ. of California Press.

NEYMAN, J., and E. L. SCOTT

(1957) On a mathematical theory of populations conceived as conglomerations of clusters. Cold Spring Harbor Symp. Quant. Biol. **22**, 109 – 120.

(1959) Stochastic models of population dynamics. Science **130**, 303 – 308.

(1964) A stochastic model of epidemics. In GURLAND (Ed.) (1964), pp. 45 – 83.

(1967) Statistical aspect of the problem of carcinogenesis. Proc. 5th Berkeley Symp. Math. Statist. Prob. Vol. IV, pp. 745 – 776. Berkeley: Univ. of California, Press.

NISSEN-MEYER, S.

(1966) Analysis of effects of antibiotics on bacteria by means of stochastic models. Biometrics **22**, 761 – 780.

NOONEY, G. C.

(1968) Age distributions. J. Theoret. Biol. **20**, 314 – 320.

NORDLING, C. O.

(1953) A new theory on the cancer-inducing mechanism. Brit. J. Cancer **7**, 68 – 72.

OHLSEN, S.

(1963) Further models for phage reproduction in a bacterium. Biometrics **19**, 441 – 449.

(1964) On estimating epidemic parameters from household data. Biometrika **51**. 511 – 512.

OHTA, T.

(1966) A theoretical study of stochastic survival of inversion chromosomes. Ph, D. Thesis, North Carolina State Univ.

OHTA, T., and M. KIMURA

(1969a) Linkage disequilibrium due to random genetic drift. Genet. Res. **13**, 47 – 55.

(1969b) Linkage disequilibrium at steady state determined by random genetic drift and recurrent mutation. Genetics **63**, 229 – 238.

OHTA, T., and K. KOJIMA

(1968) Survival probabilities of new inversions in large populations. Biometrics **24**, 501 – 516.

OLSON, J. S.
(1966) Problems of interpreting ecological compartment models. Biometrics **22**, 960. (Abstract).

ONICESCU, O., and G. MIHOC
(1937) La dépendance statistique. Chaînes et familles de chaînes discontinues. Actualités Sci. Ind. **503**. Paris: Hermann.

ORE, O.
(1966) Graphs and their uses. Yale: Random House.

PADGETT, W. J., and C. P. TSOKOS
(1970) A stochastic model for chemotherapy: Computer simulation. Math. Biosci. **9**, 119–133.

PARK, T.
(1954) Competition: an experimental and statistical study. In *Statistics and mathematics in biology*, pp. 175–195. Ames: Iowa State Univ. Press.

PARKER, R. A.
(1968) Simulation of an aquatic ecosystem. Biometrics **24**, 803–821.

PARLETT, B.
(1970) Ergodic properties of populations. I. The one sex model. Theoret. Population Biol. **1**, 191–207.

PATIL, G. P., and J. K. WANI
(1965) On certain structural properties of the logarithmic series distribution and the first type Stirling distribution. Sankhya Ser. A **27**, 271–272.

PATON, W. D. M.
(1962) Discussion on receptors. In Ciba Found. Symp. *Enzymes and drug action*, pp. 456–458. London: J & A. Churchill.

PEARSON, K.
(1912) Tuberculosis, heredity and environment. Eugenics Lab. Lecture Series **8**. London: Cambridge Univ. Press.

PEARSON, K., and J. BLAKEMAN
(1906) Mathematical contributions to the theory of evolution. XV. A mathematical theory of random migration. Drap. Co. Mem. Biom. Ser. Vol. 3, pp.1–54.

PERRIN, E. B.
(1967) Uses of stochastic models in the evaluation of population policies. II. Extension of the results by computer simulation. Proc. 5th Berkeley Symp. Math. Statist. Prob. Vol. IV, pp. 137–146. Berkeley: Univ. California Press.

PERRIN, E. B., and M. C. SHEPS
(1964) Human reproduction: a stochastic process. Biometrics **20**, 28–45.

PESKY, M., and P. TAUTU
(1971) A Markovian model of the clinical evolution of tuberculosis. Biometrics **27**, 257–258. (Abstract)

PETTIGREW, H. M., and G. H. WEISS
(1967) Epidemics with carriers: The large population approximation. J. Appl. Probability **4**, 257–263.

PICARD, PH
(1964/65) Étude analytique de l'équation de diffusion des gènes et de certaines de ses généralisations. Ann. Inst. H. Poincaré Sect. B **1**, 23–109.

(1965/66) Sur les modèles stochastiques logistiques en démographie. Ann. Inst. H. Poincaré Sect. B **2**, 151–172.

PIELOU, E. C.

(1969) An introduction to mathematical ecology. New York: Wiley.

PIKE, M. C.

(1963) Some numerical results for the queueing system $D/E_k/1$. J. Roy. Statist. Soc. Ser. B **25**, 477–488.

PODOLSKY, R. J.

(1961) The nature of the contractile mechanism in muscle. In A. M. SHANES (Ed.) *Biophysics of physiological and pharmacological actions*, pp. 461–482, Washington: Amer. Ass. for Adv. Sci.

POGGIO, G. F., and L. J. VIERNSTEIN

(1964) Time series analysis of impulse sequences of thalamic somatic sensory neurons. J. Neurophysiol. **27**, 517–545.

POLLAK, E.

(1966 a) Some effects of fluctuating offspring distributions on the survival of a gene. Biometrika **53**, 391–396.

(1966 b) On the survival of a gene in a subdivided population. J. Appl. Probability **3**, 142–155.

(1968) On random genetic drift in a subdivided population. J. Appl. Probability **5**, 314–333.

(1969) Bounds for certain branching processes. J. Appl. Probability **6**, 201–204.

POLLAK, E., and O. KEMPTHORNE

(1970) Malthusian parameters in genetic populations. I. Haploid and selfing models. Theoret. Population Biol. **1**, 315–345.

POLLARD, J. H.

(1966) On the use of the direct matrix product in analysing certain stochastic population models. Biometrika **53**, 397–415.

(1967) Hierarchical population models with Poisson recruitment. J. Appl. Probability **4**, 209–213.

(1968 a) The multi-type Galton-Watson process in a genetical context. Biometrics **24**, 147–158.

(1968 b) A note on multi-type Galton-Watson processes with random branching probabilities. Biometrika **55**, 589–590.

(1968 c) A note on the age structure of learned societies. J. Roy. Statist. Soc. Ser. A **131**, 569–578.

(1969) Continuous-time and discrete-time models of population growth. J. Roy. Statist. Soc. Ser. A **132**, 80–88.

(1970) On simple approximate calculations appropriate to populations with random growth rates. Theoret. Population Biol. **1**, 208–218.

POWELL, E. O.

(1955) Some features of the generation times of individual bacteria. Biometrika **42**, 16–44.

(1958) An outline of the pattern of bacterial generation times. J. Gen. Microbiol. **18**, 382–417.

POWELL, E. O., and F. P. ERRINGTON
(1963) Generation times of individual bacteria: some corroborative measures. J. Gen. Microbiol. **31**, 315–327.

PRABHU, N. U.
(1965 a) Stochastic processes. Basic theory and its applications. New York: Macmillan.
(1965 b) Queues and inventories. A study of their basic stochastic processes. New York: Wiley.

PRÉKOPA, A.
(1960) On the spreading process. Trans. 2nd Prague Conf. Information Theory, Statist. Decision Functions, Random Processes, pp. 521–529. Prague: Academia.

PRENDIVILLE, B. J.
(1949) Discussion *(Symposium on stochastic processes)*. J. Roy. Statist. Soc. **11**, 273.

PRESTON, K., and P. E. NORGREN
(1969) Automatic autoradiographic grain counting using the CELLSCAN^{TM} system. Ann. N. Y. Acad. Sci. **157**/1, 393–399.

PUCK, T. T., and P. I. MARCUS
(1956) Action of X-rays on mammalian cells. J. Exp. Med. **103**, 653–666.

PURI, P. S.
(1966) On the homogeneous birth-and-death process and its integral. Biometrika **53**, 61–71.
(1967 a) A class of stochastic models of response after infection in the absence of defense mechanism. Proc. 5th Berkeley Symp. Math. Statist. Prob. Vol. IV, pp. 511–535. Berkeley: Univ. California Press.
(1967 b) Some limit theorems on branching processes related to development of biological populations. Math. Biosci. **1**, 77–94.
(1968 a) Some further results on the birth-and-death process and its integral. Proc. Cambridge Philos. Soc. **64**, 141–154.
(1968 b) Interconnected birth and death processes. J. Appl. Probability **5**, 334–349.
(1968 c) A note on Gani's models on phage attachment to bacteria. Math. Biosci. **2**, 151–157.
(1969 a) Some limit theorems on branching processes and certain related processes. Sankhyā Ser. A **31**, 57–74.
(1969 b) Some new results in the mathematical theory of phage reproduction. J. Appl. Probability **6**, 493–504.

PYKE, R.
(1958) On renewal processes related to type I and type II counter models. Ann. Math. Statist. **29**, 737–754.
(1961) Markov renewal processes: Definitions and preliminary properties. Ann. Math. Statist. **32**, 1231–1242.

QUASTLER, H.
(1963) The analysis of cell population kinetics. In L. F. LAMERTON, R. J. M. FRY (Eds.) *Cell proliferation*, pp. 18–34. Oxford: Blackwell.

QUENOUILLE, M. H.
(1947) On the problem of random flights. Proc. Cambridge Philos. Soc. **43**, 581 – 582.

RAHN, O.
(1932) A chemical explanation of the variability of the growth rate. J. Gen. Physiol. **15**, 257 – 277.

RAMAKRISHNAN, A.
(1959) Probability and stochastic processes. In S. FLÜGGE (Ed.) *Handbuch der Physik*, Vol. III, Part 2, pp. 524 – 651. Berlin: Springer.

RAO, C. R.
(1965) On discrete distributions arising out of methods of ascertainment. Sankhyā Ser. A **27**, 311 – 324.

RAPOPORT, A.
(1965) A note on equivalent mathematical models. Bull. Math. Biophys. **27**, Suppl. 161 – 175.

READ, K. L. Q., and J. R. ASHFORD
(1968) A system of models for the life cycle of a biological organism. Biometrika **55**, 211 – 221.

REUTER, G. E. H.
(1961) Competition processes. Proc. 4th Berkeley Symp. Math. Statist. Prob. Vol. II, pp. 421 – 430. Berkeley: Univ. California Press.

RIDLER-ROWE, C. J.
(1967) On a stochastic model of an epidemic. J Appl. Probability **4**, 19 – 33.

RIDLEY, J. C., and M. C. SHEPS
(1966) An analytic simulation model of human reproduction with demographic and biological components. Population Studies **19**, 297 – 310.

RHODES, E. C.
(1940) Population mathematics. I, II, III. J. Roy. Statist. Soc. **103**, 61 – 89; 218 – 245; 362 – 387.

ROBERTS, F. D. K.,
(1969) Nearest neighbours in a Poisson ensemble. Biometrika **56**, 401 – 406.

ROBERTSON, A.
(1962) Selection for heterozygotes in small populations. Genetics **47**, 1291 – 1300.

ROBSON, D. S., R. F. KAHRS, and J. A. BAKER
(1967) Bounds on the mean recurrence time of subclinical epidemics in dairy herds. J. Theoret. Biol. **17**, 47 – 56.

ROSENBLATT, D.
(1957) On the graphs and asymptotic forms of finite Boolean relation matrices and stochastic matrices. Naval Res. Logist. Quart, **4**, 151 – 168.

ROSSET, E.
(1964) New tendencies in the reproduction of the population in Poland. In *Studies on fertility and social mobility*, pp. 105 – 111. Budapest: Akad. Kiadó.

ROTSCHILD, LORD
(1953) A new method of measuring sperm speeds. Nature **171**, 512 – 513.

RUBEN, H.
(1962) Some aspects of the emigration-immigration process. Ann. Math. Statist, **33**, 119—130.
(1963) The estimation of a fundamental interaction parameter in an emigration-immigration process. Ann. Math. Statist. **34**, 238—259.

RUBIN, H.
(1967) Cell growth as a function of cel' density. Proc. 5th Berkeley Symp. Math. Statist. Prob. Vol. IV, pp. 573—579. Berkeley: Univ. California Press.

RUNNENBURG, J. TH.
(1969) Limit theorems for stochastic processes occurring in studies of the light-sensitivity of the human eye. Statistica neerl. **23**, 1—17.

SAATY, T. L.
(1961) Some stochastic processes with absorbing barriers. J. Roy. Statist. Soc. Ser. B **23**, 319—334.

SACHER, G. A., and E. TRUCCO
(1962) The stochastic theory of mortality. Ann. N. Y. Acad. Sci. **96**/4, 985—1007.

SALTARELLI, M., and M. DURBIN
(1967) A semantic interpretation of kinship system. Linguistics **33**, 87—94.

SAMPFORD, M. R.
(1955) The truncated binomial distribution. Biometrika **42**, 58.

SANKOFF, D.
(1971) Duration of detectible synchrony in a binary branching process. Biometrika **58**, 77—81.

SCHACH, E., and S. SCHACH
(1970) On the variability of survival times of mice inoculated with cancer cells. Biometrische Z. **12**, 14—24.

SCHAECHTER, M., J. P. WILLIAMSON, J. R. HOOD, and A. L. KOCH
(1962) Growth, cell and nuclear division in some bacteria. J. Gen. Microbiol. **29**, 421—434.

SCHAFFER, H. E.
(1970) Survival of mutant genes as a branching process. In KOJIMA (Ed.) (1970), pp. 317—336.

SCHERBAUM, O. H.
(1962) A comparison of synchronized cell division in protozoa. J. Protozool. **9**, 61—64.

SCHERBAUM, O., and G. RASCH
(1957) Cell size distribution and single cell growth in *Tetrahymena pyriformis* GL. Acta path. microbiol. scand. **41**, 161—182.

SCHILD, H. O.
(1962) Discussion on receptors. In Ciba Found. Symp. *Enzymes and drug action*, pp. 435—439. London: J. & A. Churchill.

SCHOENFELD, R. L.
(1963) Linear network theory and tracer analysis. Ann. N. Y. Acad. Sci. **108**/1, 69—91.

SEAL, H. L.

(1945) The mathematics of a population composed of k stationary strata each recruited from the stratum below and supported at the lowest level by a uniform annual number of entrants. Biometrika **33**, 226 – 230.

SEGUNDO, J. P., D. H. PERKEL, and G. P. MOORE

(1966) Spike probability in neurons. Influence of temporal structure in the train of synaptic events. Kybernetik **3**, 67 – 82.

SELYE, H.

(1969) The pluricausal cardiopathies. Ann. N. Y. Acad. Sci. **156**/1, 195 – 206.

SENETA, E.

(1966) Quasi-stationary distributions and time-reversion in genetics. J. Roy. Statist. Soc. Ser. B **28**, 253 – 266.

(1967 a) The random walk and bacterial growth. Zastos. Mat. **9**, 135 – 147.

(1967 b) On imbedding discrete chains in continuous time. Austral. J. Statist. **9**, 1 – 7.

(1968) The stationary distribution of a branching process allowing immigration: A remark on the critical case. J. Roy. Statist. Soc. Ser. B **30**, 176 – 179.

SETHURAMAN, J.

(1965) On a characterization of three limiting types of the extreme. Sankhyā Ser. A **27**, 357 – 364.

SEVAST'JANOV, B. A.

(1958) Branching stochastic processes for particles diffusing in a bounded domain with absorbing barriers. Teor. Verojatnost. i Primenen **3**, 111 – 126 (Russian).

SEVERO, N. C.

(1967) Two theorems on solutions of differential difference equations and applications to epidemic theory. J. Appl. Probability **4**, 271 – 280.

(1969 a) Generalizations of some stochastic epidemic models. Math. Biosci. **4**, 395 – 402.

(1969 b) The probabilities of some epidemic models. Biometrika **56**, 197 – 201.

(1969 c) Right-shift processes. Proc. Nat. Acad. Sci. U.S.A. **64**, 1162 – 1164.

(1969 d) Solving non-linear problems in the theory of epidemics. Bull. Inst. Internat. Statist. **43**, Book 2, 226 – 228.

SHARPE, F. R., and A. J. LOTKA

(1911) A problem in age distribution. Phil. Mag. (Ser. 6) **21**, 435 – 438.

SHEPPARD, C. W.

(1962) Basic principles of the tracer method. New York: Wiley.

SHEPS' M. C.

(1967) Uses of stochastic models in the evaluation of population policies. I. Theory and approaches to data analysis. Proc. 5th Berkeley Symp. Math. Statist. Prob. Vol. IV, pp. 115 – 136. Berkeley: Univ. California Press.

SHEPS, M. C., D. P. DOOLITTLE, and M. L. NEW

(1969) Mammalian reproductive data fitted to a mathematical model. Biometrics **25**, 529 – 535.

SHEPS, M. C., and J. A. MENKEN

(1971) A model for studying birth rates given time dependent changes in reproductive parameters. Biometrics **27**, 325 – 343.

SHEPS, M. C., J. A. MENKEN, and A. P. RADICK
(1969) Probability models for family building: An analytical review. Demography **6**, 161–183.

SHEPS, M, C., and E. B. PERRIN
(1966) Further results from a human fertility model with a variety of pregnancy outcomes. Human Biology **38**, 180–193.

SHIKATA, M.
(1967) Effect of selections upon recombination probabilities in selfed populations. J. Theoret. Biol. **14**, 59–65.

SHIMBEL, A., and A. RAPOPORT
(1948) A statistical approach to the theory of the central nervous system. Bull. Math. Biophys. **10**, 41–55.

SHIMKIN, M. B., and M. J. POLISSAR
(1955) Some quantitative observations on the induction and growth of primary pulmonary tumors in strain A mice receiving urethane. J. Nat. Cancer Inst. **16**, 75–97.

SHIMKIN, M. B., R. WIEDER, D. MARZI, N. GUBAREGG, and V. SUNTZEFF
(1967) Lung tumors in mice receiving different schedules of urethane. Proc. 5th Berkeley Symp. Math. Statist. Prob. Vol. IV, pp. 707–719. Berkeley: Univ. California Press.

SHORTLEY, G,
(1965) A stochastic model for distributions of biological response times. Biometrics **21**, 562–582.

SHORTLEY, G., and J. R. WILKINS
(1965) Independent-action and birth and death models in experimental microbiology. Bacteriol. Rev. **29**, 102–141.

SINGER, B.
(1970) Some asymptotic results in a model of population growth. I. A class of birth and death processes. Ann. Math. Statist. **41**, 115–132.
(1971) Some asymptotic results in a model of population growth. II. Positive recurrent chains. Ann. Math. Statist. **42**, 1296–1315.

SISKEN, J. E., and L. MORASCA
(1965) Intrapopulation kinetics of the mitotic cycle. J. Cell. Biol. **25**, 179–189.

SISKIND, V.
(1965) A solution of the general stochastic epidemic. Biometrika **52**, 613–616.

SKELLAM, J. G.
(1949) The probability distribution of gene differences in relation to selection, mutation, and random extinction. Proc. Cambridge Philos. Soc. **45**, 364–367.
(1951) Random dispersal in theoretical populations. Biometrika **38**, 196–218.
(1958) On the derivation and applicability of Neyman's type A distribution. Biometrika **45**, 32–36.
(1967) Seasonal periodicity in theoretical population ecology. Proc. 5th Berkeley Symp. Math. Statist. Prob. Vol. IV, pp. 179–205. Berkeley: Univ. California Press.

SMITH, D. R., and G. K. SMITH

(1965) A statistical analysis of the continual activity of single cortical neurons in the cat unanaesthetized isolated forebrain. Biophys. J. **5**, 47–74.

SMITH, W. L.

(1958) Renewal theory and its ramifications. J. Roy. Statist. Soc. Ser. B **20**, 243–302.

(1968) Necessary conditions for almost sure extinction of a branching process with random environment. Ann. Math. Statist. **39**, 2136–2140.

SMITH, W., and W. WILKINSON

(1969) On branching processes in random environments. Ann. Math. Statist. **40**, 814–827.

SOLLBERGER, A.

(1965) Biological rhythm research. Amsterdam: Elsevier.

(1967) Biological measurements in time, with particular reference to synchronization mechanisms. Ann. N. Y. Acad. Sci. **138**/2, 561–599.

SRINIVASAN, S. K., and K. S. S. IYER

(1966) Random processes associated with random points on a line. Zastos. Mat. **8**, 221–230.

SRINIVASAN, S. K., and G. RAJAMANNAR

(1970) Counter models and dependent renewal point processes related to neuronal firing. Math. Biosci. **7**, 27–39.

SRINIVASAN, S. K., and A. RANJAN

(1970) Age-dependent stochastic models for phage reproduction. J. Appl. Probability **7**, 251–261.

SRIVASTAVA, R. C.

(1968) Estimation of the parameter in the stochastic model for phage attachment to bacteria. Ann. Math. Statist. **39**, 183–192.

STEIN, R. B.

(1967) Some models of neuronal variability. Biophys. J. **7**, 37–68.

STEINBERG, C., and F. STAHL

(1961) The clone-size distribution of mutants arising from a steady-state pool of vegetative phage. J. Theoret. Biol. **1**, 488–497.

STEINBERG, M. S.

(1962) On the mechanism of tissue reconstruction by dissociated cells. I. Population kinetics, differential adhesiveness, and the absence of directed migration. Proc. Nat. Acad. Sci. U.S.A. **48**, 1577–1582.

STEWART, P. A., H. QUASTLER, M. R. SKOUGAARD, D. R. WIMBER, M. F. WOLFSBERG, C. A. PERROTTA, B. FERBEL, and M. CARLOUGH

(1965) Four-factor model analysis of thymidine incorporation into mouse DNA and the mechanism of radiation effects. Radiation Res. **24**, 521–537.

STITELER, W. M.

1970) Measurement of spatial patterns in ecology. Ph. D. Thesis, Pennsylvania State University.

STONE, L. S.
(1955) Regeneration of the iris and lens from retina pigment cells in adult newt eyes. J. Exp. Zool. **129**, 505–534.

STUGREN, B.
(1966) Geographic variation and distribution of the Moor frog, *Rana arvalis* Nills. Ann. Zool. Fennici **3**, 23–39.

SUGIYAMA, H.
(1961) Some structural methodologies for epidemiological research of medical sciences. Bull. Inst. Internat. Statist. **38**, 137–151.

SUGIYAMA, H., G. P. MOORE, and D. H. PERKEL
(1970) Solutions for a stochastic model of neuronal spike production. Math. Biosci. **8**, 323–341.

SUMMERS, D. F., J. V. MAIZEL, and J. E. DARNELL
(1965) Evidence for virus-specific noncapsid proteins in poliovirus-infected HeLa cells. Proc. Nat. Acad. Sci. U.S.A. **54**, 505–513.

SYKES, Z. M.
(1969 a) Some stochastic versions of the matrix model for population dynamics. J. Amer. Statist. Assoc. **64**, 110–130.
(1969 b) Population projections and Markov chains. General Conf. IUSSP, London.

SZYBALSKI, W., and E. H. SZYBALSKA
(1962) Drug sensitivity as a genetic marker for human cell lines. Univ. of Michigan Medical Bull. **28**, 277–293.

TAKÁCS, L.
(1956) On a probability problem arising in the theory of counters. Proc. Cambridge Philos. Soc. **52**, 488–498.

TAKAHASHI, M.
(1968) Theoretical basis for cell cycle analysis. II. Further studies on labelled mitosis wave method. J. Theoret. Biol. **18**, 195–209.

TAKAHASHI, K., S. ISHIDA, and M. KUROKAWA
(1964) Simulation of synchronous growth. On Koch and Schaechter's model for the cell division process Ann. Inst. Statist. Math. **5**, Suppl., 21–36.

TAKASHIMA, M.
(1956) Note on evolutionary processes. Bull. Math. Statist. **7**, 18–24.

TALLIS, G. M., and M. K. LEYTON
(1969) Stochastic models of population of helminthic parasites in the definitive host. Math. Biosci. **4**, 39–48.

TATUM, E. L.
(1964) Genetic determinants. Proc. Nat. Acad. Sci. U.S.A. **51**, 908–915.

TAUTU, P.
(1968) Queues with feedback as mathematical models of some coordinated metabolic processes. 5th Intern. Congr. Cybernetic Medicine, Napoli.
(1969) The dam process in hematopoesis. Lecture at the Department of Probability and Statistics, Sheffield. (unpublished)

TAYLOR, H. M.
(1968) Some models in epidemic control. Math. Biosci. **3**, 383–398.

TEICHER, H.
(1960) On the mixture of distributions. Ann. Math. Statist. **31**, 55–73.

TEN HOOPEN, M.
(1966) Probabilistic firing of neurons considered as first passage problem. Biophys. J. **6**, 435–451.

TEN HOOPEN, M., and H. A. REUVER
(1965) Selective interaction of two recurrent processes J. Appl. Probability **2**, 286–292.
(1966) The superposition of random sequences of events. Biometrika **53**, 383–389.

THEISS, E.
(1964) Report on forecaste and international comparability of fertility. In *Studies on fertility and social mobility*, pp. 112–118. Budapest: Akad. Kiadó.

THIEBAUX, J.
(1967) Testing a Markov hypothesis with independence of intermediate states and restricted order. Biometrika **54**, 605–615.

THOMAS, E. A. C.
(1966) Mathematical models for the clustered firing of single cortical neurons. Brit. J. Math. Statist. Psychol. **19**, 151–162.

THOMAS, V. J.
(1969) A stochastic population model related to human populations. J. Roy. Statist. Soc. Ser. A **132**, 89–104.

THOMPSON, H. R.
(1955) Spatial point processes, with applications to ecology. Biometrika **42**, 102–115.
(1956) Distribution of distance to n-th nearest neighbour in a population of randomly distributed individuals. Ecology **37**, 391–394.

TILL, J. E., E. A. MCCULLOCH, and K. SIMINOVITCH
(1964) A stochastic model of stem cell proliferation based on the growth of spleen colony-forming cells. Proc. Nat. Acad. Sci. U.S.A. **51**, 29–36.

TOBIAS, C. A.
(1961) Quantitative approaches to the cell division process. Proc. 4th Nerkeley Simp. Math. Statist. Prob. Vol. IV, pp. 369–385. Berkeley: Univ. California Press.

TRUCCO, E.
(1963) On the Fokker-Planck equation in the stochastic theory of mortality. I, II. Bull. Math. Biophys. **25**, 303–323.
(1965) Mathematical models for cellular systems. The von Foerster equation I, II. Bull. Math. Biophys. **27**, 285–304; 449–471.

TUCKER, H. G.
(1961) A stochastic model for a two-stage theory of carcinogenesis. Proc. 4th Berkeley Symp. Math. Statist. Prob. Vol. IV, pp. 387–403. Berkeley: Univ. California Press.

URBANIK, K.
(1956) Remarks on the maximum quantity of bacteria in a population. Zastos. Mat. **2**, 341–348. (Polish)

USHER, M. B., and M. H. WILLIAMSON
(1970) A deterministic matrix model for handling the birth, death, and migration processes of spatially distributed populations. Biometrics **26**, 1–12.

VAJDA, S.
(1948) Introduction to a mathematical theory of the graded stationary population. Bull. Ass. Actuaires Suisses **48**, 251–273.

VERE-JONES, D.
(1966) Simple stochastic models for the release of quanta of transmitter from a nerve terminal. Austral. J. Statist. **8**, 53–63.
(1968) Some applications of probability generating functionals to the study of input-output streams. J. Roy. Statist. Soc. Ser. B **30**, 321–333.

WADDINGTON, C. H.
(1969) Paradigm for an evolutionary process. *Towards a Theoretical Biology* **2**, pp. 106–124. Edinburgh: Edinburgh Univ. Press.

WALD, G.
(1954) On the mechanism of the visual threshold and visual adaptation. Science **119**, 887–892.

WALKER, P. M. B.
(1954) The mitotic index and interphase processes. J. Exp. Biol. **31**, 8–15.

WASSER, P. G.
(1962) Relation between enzymes and cholinergic receptors. In Ciba Found. Symp. *Enzymes and drug action*, pp. 206–217. London: J. & A. Churchill.

WATERMAN, T. H., and H. J. MOROWITZ (Eds.)
(1965) Theoretical and mathematical biology. New York: Blaisdell.

WATSON, G. S.
(1965) The distribution of organisms. Biometrics **21**, 543–550.

WATTERSON, G. A.
(1961) Markov chains with absorbing states: a genetic example. Ann. Math. Statist. **32**, 716–729.
(1962) Some theoretical aspects of diffusion theory in population genetics. Ann. Math. Statist. **33**, 939–957.
(1964) The application of diffusion theory to two population genetic models of Moran. J. Appl. Probability **1**, 233–246.
(1970) On the equivalence of random mating and random union of gametes models in finite monoecious populations. Theoret. Population Biol. **1**, 233–250.

WAUGH, W. A. O'N
(1955) An age-dependent birth and death process. Biometrika **42**, 291–306.
(1958) Conditioned Markov processes. Biometrika **45**, 241–249.
(1961) Age-dependence in a stochastic model of carcinogenesis. Proc. 4th Berkeley Symp. Math. Statist. Prob. Vol. IV, pp. 405–413. Berkeley: Univ. of California Press.
(1968) Age-dependent branching processes under a condition of ultimate extinction. Biometrika **55**, 291–296.

WEISS, G. H.
(1963) Comparison of a deterministic and a stochastic model for interaction between antagonistic species. Biometrics **19**, 595–602.

(1965) On the spread of epidemics by carriers. Biometrics **21**, 481–490.

WEISS, G. H., and M. ZELEN

(1963) A stochastic model for the interpretation of clinical trials. Proc. Nat. Acad. Sci. U.S.A. **50**, 988–994.

(1965) A semi-Markov model for clinical trials. J. Appl. Probability **2**, 269–285.

WEISSNER, E. W.

(1971) Multitype branching process in random environments. J. Appl. Probability **8**, 17–31.

WHITTLE, P.

(1952) Certain nonlinear models of populations and epidemic theory. Skand. Aktuarietidskr. **14**, 211.

(1955) The outcome of a stochastic epidemic. A note on Bailey's paper. Biometrika **42**, 116.

(1964) A branching process in which individuals have variable lifetimes. Biometrika **51**, 262–264.

(1967) Nonlinear migration processes. Bull. Inst. Internat. Statist. **42**, Book 2, pp. 642–647.

(1968) Equilibrium distributions for an open migration process. J. Appl. Probability **5**, 567–571.

WIGGINS, A. D.

(1957) A statistical study of the mechanism of bacterial toxicity. Ph. D. Thesis, Univ. of California, Berkeley.

(1960) On a multicompartment migration model with chronic feeding. Biometrics **16**, 642–658.

(1961) Further aspects of a multicompartment migration model. Biometrics **17**, 508–509. (Abstract)

WILKINSON, W. E.

(1967) Branching processes in stochastic environments. Ph. D. Thesis, Univ. of North Carolina.

(1969) On calculating extinction probabilities for branching processes in random environments. J. Appl. Probability **6**, 478–492.

WILLIAMS, C. B.

(1944) Some applications of the logarithmic series and the index of diversity to ecological problems. J. Ecol. **32**, 1–44.

WILLIAMS, E. J.

(1961 a) The growth and age-distribution of a population of insects under uniform conditions. Biometrics **17**, 349–358.

(1961 b) The distribution of larvae of randomly moving insects. Austral. J. Biol. Sci. **14**, 598–604.

WILLIAMS, T.

(1965 a) The simple stochastic epidemic curve for large populations of susceptibles. Biometrika **52**, 571–579.

(1965 b) The basic birth-and-death model for microbial infections. J. Roy. Statist. Soc. Ser. B **27**, 338–360.

(1965 c) The distribution of response times in a birth-death process. Biometrika **52,** 581 – 585.

(1969) The distribution of inanimate marks over a non-homogeneous birth-death process. Biometrika **56,** 225 – 227.

WINSTON, J. A.

(1970) Survival of a mutation under mixed positive assortative and random mating. I. One locus, two alleles without dominance. Biometrics **26,** 433 – 450.

WISHART, D. M. G.

(1956) A queueing system with χ^2 service time distribution. Ann. Math. Statist. **27,** 768 – 779.

WODINSKY, I., J. SWINIARSKI, and C. J. KENSLER

(1967) Spleen colony studies of leukemia L 1210. I. Growth kinetics of lymphocytic L 1210 cells in vivo as determined by spleen colony assay. Cancer Chemother. Rep. **51,** 415 – 421.

WOODGER, J. H.

(1962) Biology and the axiomatic method. Ann. N. Y. Acad. Sci. **96**/4, 1093 – 1104.

(1965) Theorems of random evolution. Bull. Math. Biophys. **27,** Suppl., 145 – 150.

WRIGHT, S.

(1929) Fisher's theory of dominance. Amer. Naturalist **63,** 274 – 279.

(1931) Evolution in Mendelian populations. Genetics **16,** 97 – 159.

(1937) The distribution of gene frequencies in populations. Proc. Nat. Acad. Sci. U.S.A. **23,** 307 – 320.

(1940) Breeding structure of populations in relation to speciation. Amer. Nat. **74,** 232 – 248.

(1942) Statistical genetics and evolution. Bull. Amer. Math. Soc. **48,** 223 – 246.

(1948) On the role of directed and random changes in gene frequency in the genetics of populations. Evolution **2,** 279 – 294.

(1964) Stochastic processes in evolution. In GURLAND (Ed.) (1964), pp. 199 – 241.

(1970) Random drift and the shifting balance theory of evolution. In KOJIMA (Ed.) (1970), pp. 1 – 31.

YAMAMOTO, S.

(1961) On the homogeneous birth and death process with an absorbing barrier. Bull. Math. Statist. **10,** 45 – 56.

YANG, G. L.

(1968) Contagion in stochastic models for epidemics. Ann. Math. Statist. **39,** 1863 – 1889.

YASSKY, D.

(1962) A model for the kinetics of phage attachment to bacteria in suspension Biometrics **18,** 185 – 191.

YOFFEY, J. M.

(1957) Cellular equilibria in blood and blood-forming tissues. Brookhaven Symp. in Biology **10,** 1 – 25. Upton: Brookhaven Nat. Lab.

YULE, G. U.

(1924) A mathematical theory of evolution based on the conclusions of Dr J. C. Willis, FRS. Philos. Trans. Roy. Ser. London, Ser. **B 213,** 21 – 87.

ZAHL, S.
(1955) A Markov process model for follow-up studies. Human Biology, **25**, 90–120.
ZAMENHOF, S.
(1963) Mutations. Amer. J. Med. **34**, 609–626.
ZELEN, M., and M. FEINLEIB
(1969 a) On the theory of screening for chronic diseases. Biometrika **56**, 601–614.
(1969 b) Estimating the mean forward recurrence time associated with early detection of chronic disease. Bull. Inst. Internat. Statist. **43**, Book 2, pp. 235–237.
ZUCKERKANDL, E., and L. PAULING
(1962) Molecular disease, evolution, and genetic heterogeneity. In M. KASHA, B. PULLMAN (Eds.) *Horizons in biochemistry*, pp. 189–225, New York: Academic Press.

Notation index

$N = \{0, 1, 2, \ldots, n, \ldots\}$
$N^* = \{1, 2, \ldots, n, \ldots\}$
$-N = \{\ldots, n, \ldots, -2, -1, 0\}$
$Z = -N \cup N^*$
R = the set of all real number
Rez = real part of the complex number z
$\mathbf{A}'$ = the transpose of the matrix $\mathbf{A}$
E = expectation
D = variance
i = the imaginary unit $\sqrt{-1}$
a.s. = almost surely
iff = if and only if
◇ = end of a proof

Subject index

The contents should be also consulted for subject matter

Author index

For a complete information the bibliography should be also consulted.